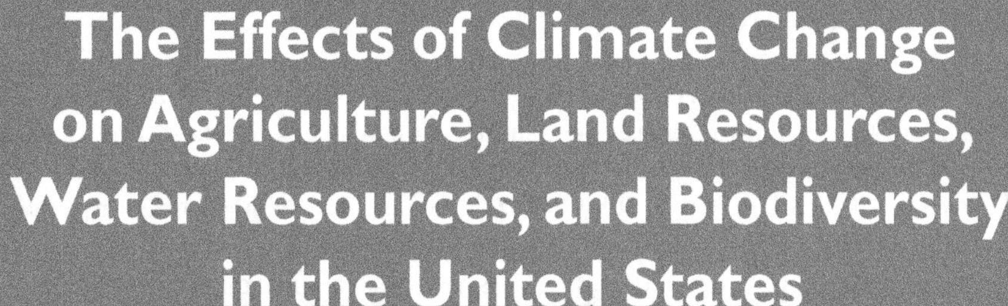

The Effects of Climate Change on Agriculture, Land Resources, Water Resources, and Biodiversity in the United States

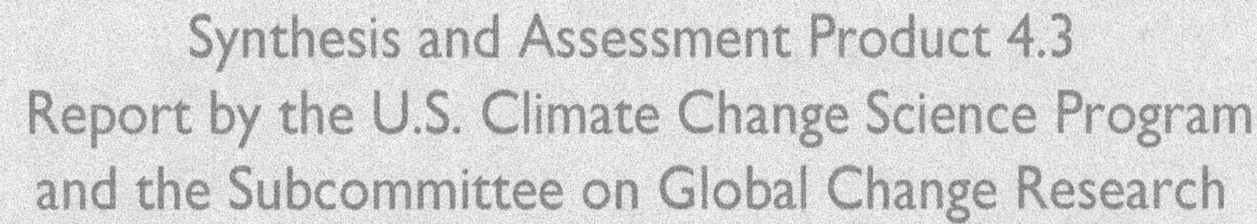

Synthesis and Assessment Product 4.3
Report by the U.S. Climate Change Science Program
and the Subcommittee on Global Change Research

CONVENING LEAD AUTHORS:
Peter Backlund, Anthony Janetos, and David Schimel

MANAGING EDITOR:
Margaret Walsh

June 2008

Members of Congress:

On behalf of the National Science and Technology Council, the U.S. Climate Change Science Program (CCSP) is pleased to transmit to the President and the Congress this Synthesis and Assessment Product (SAP), *The Effects of Climate Change on Agriculture, Biodiversity, Land, and Water Resources in the United States*. This is part of a series of 21 SAPs produced by the CCSP aimed at providing current assessments of climate change science to inform public debate, policy, and operational decisions. These reports are also intended to help the CCSP develop future program research priorities. This SAP is issued pursuant to Section 106 of the Global Change Research Act of 1990 (Public Law 101-606).

The CCSP's guiding vision is to provide the Nation and the global community with the science-based knowledge needed to manage the risks and capture the opportunities associated with climate and related environmental changes. The SAPs are important steps toward achieving that vision and help to translate the CCSP's extensive observational and research database into informational tools that directly address key questions being asked of the research community.

This SAP assesses the effects of climate change on U.S. land resources, water resources, agriculture, and biodiversity. It was developed with broad scientific input and in accordance with the Guidelines for Producing CCSP SAPs, the Federal Advisory Committee Act, the Information Quality Act, Section 515 of the Treasury and General Government Appropriations Act for fiscal year 2001 (Public Law 106-554), and the guidelines issued by the Department of Agriculture to Section 515.

We commend the report's authors for both the thorough nature of their work and their adherence to an inclusive review process.

Sincerely,

Carlos M. Gutierrez
Secretary of Commerce
Chair, Committee on
Climate Change Science
and Technology Integration

Samuel W. Bodman
Secretary of Energy
Vice Chair, Committee on
Climate Change Science
and Technology Integration

John H. Marburger III
Director, Office of Science
and Technology Policy
Executive Director, Committee
on Climate Change Science and
Technology Integration

TABLE OF CONTENTS

CHAPTER

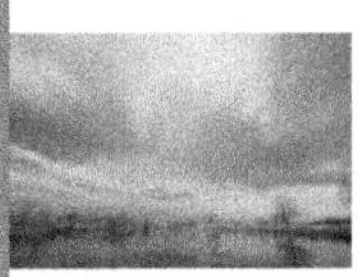

RECOMMENDED CITATIONS

For the Report as a Whole:

CCSP, 2008: *The effects of climate change on agriculture, land resources, water resources, and biodiversity.* A Report by the U.S. Climate Change Science Program and the Subcommittee on Global Change Research. **P. Backlund, A. Janetos, D. Schimel, J. Hatfield, K. Boote, P. Fay, L. Hahn, C. Izaurralde, B.A. Kimball, T. Mader, J. Morgan, D. Ort, W. Polley, A. Thomson, D. Wolfe, M. Ryan, S. Archer, R. Birdsey, C. Dahm, L. Heath, J. Hicke, D. Hollinger, T. Huxman, G. Okin, R. Oren, J. Randerson, W. Schlesinger, D. Lettenmaier, D. Major, L. Poff, S. Running, L. Hansen, D. Inouye, B.P. Kelly, L Meyerson, B. Peterson, R. Shaw**. U.S. Environmental Protection Agency, Washington, DC., USA, **362** pp

For the Executive Summary:

Backlund, P., A. Janetos, D.S. Schimel, J. Hatfield, M. Ryan, S. Archer, and D. Lettenmaier, 2008. **Executive Summary**. In: *The effects of climate change on agriculture, land resources, water resources, and biodiversity.* A Report by the U.S. Climate Change Science Program and the Subcommittee on Global Change Research. Washington, DC., USA, **362** pp

For Chapter 1:

Backlund, P., D. Schimel, A. Janetos, J. Hatfield, M. Ryan, S. Archer, and D. Lettenmaier, 2008. **Introduction**. In: *The effects of climate change on agriculture, land resources, water resources, and biodiversity.* A Report by the U.S. Climate Change Science Program and the Subcommittee on Global Change Research. Washington, DC., USA, **362** pp

For Chapter 2:

Hatfield, J., K. Boote, P. Fay, L. Hahn, C. Izaurralde, B.A. Kimball, T. Mader, J. Morgan, D. Ort, W. Polley, A. Thomson, and D. Wolfe, 2008. **Agriculture**. In: *The effects of climate change on agriculture, land resources, water resources, and biodiversity.* A Report by the U.S. Climate Change Science Program and the Subcommittee on Global Change Research. Washington, DC., USA, **362** pp

For Chapter 3:

Ryan, M., S. Archer, R. Birdsey, C. Dahm, L. Heath, J. Hicke, D. Hollinger, T. Huxman, G. Okin, R. Oren, J. Randerson, and W. Schlesinger, 2008. **Land Resources**. In: *The effects of climate change on agriculture, land resources, water resources, and biodiversity.* A Report by the U.S. Climate Change Science Program and the Subcommittee on Global Change Research. Washington, DC., USA, **362** pp

For Chapter 4:

Lettenmaier, D., D. Major, L. Poff, and S. Running, 2008. **Water Resources**. In: *The effects of climate change on agriculture, land resources, water resources, and biodiversity.* A Report by the U.S. Climate Change Science Program and the subcommittee on Global change Research. Washington, DC., USA, **362** pp

For Chapter 5:

Janetos, A., L. Hansen, D. Inouye, B.P. Kelly, L. Meyerson, B. Peterson, and R. Shaw, 2008. **Biodiversity**. In: *The effects of climate change on agriculture, land resources, water resources, and biodiversity.* A Report by the U.S. Climate Change Science Program and the Subcommittee on Global Change Research. Washington, DC., USA, **362** pp

For Chapter 6:

Schimel, D., A. Janetos, P. Backlund, J. Hatfield, M. Ryan, S. Archer, D. Lettenmaier, 2008. **Synthesis**. In: *The effects of climate change on agriculture, land resources, water resources, and biodiversity.* A Report by the U.S. Climate Change Science Program and the subcommittee on Global change Research. Washington, DC., USA, **362** pp

AUTHOR TEAM FOR THIS REPORT

Executive Summary

Lead Authors: Peter Backlund, NCAR; Anthony Janetos, PNNL/Univ. Maryland; David Schimel, National Ecological Observatory Network
Contributing Authors: Jerry Hatfield USDA ARS; Mike Ryan, USDA Forest Service; Steven Archer, Univ. Arizona; Dennis Lettenmaier, Univ. Washington

Chapter 1

Lead Authors: Peter Backlund, NCAR; David Schimel, National Ecological Observatory Network; Anthony Janetos, PNNL/Univ. Maryland
Contributing Authors: Jerry Hatfield USDA ARS; Mike Ryan, USDA Forest Service; Steven Archer, Univ. Arizona; Dennis Lettenmaier, Univ. Washington

Chapter 2

Lead Author: Jerry Hatfield USDA ARS
Contributing Authors: Kenneth Boote, Univ. Florida; Philip Fay, USDA ARS; Leroy Hahn, Univ. Nebraska; César Izaurralde, PNNL/Univ. Maryland; Bruce A. Kimball, USDA ARS; Terry Mader, Univ. Nebraska-Lincoln; Jack Morgan, USDA ARS; Donald Ort, Univ. Illinois; Wayne Polley, USDA ARS; Allison Thomson, PNNL/Univ. Maryland; David Wolfe, Cornell Univ.

Chapter 3

Lead Authors: Mike Ryan, USDA Forest Service; Steven Archer, Univ. Arizona
Contributing Authors: Richard Birdsey, USDA Forest Service; Cliff Dahm, Univ. New Mexico; Linda Heath, USDA Forest Service; Jeff Hicke, Univ. Idaho; David Hollinger, USDA Forest Service; Travis Huxman, Univ. Arizona; Gregory Okin, Univ. California-Los Angeles; Ram Oren, Duke Univ.; James Randerson, Univ. California-Irvine; William Schlesinger, Duke Univ.

Chapter 4

Lead Author: Dennis Lettenmaier, Univ. Washington
Contributing Authors: David Major, Columbia Univ.; Leroy Poff, Colo. State Univ.; Steve Running, Univ. Montana

Chapter 5

Lead Author: Anthony Janetos, PNNL/Univ. Maryland
Contributing Authors: Lara Hansen, World Wildlife Fund Intl.; David Inouye, Univ. Maryland; Brendan P. Kelly, Univ. Alaska and National Science Foundation; Laura Meyerson, Univ. Rhode Island; Bill Peterson, NOAA; Rebecca Shaw, The Nature Conservancy

Chapter 6

Lead Authors: David Schimel, National Ecological Observatory Network; Anthony Janetos, PNNL/Univ. Maryland; Peter Backlund, NCAR;
Contributing Authors: Jerry Hatfield USDA ARS; Mike Ryan, USDA Forest Service; Steven Archer, Univ. Arizona; Dennis Lettenmaier, Univ. Washington

ACKNOWLEDGEMENTS

In addition to the authors, many people made important contributions to this document.

Special thanks are due to Margaret Walsh of the USDA Climate Change Program Office. Her contributions to the report and the successful completion of this project are too numerous to document.

We are very appreciative of the excellent support from the project team at the University Corporation for Atmospheric Research (UCAR) and the National Center for Atmospheric Research (NCAR). From NCAR, Carol Park assisted with editing and research, and coordinated all aspects of the project; Greg Guibert and Rachel Hauser assisted with editing, technical writing, and research; Brian Bevirt contributed to editing and layout; and Kristin Conrad, Steve Aulenbach, and Erica Markum provided web support. From the UCAR Communications Office, Nicole Gordon and Lucy Warner provided copy editing services. Michael Shibao of Smudgeworks provided graphic design services.

We are also grateful to the members of the Committee for the Expert Review of Synthesis and Assessment Product 4.3 (CERSAP) who oversaw the project on behalf of USDA and provided helpful guidance and insightful comments: Thomas Lovejoy (Heinz Center for Science, Economics, and the Environment), J. Roy Black (Michigan State University), David Breshears (University of Arizona), Glenn Guntenspergen (USGS), Brian Helmuth (University of South Carolina), Frank Mitloehner (UC Davis), Harold Mooney (Stanford University), Dennis Ojima (Heinz Center), Charles Rice (Kansas State University), William Salas (Applied Geosciences LLC), William Sommers (George Mason University), Soroosh Sorooshian (UC Irvine), Eugene Takle (Iowa State University), and Carol Wessman (University of Colorado).

Lawrence Buja (NCAR Climate and Global Dynamics Division (CGD) and Julie Arblaster (NCAR/CGD and the Australian Bureau of Meteorology) created and provided a series of figures showing projections of future climate conditions for the Introduction and Context chapter. Kathy Hibbard (NCAR/CGD) provided expert scientific input and review for the Biodiversity chapter. Timothy R. Green (Agricultural Engineer, USDA Agricultural Research Division) provided expert scientific input and review for aspects of the Water chapter. Chris Milly (USGS Continental Water, Climate, and Earth-System Dynamics Project), and Harry Lins (USGS Surface-water Program) also contributed to this chapter by providing figures 4.10 and 4.7, respectively. Figures 4.1 through 4.4 were prepared with assistance from Jennifer C. Adam (Washington State University). Authors in the Arid Land section of the Land Resources Chapter benefited from information generated by the Jornada and Sevilleta Long Term Ecological Research programs, and New Mexico's Experimental Program to Stimulate Competitive Research (EPSCoR). Julio Betancourt (USGS) provided valuable discussions on exotic species invasions. Craig Allen (USGS) provided the thumbnail image of Bandelier National Monument.

Finally, we wish to thank everyone who took the time to read and provide comments on this document during its various stages of review.

ABSTRACT

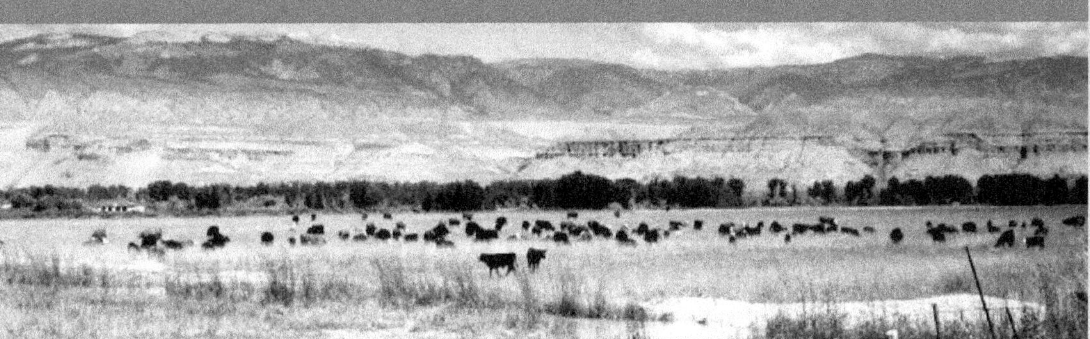

This report provides an assessment of the effects of climate change on U.S. agriculture, land resources, water resources, and biodiversity. It is one of a series of 21 Synthesis and Assessment Products (SAP) that are being produced under the auspices of the U.S. Climate Change Science Program (CCSP).

This SAP builds on an extensive scientific literature and series of recent assessments of the historical and potential impacts of climate change and climate variability on managed and unmanaged ecosystems and their constituent biota and processes. It discusses the nation's ability to identify, observe, and monitor the stresses that influence agriculture, land resources, water resources, and biodiversity, and evaluates the relative importance of these stresses and how they are likely to change in the future. It identifies changes in resource conditions that are now being observed, and examines whether these changes can be attributed in whole or part to climate change. The general time horizon for this report is from the recent past through the period 2030-2050, although longer-term results out to 2100 are also considered.

There is robust scientific consensus that human-induced climate change is occurring. Records of temperature and precipitation in the United States show trends consistent with the current state of global-scale understanding and observations of change. Observations also show that climate change is currently impacting the nation's ecosystems and services in significant ways, and those alterations are very likely to accelerate in the future, in some cases dramatically. Current observational capabilities are considered inadequate to fully understand and address the future scope and rate of change in all ecological sectors. Additionally, the complex interactions between change agents such as climate, land use alteration, and species invasion create dynamics that confound simple causal relationships and will severely complicate the development and assessment of mitigation and adaptation strategies.

Even under the most optimistic CO_2 emission scenarios, important changes in sea level, regional and super-regional temperatures, and precipitation patterns will have profound effects. Management of water resources will become more challenging. Increased incidence of disturbances such as forest fires, insect outbreaks, severe storms, and drought will command public attention and place increasing demands on management resources. Ecosystems are likely to be pushed increasingly into alternate states with the possible breakdown of traditional species relationships, such as pollinator/plant and predator/prey interactions, adding additional stresses and potential for system failures. Some agricultural and forest systems may experience near-term productivity increases, but over the long term, many such systems are likely to experience overall decreases in productivity that could result in economic losses, diminished ecosystem services, and the need for new, and in many cases significant, changes to management regimes.

Lead Authors: Peter Backlund, NCAR; Anthony Janetos, PNNL/Univ. Maryland; David Schimel, National Ecological Observatory Network
Contributing Authors: J. Hatfield, USDA ARS; M. Ryan, USDA Forest Service; S. Archer, Univ. Arizona; D. Lettenmaier, Univ. Washington

1 INTRODUCTION AND BACKGROUND

This report is an assessment of the effects of climate change on U.S. land resources, water resources, agriculture, and biodiversity. It is one of a series of 21 Synthesis and Assessment Products being produced under the auspices of the U.S. Climate Change Science Program (CCSP), which coordinates the climate change research activities of U.S. government agencies. The lead sponsor of this particular assessment product is the U.S. Department of Agriculture (USDA). The project was led and coordinated by the National Center for Atmospheric Research (NCAR).

This assessment is based on extensive review of the relevant scientific literature and measurements and data collected and published by U.S. government agencies. The team of authors includes experts in the fields of agriculture, biodiversity, and land and water resources – scientists and researchers from universities, national laboratories, non-government organizations, and government agencies. To generate this assessment of the effects of climate and climate change, the authors conducted an exhaustive review, analysis, and synthesis of the scientific literature, considering more than 1,000 separate publications.

Scope

The CCSP agencies agreed on the following set of topics for this assessment. Descriptions of the major findings in each of these sectors can be found in Section 4 of this Executive Summary.

- Agriculture: (a) cropping systems, (b) pasture and grazing lands, and (c) animal management
- Land Resources: (a) forests and (b) arid lands
- Water Resources: (a) quantity, availability, and accessibility and (b) quality
- Biodiversity: (a) species diversity and (b) rare and sensitive ecosystems

The CCSP also agreed on a set of questions to guide the assessment process. Answers to these questions can be found in Section 3 of this summary:

- What factors influencing agriculture, land resources, water resources, and biodiversity in the United States are sensitive to climate and climate change?
- How could changes in climate exacerbate or ameliorate stresses on agriculture, land resources, water resources, and biodiversity? What are the indicators of these stresses?
- What current and potential observation systems could be used to monitor these indicators?
- Can observation systems detect changes in agriculture, land resources, water resources, and biodiversity that are caused by climate change, as opposed to being driven by other causes?

Our charge from the CCSP was to address the specific topics and questions from the prospectus. This had several important consequences for this report. We were asked not to make recommendations and we have adhered to this request. Our document is not a plan for scientific or agency action, but rather an assessment and analysis of current scientific understanding of the topics defined by the CCSP. In addition, we were asked not to define and examine options for adapting to climate change impacts. This topic is addressed in a separate CCSP Synthesis and Assessment Product. Our authors view adaptation as a very important issue and recognize that adaptation options will certainly affect the ultimate severity of many climate change impacts. Our findings and conclusions are relevant to informed assessment of adaptation options, but we have not attempted that task in this report.

Time Horizon

Many studies of climate change have focused on the next 100 years. Model projections out to 2100 have become the de facto standard, as in the assessment reports produced by the Intergovernmental Panel on Climate Change (IPCC). This report has benefited greatly from such literature, but our main focus is on the recent past and the nearer-term future – the next 25 to 50 years. This period is within the planning horizon of many natural resources managers. Furthermore, the climate change that will occur during this period is relatively well understood. Much of this change will be caused by greenhouse gas emissions that have already happened. It is thus partially independent of current or planned emissions control measures and the large scenario uncertainty that affects longer-term projections. We report some results out to 100 years to frame our assessment, but we emphasize the coming decades.

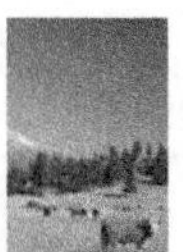

... our main focus is on the recent past and the nearer-term future – the next 25 to 50 years. This period is within the planning horizon of many natural resources managers. Furthermore, the climate change that will occur during this period is relatively well understood.

Ascribing Confidence to Findings

The authors have endeavored to use consistent terms, agreed to by the CCSP agencies, to describe their confidence in the findings and conclusions in this report, particularly when these involve projections of future conditions and accumulation of information from multiple sources. The use of these terms represents the judgment of the authors of this document; much of the underlying literature does not use such a lexicon and we have not retroactively applied this terminology to previous studies by other authors.

Climate Context

There is a robust scientific consensus that human-induced climate change is occurring. The Fourth Assessment Report (AR4) of the IPCC, the most comprehensive and up-to-date scientific assessment of this issue, states with "very high confidence" that human activities, such as fossil fuel burning and deforestation, have altered the global climate. During the 20th century, the global average surface temperature increased by about 0.6°C and global sea level increased by about 15 to 20 cm. Global precipitation over land increased about two percent during this same period. Looking ahead, human influences will continue to change Earth's climate throughout the 21st century. The IPCC AR4 projects that the global average temperature will rise another 1.1 to 5.4°C by 2100, depending on how much the atmospheric concentrations of greenhouse gases increase during this time. This temperature rise will result in continued increases in sea level and overall rainfall, changes in rainfall patterns and timing, and decline in snow cover, land ice, and sea ice extent. It is very likely that the Earth will experience a faster rate of climate change in the 21st century than seen in the last 10,000 years.

The United States warmed and became wetter overall during the 20th century, with changes varying by region. Parts of the South have cooled, while northern regions have warmed – Alaskan temperatures have increased by 2 to 4°C (more than four times the global average). Much of the eastern and southern United States now receive more precipitation than 100 years ago, while other areas, especially in the Southwest, receive less. The frequency and duration of heat waves has increased, there have been large declines in summer sea ice in the Arctic, and there is some evidence of increased frequency of heavy rainfalls. Observational and modeling results documented in the IPCC AR4 indicate that these trends are very likely to continue. Temperatures in the United States are very likely to increase by another 1°C to more than 4°C. The West and Southwest are likely to become drier, while the eastern United States is likely to experience increased rainfall. Heat waves are very likely to be hotter, longer, and more frequent, and heavy rainfall is likely to become more frequent.

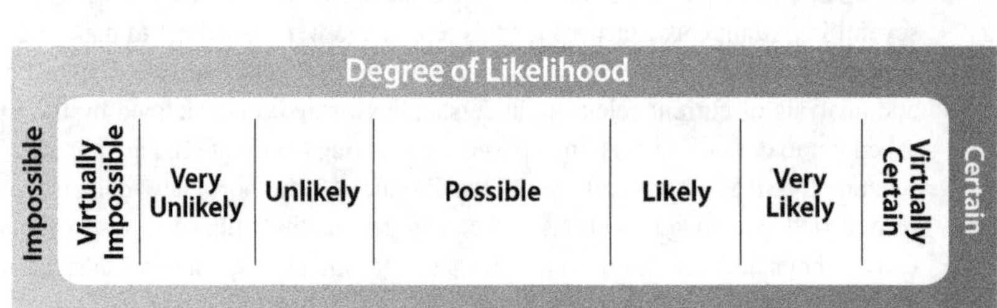

Figure 1 Language for Describing Confidence in Findings

2 OVERARCHING CONCLUSIONS

Climate changes – temperature increases, increasing CO_2 levels, and altered patterns of precipitation – are already affecting U.S. water resources, agriculture, land resources, and biodiversity *(very likely).* The literature reviewed for this assessment documents many examples of changes in these resources that are the direct result of variability and changes in the climate system, even after accounting for other factors. The number and frequency of forest fires and insect outbreaks are increasing in the interior West, the Southwest, and Alaska. Precipitation, streamflow, and stream temperatures are increasing in most of the continental United States. The western United States is experiencing reduced snowpack and earlier peaks in spring runoff. The growth of many crops and weeds is being stimulated. Migration of plant and animal species is changing the composition and structure of arid, polar, aquatic, coastal, and other ecosystems.

Climate change will continue to have significant effects on these resources over the next few decades and beyond *(very likely).* Warming is very likely to continue in the United States during the next 25 to 50 years, regardless of reductions in greenhouse gas emissions, due to emissions that have already occurred. U.S. ecosystems and natural resources are already being affected by climate system changes and variability. It is very likely that the magnitude and frequency of ecosystem changes will continue to increase during this period, and it is possible that they will accelerate. As temperature rises, crops will increasingly experience temperatures above the optimum for their reproductive development, and animal production of meat or dairy products will be impacted by temperature extremes. Management of Western reservoir systems is very likely to become more challenging as runoff patterns continue to change. Arid areas are very likely to experience increases erosion and fire risk. In arid ecosystems that have not coevolved with a fire cycle, the probability of loss of iconic, charismatic megaflora such as Saguaro cacti and Joshua trees will greatly increase.

Many other stresses and disturbances are also affecting these resources *(very likely).* For many of the changes documented in this assessment, there are multiple environmental drivers – land use change, nitrogen cycle changes, point and nonpoint source pollution, wildfires, invasive species – that are also changing. Atmospheric deposition of biologically available nitrogen compounds continues to be an important issue, along with persistent ozone pollution in many parts of the country. It is very likely that these additional atmospheric effects cause biological and ecological changes that interact with changes in the physical climate system. In addition, land cover and land use patterns are changing, e.g., the increasing fragmentation of U.S. forests as exurban development spreads to previously undeveloped areas, further raising fire risk and compounding the effects of summer drought, pests, and warmer winters. There are several dramatic examples of extensive spread of invasive species throughout rangeland and semiarid ecosystems in western states, and indeed throughout the United States. It is likely that the spread of these invasive species, which often change ecosystem processes, will exacerbate the risks from climate change alone. For example, in some cases invasive species increase fire risk and decrease forage quality.

Climate change impacts on ecosystems will affect the services that ecosystems provide, such as cleaning water and removing carbon from the atmosphere *(very likely)*, but we do not yet possess sufficient understanding to project the timing, magnitude, and consequences of many of these effects. One of the main reasons to assess changes in ecosystems is to understand the consequences of those changes for the delivery of services that our society values. There are many analyses of the impacts of climate change on individual species and ecosystems in the scientific literature, but there is not yet adequate integrated analysis of how climate change could affect ecosystem services. A comprehensive understanding of impacts on these services will only be possible through quantification of anticipated alterations in ecosystem function and productivity. As described by the Millennium Ecosystem Assessment, some products of ecosystems, such as food and fiber, are priced and traded in markets.

Climate changes – temperature increases, increasing CO_2 levels, and altered patterns of precipitation – are already affecting U.S. water resources, agriculture, land resources, and biodiversity.

Others, such as carbon sequestration capacity, are only beginning to be understood and traded in markets. Still others, such as the regulation of water quality and quantity and the maintenance of soil fertility, while not priced and traded, are valuable nonetheless. Although these points are recognized and accepted in the scientific literature and increasingly among decision makers, there is no analysis specifically devoted to understanding changes in ecosystem services in the United States from climate change and associated stresses. It is possible to make some generalizations from the literature on the physical changes in ecosystems, but interpreting what these changes mean for services provided by ecosystems is very challenging and can only be done for a limited number of cases. This is a significant gap in our knowledge base.

Existing monitoring systems, while useful for many purposes, are not optimized for detecting the impacts of climate change on ecosystems. There are many operational and research monitoring systems in the United States that are useful for studying the consequences of climate change on ecosystems and natural resources. These range from the resource- and species-specific monitoring systems that land-management agencies depend on to research networks, such as the Long-Term Ecological Research (LTER) sites, which the scientific community uses to understand ecosystem processes. All of the existing monitoring systems, however, have been put in place for other reasons, and none have been optimized specifically for detecting the effects and consequences of climate change. As a result, it is likely that only the largest and most visible consequences of climate change are being detected. In some cases, marginal changes and improvements to existing observing efforts, such as USDA snow and soil moisture measurement programs, could provide valuable new data detection of climate impacts. But more refined analysis and/or monitoring systems designed specifically for detecting climate change effects would provide more detailed and complete information and probably capture a range of more subtle impacts. Such systems, in turn, might lead to early-warning systems and more accurate forecasts of potential future changes. But it must be emphasized that improved observations, while needed, are not sufficient for improving understanding of ecological impacts of climate change. Ongoing, integrated and systematic analysis of existing and new observations could enable forecasting of ecological change, thus garnering greater value from observational activities, and contribute to more effective evaluation of measurement needs. This issue is addressed in greater detail in Section 3.

Existing monitoring systems, while useful for many purposes, are not optimized for detecting the impacts of climate change on ecosystems.

3 KEY QUESTIONS AND ANSWERS

This section presents a set of answers to the guiding questions posed by the CCSP agencies, derived from the longer chapters that follow this Executive Summary.

What factors influencing agriculture, land resources, water resources, and biodiversity in the United States are sensitive to climate and climate change? Climate change affects average temperatures and temperature extremes; timing and geographical patterns of precipitation; snowmelt, runoff, evaporation, and soil moisture; the frequency of disturbances, such as drought, insect and disease outbreaks, severe storms, and forest fires; atmospheric composition and air quality; and patterns of human settlement and land use change. Thus, climate change leads to myriad direct and indirect effects on U.S. ecosystems. Warming temperatures have led to effects as diverse as altered timing of bird migrations, increased evaporation, and longer growing seasons for wild and domestic plant species. Increased temperatures often lead to a complex mix of effects. Warmer summer temperatures in the western United States have led to longer forest growing seasons but have also increased summer drought stress, vulnerability to insect pests, and fire hazard. Changes to precipitation and the size of storms affect plant-available moisture, snowpack and snowmelt, streamflow, flood hazard, and water quality.

How could changes in climate exacerbate or ameliorate stresses on agriculture, land resources, water resources, and biodiversity? What are the indicators of these stresses? Ecosystems and their services (land and water resources, agriculture, biodiversity) experience a wide range of stresses, including pests and pathogens, invasive species, air pollution, extreme events, wildfires and floods. Climate change can cause or exacerbate direct stress through high temperatures, reduced water availability, and altered frequency of extreme events and severe storms. It can ameliorate stress through warmer springs and longer growing seasons, which, assuming adequate moisture, can increase agricultural and forest productivity. Climate change can also modify the frequency and severity of stresses. For example, increased minimum temperatures and warmer springs extend the range and lifetime of many pests that stress trees and crops. Higher temperatures and/or decreased precipitation increase drought stress on wild and crop plants, animals and humans. Reduced water availability can lead to increased withdrawals from rivers, reservoirs, and groundwater, with consequent effects on water quality, stream ecosystems, and human health.

What current and potential observation systems could be used to monitor these indicators? A wide range of observing systems within the United States provides information on environmental stress and ecological responses. Key systems include National Aeronautics and Space Administration (NASA) research satellites, operational satellites and ground-based observing networks from the National Oceanic and Atmospheric Administration (NOAA) in the Department of Commerce, Department of Agriculture (USDA) forest and agricultural survey and inventory systems, Department of Interior/U.S. Geological Survey (USGS) stream gauge networks, Environmental Protection Agency (EPA) and state-supported water quality observing systems, the Department of Energy (DOE) Ameriflux network, and the LTER network and the proposed National Ecological Observing Network (NEON) sponsored by the National Science Foundation (NSF). However, many key biological and physical indicators are not currently monitored, are monitored haphazardly or with incomplete spatial coverage, or are monitored only in some regions. In addition, the information from these disparate networks is not well integrated. Almost all of the networks were originally instituted for specific purposes unrelated to climate change and cannot necessarily be adapted to address these new questions.

Climate change presents new challenges for operational management. Understanding climate impacts requires monitoring both many aspects of climate and a wide range of biological and physical responses. Putting climate change impacts in the context of multiple stresses and forecasting future services requires an integrated analysis. Beyond the problems of integrating the data sets, the nation has limited operational capability for integrated ecological monitoring, analyses, and forecasting. A few centers exist, aimed at specific questions and/or regions, but no coordinating agency or center has the mission to conduct integrated environmental analysis and assessment by pulling this information together.

Warming temperatures have led to effects as diverse as altered timing of bird migrations, increased evaporation, and longer growing seasons for wild and domestic plant species.

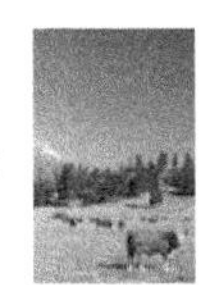

Operational weather and climate forecasting provides an analogy. Weather-relevant observations are collected in many ways, ranging from surface observations through radiosondes to operational and research satellites. These data are used at a handful of university, federal, and private centers as the basis for analysis, understanding, and forecasting of weather through highly integrative analyses blending data and models. This operational activity requires substantial infrastructure and depends on federal, university, and private sector research for continual improvement. By contrast, no such integrative analysis of comprehensive ecological information is carried out, although the scientific understanding and societal needs have probably reached the level where an integrative and operational approach is both feasible and desirable.

Can observation systems detect changes in agriculture, land resources, water resources, and biodiversity that are caused by climate change, as opposed to being driven by other causes? In general, the current suite of observing systems is reasonably able overall to monitor ecosystem change and health in the United States, but neither the observing systems nor the current state of scientific understanding is adequate to rigorously quantify climate contributions to ecological change and separate these from other influences. Monitoring systems for measuring long-term response of **agriculture** to climate and other stresses are numerous, but integration across these systems is limited. There is no coordinated national network for monitoring changes in **land resources** associated with climate change, most disturbances, such as storms, insects, and diseases, and changes in land cover/land use. No aspect of the current hydrologic observing system was designed specifically to detect climate change or its effects on **water resources**. The monitoring systems that have been used to evaluate the relationship between changes in the physical climate system and **biological diversity** were likewise not designed with climate variability or change in mind.

Climate change is likely to lead to a northern migration of weeds.

So for the moment, there is no viable alternative to using the existing systems for identifying climate change and its impacts on U.S. agriculture, land resources, water resources, and biodiversity, even though these systems were not originally designed for this purpose. There has obviously been some considerable success so far in doing so, but there is limited confidence that the existing systems provide a true early warning system capable of identifying potential impacts in advance. The authors of this report also have very limited confidence in the ability of current observation and monitoring systems to provide the information needed to evaluate the effectiveness of actions that are taken to mitigate or adapt to climate change impacts. Furthermore, we emphasize that improvements in observations and monitoring of ecosystems, while desirable, are not sufficient by themselves for increasing our understanding of climate change impacts. Experiments that directly manipulate climate and observe impacts are critical for developing more detailed information on the interactions of climate and ecosystems, attributing impacts to climate, differentiating climate impacts from other stresses, and designing and evaluating response strategies. Much of our understanding of the direct effects of temperature, elevated CO_2, ozone, precipitation, and nitrogen deposition has come from manipulative experiments. Institutional support for such experiments is a concern.

4 SECTORAL FINDINGS

Agriculture

The broad subtopics considered in this section are cropping systems, pasture and grazing lands, and animal management. The many U.S. crops and livestock varieties (valued at about $200 billion in 2002) are grown in diverse climates, regions, and soils. No matter the region, however, weather and climate factors such as temperature, precipitation, CO_2 concentrations, and water availability directly impact the health and well-being of plants, pasture, rangeland, and livestock. For any agricultural commodity, variation in yield between years is related to growing-season weather; weather also influences insects, disease, and weeds, which in turn affect agricultural production.

- With increased CO_2 and temperature, the life cycle of grain and oilseed crops will likely progress more rapidly. But, as temperature rises, these crops will increasingly begin to experience failure, especially if climate variability increases and precipitation lessens or becomes more variable.

- The marketable yield of many horticultural crops – e.g., tomatoes, onions, fruits – is very likely to be more sensitive to climate change than grain and oilseed crops.

- Climate change is likely to lead to a northern migration of weeds. Many weeds respond more positively to increasing CO_2 than most cash crops, particularly C3 "invasive" weeds. Recent research also suggests that glyphosate, the most widely used herbicide in the United States, loses its efficacy on weeds grown at the increased CO_2 levels likely in the coming decades.

- Disease pressure on crops and domestic animals will likely increase with earlier springs and warmer winters, which will allow proliferation and higher survival rates of pathogens and parasites. Regional variation in warming and changes in rainfall will also affect spatial and temporal distribution of disease.

- Projected increases in temperature and a lengthening of the growing season will likely extend forage production into late fall and early spring, thereby decreasing need for winter season forage reserves. However, these benefits will very likely be affected by regional variations in water availability.

- Climate change-induced shifts in plant species are already under way in rangelands. Establishment of perennial herbaceous species is reducing soil water availability early in the growing season. Shifts in plant productivity and type will likely also have significant impact on livestock operations.

- Higher temperatures will very likely reduce livestock production during the summer season, but these losses will very likely be partially offset by warmer temperatures during the winter season. For ruminants, current management systems generally do not provide shelter to buffer the adverse effects of changing climate; such protection is more frequently available for non-ruminants (e.g., swine and poultry).

- Monitoring systems for measuring long-term response of agricultural lands are numerous, but integration across these systems is limited. Existing state-and-transition models could be expanded to incorporate knowledge of how agricultural lands and products respond to global change; integration of such models with existing monitoring efforts and plant developmental data bases could provide cost-effective strategies that both enhance knowledge of regional climate change impacts and offer ecosystem management options. In addition, at present, there are no easy and reliable means to accurately ascertain the mineral and carbon state of agricultural lands, particularly over large areas; a fairly low-cost method of monitoring biogeochemical response to global change would be to sample ecologically important target species in different ecosystems.

Land Resources

The broad subtopics considered in this section are forest lands and arid lands. Climate strongly influences forest productivity, species composition, and the frequency and magnitude of disturbances that impact forests. The effect of climate change on disturbances such as forest fire, insect outbreaks, storms, and severe drought will command public attention and place increasing demands on management resources. Disturbance and land use will control the response of arid lands to climate change. Many plants and animals in arid ecosystems are near their physiological limits for tolerating temperature and water stress and even slight changes in stress will have significant consequences. In the near term, fire effects will trump climate effects on ecosystem structure and function.

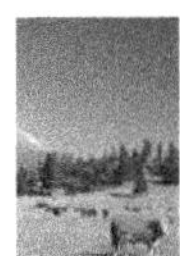

- Climate change has very likely increased the size and number of forest fires, insect outbreaks, and tree mortality in the interior West, the Southwest, and Alaska, and will continue to do so.

- Rising CO_2 will very likely increase photosynthesis for forests, but this increase will likely only enhance wood production in young forests on fertile soils.

- Nitrogen deposition and warmer temperatures have very likely increased forest growth where adequate water is available and will continue to do so in the near future.

- The combined effects of rising temperatures and CO_2, nitrogen deposition, ozone, and forest disturbance on soil processes and soil carbon storage remains unclear.

- Higher temperatures, increased drought, and more intense thunderstorms will very likely increase erosion and promote invasion of exotic grass species in arid lands.

- Climate change in arid lands will create physical conditions conducive to wildfire, and the proliferation of exotic grasses will provide fuel, thus causing fire frequencies to increase in a self-reinforcing fashion.

Climate change has very likely increased the size and number of forest fires, insect outbreaks, and tree mortality in the interior West, the Southwest, and Alaska, and will continue to do so.

- In arid regions where ecosystems have not coevolved with a fire cycle, the probability of loss of iconic, charismatic megaflora such as saguaro cacti and Joshua trees is very likely.

- Arid lands very likely do not have a large capacity to absorb CO_2 from the atmosphere and will likely lose carbon as climate-induced disturbance increases.

- River and riparian ecosystems in arid lands will very likely be negatively impacted by decreased streamflow, increased water removal, and greater competition from non-native species.

- Changes in temperature and precipitation will very likely decrease the cover of vegetation that protects the ground surface from wind and water erosion.

- Current observing systems do not easily lend themselves to monitoring change associated with disturbance and alteration of land cover and land use, and distinguishing such changes from those driven by climate change. Adequately distinguishing climate change influences is aided by the collection of data at certain spatial and temporal resolutions, as well as supporting ground truth measurements.

Stream temperatures are likely to increase as the climate warms, and are very likely to have both direct and indirect effects on aquatic ecosystems.

Water Resources

The broad subtopics considered in this section are water quantity and water quality. Plants, animals, natural and managed ecosystems, and human settlements are susceptible to variations in the storage, fluxes, and quality of water, all of which are sensitive to climate change. The effects of climate on the nation's water storage capabilities and hydrologic functions will have significant implications for water management and planning as variability in natural processes increases. Although U.S. water management practices are generally quite advanced, particularly in the West, the reliance on past conditions as the foundation for current and future planning and practice will no longer be tenable as climate change and variability increasingly create conditions well outside of historical parameters and erode predictability.

- Most of the United States experienced increases in precipitation and streamflow and decreases in drought during the second half of the 20th century. It is likely that these trends are due to a combination of decadal-scale variability and long-term change.

- Consistent with streamflow and precipitation observations, most of the continental United States experienced reductions in drought severity and duration over the 20th century. However, there is some indication of increased drought severity and duration in the western and southwestern United States.

- There is a trend toward reduced mountain snowpack and earlier spring snowmelt runoff peaks across much of the western United States. This trend is very likely attributable at least in part to long-term warming, although some part may have been played by decadal-scale variability, including a shift in the phase of the Pacific Decadal Oscillation in the late 1970s. Where earlier snowmelt peaks and reduced summer and fall low flows have already been detected, continuing shifts in this direction are very likely and may have substantial impacts on the performance of reservoir systems.

- Water quality is sensitive to both increased water temperatures and changes in precipitation. However, most water quality changes observed so far across the continental United States are likely attributable to causes other than climate change.

- Stream temperatures are likely to increase as the climate warms, and are very likely to have both direct and indirect effects on aquatic ecosystems. Changes in temperature will be most evident during low flow periods, when they are of greatest concern. Stream temperature increases have already begun to be detected across some of the United States, although a comprehensive analysis similar to those reviewed for streamflow trends has yet to be conducted.

- A suite of climate simulations conducted for the IPCC AR4 show that the United States may experience increased runoff in eastern regions, gradually transitioning to little change in the Missouri and lower Mississippi, to substantial decreases in annual runoff in the interior of the west (Colorado and Great Basin).

- Trends toward increased water use efficiency are likely to continue in the coming decades. Pressures for reallocation of water will be greatest in areas of highest population growth, such as the Southwest. Declining per capita (and, for some cases, total) water consumption will help mitigate the impacts of climate change on water resources.

- Essentially no aspect of the current hydrologic observing system was designed specifically to detect climate change or its effects on water resources. Recent efforts have the potential to make improvements, although many systems remain technologically obsolete, incompatible, and/or have significant data collection gaps in their operational and maintenance structures. As a result, many of the data are fragmented, poorly integrated, and unable to meet the predictive challenges of a rapidly changing climate.

Biodiversity

The broad subtopics considered in this section are species diversity and rare and sensitive ecosystems. Biodiversity, the variation of life at the genetic, species, and ecosystem levels of biological organization, is the fundamental building block of the services that ecosystems deliver to human societies. It is intrinsically important both because of its contribution to the functioning of ecosystems, and because it is difficult or impossible to recover or replace, once it is eroded. Climate change is affecting U.S. biodiversity and ecosystems, including changes in growing season, phenology, primary production, and species distributions and diversity. It is very likely that climate change will increase in importance as a driver for changes in biodiversity over the next several decades, although for most ecosystems it is not currently the largest driver of change.

- There has been a significant lengthening of the growing season and increase in net primary productivity (NPP) in the higher latitudes of North America. Over the last 19 years, global satellite data indicate an earlier onset of spring across the temperate latitudes by 10 to 14 days.

- In an analysis of 866 peer-reviewed papers exploring the ecological consequences of climate change, nearly 60 percent of the 1598 species studied exhibited shifts in their distributions and/or phenologies over the 20- and 140-year time frame. Analyses of field-based phenological responses have reported shifts as great as 5.1 days per decade, with an average of 2.3 days per decade across all species.

- Subtropical and tropical corals in shallow waters have already suffered major bleaching events that are clearly driven by increases in sea surface temperatures. Increases in ocean acidity, which are a direct consequence of increases in atmospheric carbon dioxide, are calculated to have the potential for serious negative consequences for corals.

- The rapid rates of warming in the Arctic observed in recent decades, and projected for at least the next century, are dramatically reducing the snow and ice covers that provide denning and foraging habitat for polar bears.

- There are other possible, and even probable, impacts and changes in biodiversity (e.g., disruption of the relationships between pollinators, such as bees, and flowering plants), for which we do not yet have a substantial observational database. However, we cannot conclude that the lack of complete observations is evidence that changes are not occurring.

- It is difficult to pinpoint changes in ecosystem services that are specifically related to changes in biological diversity in the United States. A specific assessment of changes in ecosystem services for the United States as a consequence of changes in climate or other drivers of change has not been done.

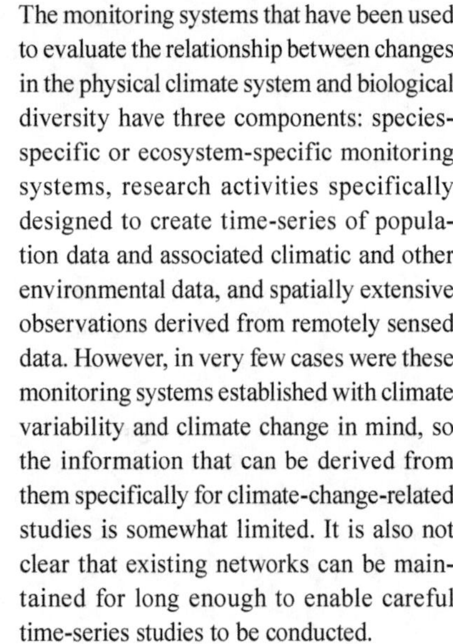

- The monitoring systems that have been used to evaluate the relationship between changes in the physical climate system and biological diversity have three components: species-specific or ecosystem-specific monitoring systems, research activities specifically designed to create time-series of population data and associated climatic and other environmental data, and spatially extensive observations derived from remotely sensed data. However, in very few cases were these monitoring systems established with climate variability and climate change in mind, so the information that can be derived from them specifically for climate-change-related studies is somewhat limited. It is also not clear that existing networks can be maintained for long enough to enable careful time-series studies to be conducted.

It is also not clear that existing networks can be maintained for long enough to enable careful time-series studies to be conducted.

Introduction

Lead Authors: Peter Backlund, NCAR; Anthony Janetos, PNNL/ Univ. Maryland; David Schimel, National Ecological Observatory Network

Contributing Authors: J. Hatfield, USDA ARS; M. Ryan, USDA Forest Service; S. Archer, Univ. Arizona; D. Lettenmaier, Univ. Washington

This report is an assessment of the effects of climate change on U.S. land resources, water resources, agriculture, and biodiversity. It is based on extensive examination of the relevant scientific literature, and is one of a series of 21 Synthesis and Assessment Products that are being produced under the auspices of the U.S. Climate Change Science Program (CCSP). The lead sponsor of this particular assessment product is the U.S. Department of Agriculture.

The purpose of this assessment and more broadly, of all the CCSP Scientific Assessment Products (SAPs) is to integrate existing scientific knowledge on issues and questions related to climate change that are important to policy and decision makers. The assessments are meant to support informed discussion and decision makers by a wide audience of potential stakeholders, including, for example, federal and state land managers, private citizens, private industry, and non-governmental organizations. The scientific research community is also an important stakeholder, as an additionally important feature of the SAPs is to inform decision making about the future directions and priorities of the federal scientific research programs by pointing out where there are important knowledge gaps. It is a goal of the SAPs that they not only be useful and informative scientific documents, but that they are also accessible and understandable to a more general, well-informed public audience. The team of authors was selected by the agencies after asking for public comment, and it includes scientists and researchers from universities, non-governmental organizations, and government agencies, coordinated by the National Center for Atmospheric Research (NCAR). The team has reviewed hundreds of peer-reviewed papers, guided by a prospectus agreed upon by the CCSP agencies (see Appendix C).

Intent of this Report

Strong scientific consensus highlights that anthropogenic effects of climate change are already occurring and will be substantial (IPCC). A recent U.S. government analysis (GAO) shows that that U.S. land management agencies are not prepared to address this issue. This analysis also highlights the need for assessment of climate change impacts on U.S. natural resources and assessment of monitoring systems needed to provide information to support effective decision making about mitigation and adaptation in periods of potentially rapid change. This report addresses this issue by providing an assessment specific to U.S. natural resources in agriculture, land resources, water resources, and biodiversity, and by assessing the ability of existing monitoring systems to aid decision making. The report documents that (1) numerous, substantial impacts of climate change on U.S. natural resources are already occurring, (2) that these are likely to become exacerbated as warming progresses, and (3) that existing monitoring systems are insufficient to address this issue.

This report is an assessment of the effects of climate change on U.S. land resources, water resources, agriculture, and biodiversity.

Strong scientific consensus highlights that anthropogenic effects of climate change are already occurring and will be substantial (IPCC).

Scope of this Report

The overall scope of the report has been determined by agreement among the CCSP agencies. Important features of the scope include the topics to be addressed:

Agriculture
- Cropping systems
- Pasture and grazing lands
- Animal management

Land Resources
- Forests
- Arid lands

Water Resources
- Quantity, availability, and accessibility
- Quality

Biodiversity
- Species diversity
- Rare and sensitive ecosystems

Equally important are the elements of the climate change problem that are not addressed by this report. Some key issues, such as climate impacts on freshwater ecosystems, did not receive extensive attention. This is mainly due to timing and length constraints – it does not represent a judgment on the part of the authors that such impacts are not important. In addition, while the report was specifically asked to address issues of climate impacts, it was not asked to address the challenge of what adaptation and management strategies exist, their potential effectiveness, and potential costs. While these topics are acknowledged to be important in the scientific literature (Parsons et al.; Granger Morgan et al.; U.S. National Assessment), they are the subject of another of the CCSP Synthesis and Assessment Products (4.4). Nevertheless, the information synthesized in this report is meant to be of use to stakeholders concerned with planning, undertaking, and evaluating the effectiveness of adaptation options.

This report also deals almost exclusively with biological, ecological, and physical impacts of climate change. With the exception of some information in agricultural systems, market impacts on natural resources are not discussed, nor are the potential costs or benefits of changes in the management of natural resources. We recognize that this leaves an incomplete picture of the overall impacts of climate change on those resources that the nation considers significant. Again, however, further consideration of economic effects requires a firm foundation in understanding the biological, ecological, and physical impacts.

Guiding Questions for this Report

This synthesis and assessment report builds on an extensive scientific literature and series of recent assessments of the historical and potential impacts of climate change and climate variability on managed and unmanaged ecosystems and their constituent biota and processes. It discusses the nation's ability to identify, observe, and monitor the stresses that influence agriculture, land resources, water resources, and biodiversity, and evaluates the relative importance of these stresses and how they are likely to change in the future. It identifies changes in resource conditions that are now being observed, and examines whether these changes can be attributed in whole or part to climate change. It also highlights changes in resource conditions that recent scientific studies suggest are most likely to occur in response to climate change, and when and where to look for these changes. The assessment is guided by five overarching questions:

What factors influencing agriculture, land resources, water resources, and biodiversity in the United States are sensitive to climate and climate change?

How could changes in climate exacerbate or ameliorate stresses on agriculture, land resources, water resources, and biodiversity?

What are the indicators of these stresses?

What current and potential observation systems could be used to monitor these indicators?

Can observation systems detect changes in agriculture, land resources, water resources, and biodiversity that are caused by climate change, as opposed to being driven by other causal activities?

Ascribing Confidence to Findings

The authors of this document have used language agreed to by the CCSP agencies to describe their confidence in findings that project future climate changes and impacts, as shown in Figure 1.1. The intent is to use a limited set of terms in a systematic and consistent fashion to communicate clearly with readers. The use of these terms represents the qualitative judgment of the authors of this document; much of the underlying literature does not use such a lexicon. Unless explicitly describing a formal statistical analysis, the use of these terms by the authors of this assessment should be treated as a statement of their expert judgment in the confidence of our findings and conclusions. There are cases where we have not applied the agreed terminology because we felt it was not an accurate representation of work published by others.

Time Horizon for this Report

Climate change is a long-term issue and will affect the world for the foreseeable future. Many studies of climate change have focused on the next 100 years and model projections out to 2100 have become the de facto standard, as reported in the assessment reports produced by the Intergovernmental Panel on Climate Change (IPCC) and many other documents. In this report, however, the focus is on the mid-term future. Key results are reported out to 100 years to frame the report, but the emphasis is on the next 25-50 years.

This mid-term focus is chosen for several reasons. First, for many natural resources, planning and management activities already address these time scales through the development of long-lived infrastructure, forest rotations, and other significant investments. Second, we will experience significant warming from greenhouse gas emissions that have already occurred, regardless of the effectiveness of any emissions reduction activities. And most emission scenarios for the next few decades do not significantly diverge from each other because it will take decades to make major changes in energy infrastructure in the U.S. and other nations. As a result, high- and low-emission scenarios only begin to separate strongly in the 2030s-2050s. As emissions diverge, so do climate projections, and uncertainty about future climates rapidly becomes more pronounced. Averaging over climate models, a rate of a few tenths of a degree per decade can be assumed likely for the next two to four decades.

Climate change is a long-term issue and will affect the world for the foreseeable future.

Global Climate Context

There is a robust scientific consensus that human-induced climate change is occurring. The recently released Fourth Assessment Report of the IPCC (IPCC AR4) states with "very high confidence," that human activity has caused the global climate to warm. Many well-documented observations show that fossil fuel burning, deforestation, and other industrial processes are rapidly increasing the atmospheric concentrations of CO_2 and other greenhouse gases. The IPCC report describes an increasing body of observations and modeling results, summarized below, which show that these changes in atmospheric composition are changing the global climate and beginning to affect terrestrial and marine ecosystems.

The global-average surface temperature increased by about 0.6°C over the 20th century. Global sea level increased by about 15-20 cm during this period.

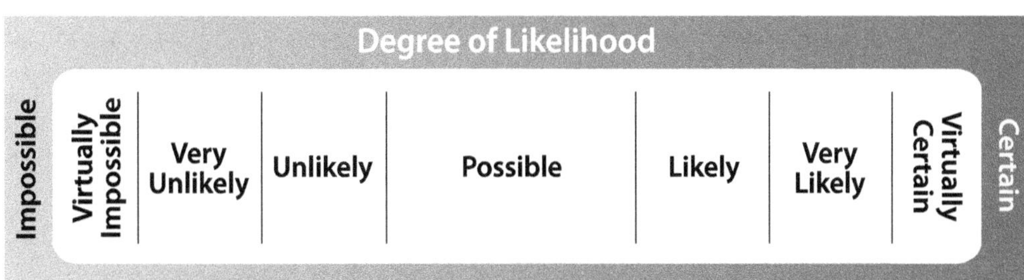

Figure 1.1 Language for discussing confidence in findings.

Observations since 1961 show that the average temperature of the global ocean has increased to depths of at least 3,000 meters, and that the ocean has been absorbing more than 80 percent of the heat added to the climate system.

Long-term temperature records from ice sheets, glaciers, lake sediments, corals, tree rings, and historical documents show that 1995-2004 was the warmest decade worldwide in the last 1-2,000 years. Nine of the 10 warmest years on record occurred since 1996.

Global precipitation over land increased about 2 percent over the last century, with considerable variability by region (Northern Hemisphere precipitation increased by about 5 to 10 percent during this time, while West Africa and other areas experienced decreases).

Mountain glaciers are melting worldwide, Greenland's ice sheet is melting, the extent and thickness of Arctic sea ice is declining, and lakes and rivers freeze later in the fall and melt earlier in the spring. The growing season

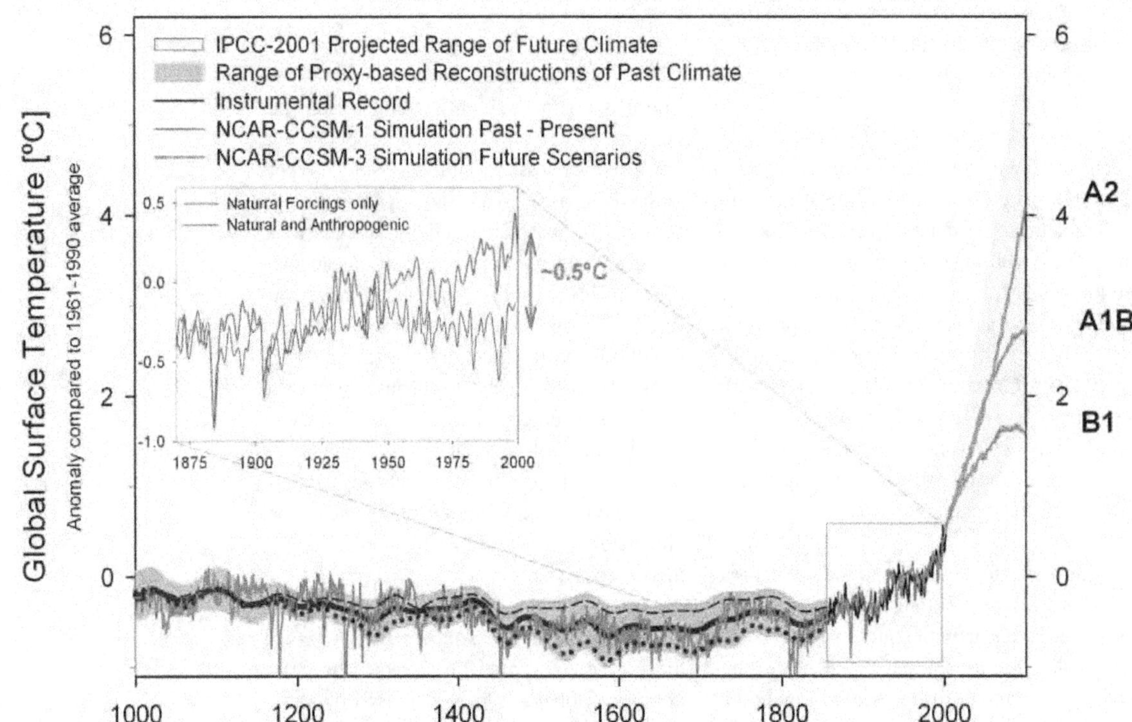

Figure 1.2 Temperatures of the Last Millennium and the Next Century. The effects of historical reconstructions of solar variability and volcanic eruptions were modeled using an NCAR climate model and compared to several reconstructions of past temperatures. The model reproduces many temperature variations of the past 1,000 years, and shows that solar and volcanic forcing has been a considerable impact on past climate. When only 20th century solar and volcanic data are used, the model fails to reproduce the recent warming, but captures it well when greenhouse gases are included.

in the Northern Hemisphere has lengthened by about 1 to 4 days per decade in the last 40 years, especially at high latitudes.

The ranges of migrating birds, and some fish and insect species are changing. Tropical regions are losing animal species, especially amphibians, to warming and drying.

Although much (but not all) of recent increases have been in nighttime maximum temperatures rather than daytime maxima, the expectation for the future is that daytime temperatures will become increasingly responsible for higher overall average temperatures.

Change and variability are persistent features of climate, and the anthropogenic climate change now occurring follows millennia of strictly natural climate changes and variability. Paleoclimate records, including natural archives in tree rings, corals, and glacial ice, now show that the climate of the last millennium has varied significantly with hemispheric-to-global changes in temperature and precipitation resulting from the effects of the sun, volcanoes, and the climate system's natural variability (Ammann et al. 2007). The anthropogenic changes now being observed are superimposed on this longer-term, ongoing variability, some of which can be reproduced by today's advanced climate models. Importantly, the model that captures the past thousand years of global temperature patterns successfully (Figure 1.2) using only solar and volcanic inputs does not accurately simulate the 20th century's actual, observed climate unless greenhouse gases are factored in (Ammann et al. 2007).

It is also clear that human influences will continue to alter Earth's climate throughout the 21st century. The IPCC AR4 describes a large body of modeling results, which show that changes in atmospheric composition will result in further increases in global average temperature and sea level, and continued declines in snow cover, land ice, and sea ice extent. Global average rainfall, variability of rainfall, and heavy rainfall events are projected to increase. Heat waves in Europe, North America, and other regions will become more intense, more frequent, and longer lasting. It is very likely that the rate of climate change in the 21st century will be faster than that seen in the last 10,000 years. The IPCC AR4 contains projections of the temperature increases that would result from a variety of different emissions scenarios:

If atmospheric concentration of CO_2 increases to about 550 parts per million (ppm), global average surface temperature would likely increase by about 1.1-2.9°C by 2100.

If atmospheric concentration of CO_2 increases to about 700 ppm, global average surface temperature would likely increase about 1.7-4.4°C by 2100.

If atmospheric concentration of CO_2 increases to about 800 ppm, global average surface temperature would likely increase about 2.0-5.4°C by 2100.

Even if atmospheric concentration of CO_2 were stabilized at today's concentrations of about 380 ppm, global average surface temperatures would likely continue to increase by another 0.3–0.9°C by 2100.

U.S. Climate Context

Records of temperature and precipitation in the United States show trends that are consistent with the global-scale changes discussed above. The United States has warmed significantly overall, but change varies by region (Figure 1.3). Some parts of the United States have cooled, but Alaska and other northern regions have warmed significantly. Much of the eastern and southern U.S. now receive more precipitation than 100 years ago, while other areas, especially in the Southwest, now receive less (Figure 1.4).

The United States has warmed significantly overall, but change varies by region.

The scenarios of global temperature change discussed in the global climate context section above would result in large changes in U.S. temperatures and precipitation, with considerable variation by region. Figure 1.5, which is based on multiple model simulations, show how IPCC global scenario A1B, generally considered a moderate emissions growth scenario, would affect U.S. temperatures and precipitation by 2030. The projected temperature increases range

Extreme climate conditions, such as droughts, heavy rainfall, snow events, and heat waves affect individual species and ecosystems structure and function.

from approximately 1°C in the southeastern United States, to more than 2°C in Alaska and northern Canada, with other parts of North America having intermediate values.

Although precipitation increases are anticipated for large areas of the U.S., it is important to note this does not necessarily translate into more available moisture for biological and ecological processes. Higher temperatures increase evapotranspirative losses to the atmosphere, and the relative balance of the two factors on average in the U.S. leads to less moisture in soils and surface waters for organisms or ecosystems to utilize both now and in the future.

The average temperature and precipitation are not the only factors that affect ecosystems. Extreme climate conditions, such as droughts, heavy rainfall, snow events, and heat waves affect individual species and ecosystems structure and function. Change in the incidence of extreme events could thus have major impacts on U.S. ecosystems and must be considered when assessing vulnerability to and impacts of climate change. Figure 1.6 shows how the IPCC A1B scenario will change the incidence of heat waves and warm nights by approximately 2030. Figure 1.7 shows projected changes in frost days and growing season.

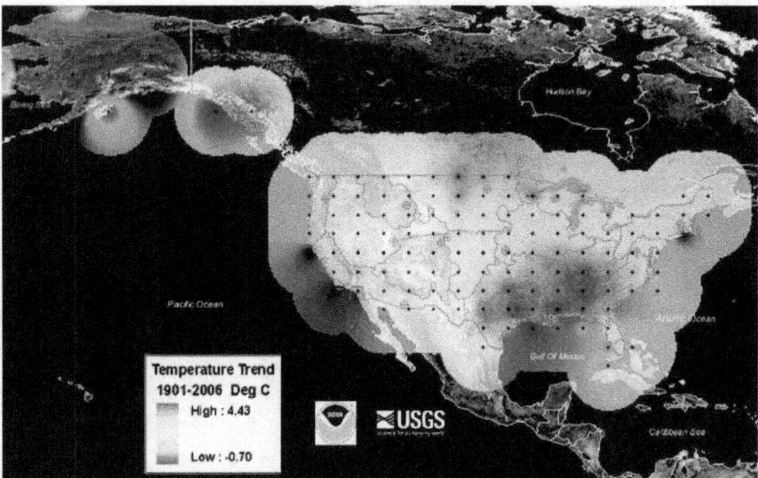

Figure 1.3 Mapped trends in temperature across the lower 48 states and Alaska. These data, which show the regional pattern of U.S. warming, are averaged from weather stations across the country using stations that have as complete, consistent, and high quality records as can be found. Courtesy of NOAA's National Climate Data Center and the U.S. Geological Survey.

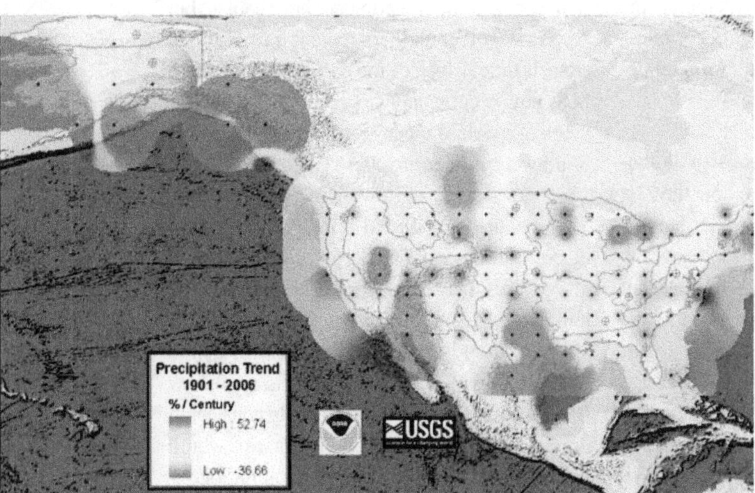

Figure 1.4 Precipitation changes over the past century from the same weather stations as for temperature. The changes are shown as percentage changes from the long-term average. Courtesy of NOAA's National Climate Data Center and the U.S. Geological Survey.

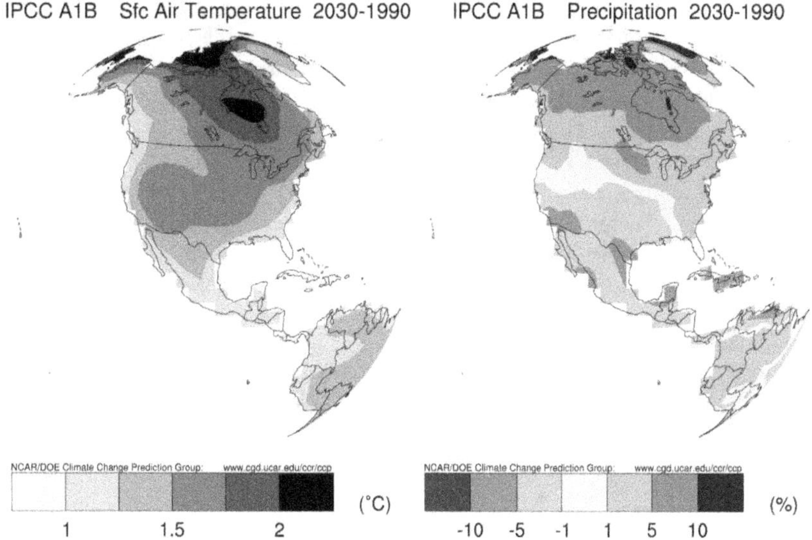

Figure 1.5 U.S. Temperature and Precipitation Changes by 2030. This figure shows how U.S. temperatures and precipitation would change by 2030 under IPCC emissions scenario A1B, which would increase the atmospheric concentration of greenhouse gases to about 700 parts per million by 2100 (this is roughly double the pre-industrial level). The changes are shown as the difference between two 20-year averages (2020-2040 minus 1980-1999). These results are based on simulations from nine different climate models from the IPCC AR4 multi-model ensemble. The simulations were created on supercomputers at research centers in France, Japan, Russia, and the United States. Adapted by Lawrence Buja and Julie Arblaster from Tebaldi et al. 2006: Climatic Change, Going to the extremes; An intercomparison of model-simulated historical and future changes in extreme events, *Climatic Change*, 79:185-211.

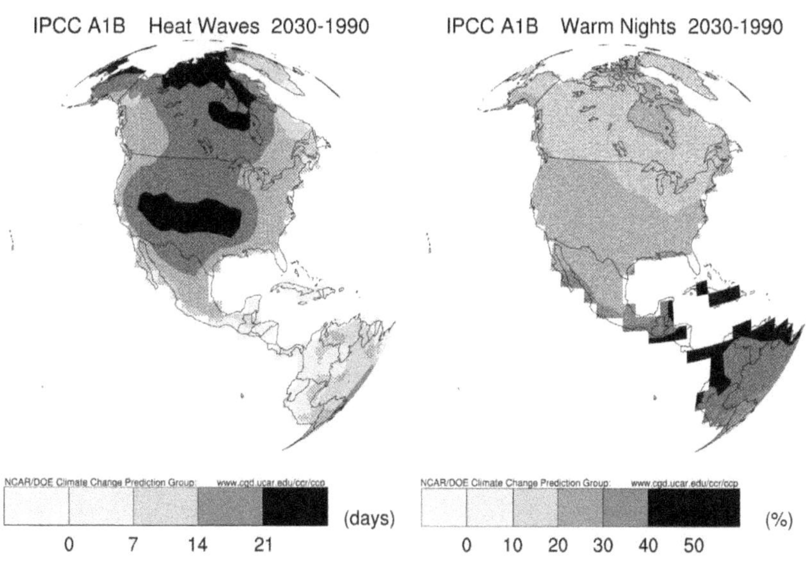

Figure 1.6 Simulated U.S. Heat Wave Days and Warm Nights in 2030. The left panel shows the projected change in number of heat wave days (days with maximum temperature higher by at least 5°C (with respect to the climatological norm)). The right panel shows changes in warm nights (percent of times when minimum temperature is above the 90th percentile of the climatological distribution for that day). Both panels show results for IPCC emissions scenario A1B, which would increase the atmospheric concentration of greenhouse gases to about 700 parts per million by 2100 (this is roughly double the pre-industrial level). The changes are shown as the difference between two 20-year averages (2020-2040 minus 1980-1999). Shading indicates areas of high inter-model agreement. These results are based on simulations from nine different climate models from the IPCC AR4 multi-model ensemble. The simulations were created on supercomputers at research centers in France, Japan, Russia, and the United States. Adapted by Lawrence Buja and Julie Arblaster from Tebaldi et al. 2006: *Climatic Change*, Going to the extremes; An intercomparison of model-simulated historical and future changes in extreme events, *Climatic Change*, 79:185-211.

Ecological and Biological Context

Climate variability and change have many impacts on terrestrial and marine ecosystems. Ecosystem responses to climate have implications for sustainability, biodiversity, and the ecosystem goods and services available to society. Some of these impacts affect the biological systems only, but some create further feedbacks to the climate system through greenhouse gas fluxes, albedo changes, and other processes.

Much research on terrestrial ecosystems and climate change has focused on their role as carbon sources or sinks. The observation that atmospheric CO_2 was increasing more slowly than expected from fossil fuel use and ocean uptake led to the speculation of a "missing sink," and the conclusion that increased plant photosynthesis was due to elevated atmospheric CO_2 (Gifford et al. 1994). It is now evident that several mechanisms, and not just CO_2 fertilization, contribute to the 'missing sink' (Field et al. 2007). These mechanisms include recovery from historic land use, fertilizing effects of nitrogen in the environment, expansion of woody vegetation ranges, storage of carbon in landfills and other depositional sites, and sequestration in long-lived timber products (Schimel et al. 2001).

Responses of photosynthesis and other processes that contribute to overall plant growth to warming are nonlinear. Each process (e.g.,

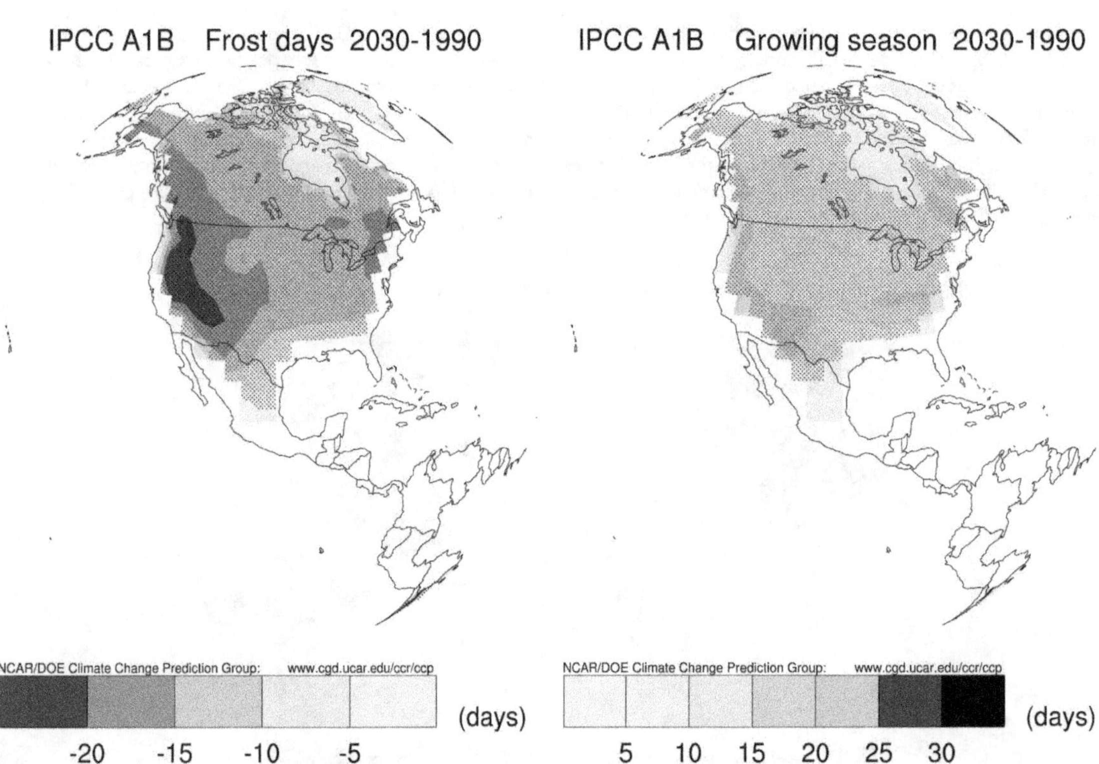

Figure 1.7 Changes in U.S. Frost days and Growing season by 2030. This figure shows decreases in frost days and increases in growing season length that would occur by about 2030 if the world follows IPCC emissions scenario A1B, which would increase the atmospheric concentration of greenhouse gases to about 700 parts per million by 2100 (this is roughly double the pre-industrial level). The changes are shown as the difference between two 20-year averages (2020-2040 minus 1980-1999). Shading indicates areas of high inter-model agreement. These results are based on simulations from nine different climate models from the IPCC AR4 multi-model ensemble. The simulations were created on supercomputers at research centers in France, Japan, Russia, and the United States. Adapted by Lawrence Buja and Julie Arblaster from Tebaldi et al. 2006: Climatic Change, Going to the extremes; An intercomparison of model-simulated historical and future changes in extreme events, *Climatic Change*, 79:185-211.

photosynthesis, respiration) typically has its own optimal response to temperature, which then decreases as temperatures change either below or above that optimum. The response of plants from different ecosystems is usually adapted to local conditions. Extreme hot and cold events affect photosynthesis and growth and may reduce carbon uptake or even cause mortality. Warming can lead to either increased or decreased plant growth, depending on the balance of the response of the individual processes.

Comprehensive analyses show that climate change is already causing the shift of many species to higher latitudes and/or altitudes, as well as changes in phenology. Not all species can successfully adjust, and some models suggest that biomes that are shifting in a warm, high-CO_2 world lose an average of a tenth of their biota.

Climate will affect ecosystems through fire, pest outbreaks, diseases, and extreme weather, as well as through changes to photosynthesis and other physiological processes. Disturbance regimes are a major control of climate-biome patterns. Fire-prone ecosystems cover about half the land area where forests would be expected, based on climate alone, and lead to grasslands and savannas in some of these areas. Plant pathogens, and insect defoliators are pervasive as well, and annually affect more than 40 times the acreage of forests in the United States damaged by fire. Disturbance modifies the climatic conditions where a vegetation type can exist.

While much of the ecosystems and climate change literature focuses on plants and soil processes, significant impacts on animal species are also known. A substantial literature documents impacts on the timing of bird migrations, on the latitudinal and elevational ranges of species and on more complex interactions between species, e.g., when predator and prey species respond to climate differently, breaking their relationships (Parmesan and Yohe 2003). The seasonality of animal processes may also respond to changes in climate, and this effect can have dramatic consequences, as occurs, for example, with changes in insect pest or pathogen-plant host interactions. Domestic animals also respond significantly to climate, both through direct physiological impacts on livestock, and through more complex effects of climate on livestock and their habitats.

Marine and coastal ecosystems are similarly sensitive in general to variability and change in the physical climate system, and in some cases directly to atmospheric concentrations of carbon dioxide. Fish populations in major large marine biomes are known to shift their geographic ranges in response to specific modes of climate variation, such as the Pacific Decadal Oscillation and the North Atlantic Oscillation, and there have been shifts in geographical range of some fish species in response to surface water warming over the past several decades on both West and East coasts of North America. Subtropical and tropical corals in shallow waters have already suffered major bleaching events that are clearly driven by increases in sea surface temperatures, and increases in ocean acidity, which are a direct consequence of increases in atmospheric carbon dioxide, are calculated to have the potential for serious negative consequences for corals.

Not all species can successfully adjust, and some models suggest that biomes that are shifting in a warm, high-CO_2 world lose an average of a tenth of their biota.

Many studies on climate impacts on ecosystems look specifically at impacts only of variation and change in the physical climate system and CO_2 concentrations. But there are many factors that affect the distribution, complexity, make-up, and performance of ecosystems. Disturbance, pests, invasive species, deforestation, human management practices, overfishing, etc., are powerful influences on ecosystems. Climate change impacts are but one of many such features, and need to be considered in this broader context.

Attribution of Ecosystem Changes

It is important to note that the changes due to climate change occur against a background of rapid changes in other factors affecting ecosystems. These include changing patterns of land management, intensification of land use and exurban development, new management practices (e.g., biofuel production), species invasions and changing air quality (Lodge et al. 2006). Because many factors are affecting ecosystems simultaneously, it is difficult and in some cases impossible to factor out the magnitude of each impact separately. In a system affected by, for example, temperature, ozone, and changing precipitation, assigning a percentage of an observed change to each factor is generally impossible. Research on improving techniques for separating influences is ongoing, but in some cases drivers of change interact with each other, making the combined effects different from the sum of the separate effects. Scientific concern about such multiple stresses is rising rapidly.

The changes that are likely to occur will continue to have significant effects on the ecosystems of the United States, and the services those ecosystems provide.

Summary

The changes in temperature and precipitation over the past century now form a persistent pattern and show features consistent with the scientific understanding of climate change. For example, scientists expect larger changes near the poles than near the equator. This pattern can be seen in the dramatically higher rates of warming in Alaska compared to the rest of the country. Most of the warming is concentrated in the last decades of the century. Prior to that, large natural variations due to solar and volcanic effects were comparable in magnitude to the then-lower greenhouse gas effects. These natural swings sometimes enhanced and sometimes hid the effects of greenhouse gases. The warming due to greenhouse gases is now quite large and the "signal" of the greenhouse warming has more clearly emerged from the "noise" of the planet's natural variations. The effects of greenhouse gases have slowly accumulated, but in the past few years, their effects have become evident. Recent data show clearly both the trends in climate, and climate's effects on many aspects of the nation's ecology.

The changes that are likely to occur will continue to have significant effects on the ecosystems of the United States, and the services those ecosystems provide. The balance of this report will document some of the observed historical changes and provide insights into how the continuing changes may affect the nation's ecosystems.

Agriculture

Lead Author: J. L. Hatfield, USDA ARS

Contributing Authors:

Cropland Response: K.J. Boote, B.A. Kimball, D.W. Wolfe, D.R. Ort
Pastureland: R.C. Izaurralde, A.M. Thomson
Rangeland: J.A. Morgan, H.W. Polley, P.A. Fay
Animal Management: T.L. Mader, G.L. Hahn

2.1 INTRODUCTION

This synthesis and assessment report builds on an extensive scientific literature and series of recent assessments of the historical and potential impacts of climate change and climate variability on managed and unmanaged ecosystems and their constituent biota and processes. It identifies changes in resource conditions that are now being observed, and examines whether these changes can be attributed in whole or part to climate change. It also highlights changes in resource conditions that recent scientific studies suggest are most likely to occur in response to climate change, and when and where to look for these changes. As outlined in the Climate Change Science Program (CCSP) Synthesis and Assessment Product 4.3 (SAP 4.3) prospectus, this chapter will specifically address climate-related issues in cropping systems, pasture and grazing lands, and animal management.

In this chapter the focus is on the near-term future. In some cases, key results are reported out to 100 years to provide a larger context but the emphasis is on the next 25-50 years. This nearer term focus is chosen for two reasons. First, for many natural resources, planning and management activities already address these time scales through the development of long-lived infrastructure, plant species rotation, and other significant investments. Second, climate projections are relatively certain over the next few decades. Emission scenarios for the next few decades do not diverge from each other significantly because of the "inertia" of the energy system. Most projections of greenhouse gas emissions assume that it will take decades to make major changes in the energy infrastructure, and only begin to diverge rapidly after several decades have passed (30-50 years).

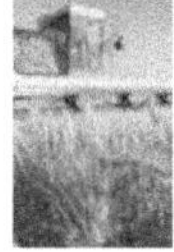

To average consumers, U.S. agricultural production seems uncomplicated – they see only the staples that end up on grocery store shelves. The reality, however, is far from simple. Valued at $200 billion in 2002, agriculture includes a wide range of plant and animal production systems (Figure 2.1).

The United States Department of Agriculture (USDA) classifies 116 plant commodity groups as agricultural products, as well as four livestock groupings (beef cattle, dairy, poultry, swine) and products derived from animal production, e.g., cheese or eggs. Of these commodities, 52 percent of the total sales value is generated from livestock, 21 percent from fruit and nuts, 20 percent from grain and oilseed, two percent from cotton, and five percent from other commodity production, not including pastureland or rangeland production (Figure 2.2).

The many U.S. crops and livestock varieties are grown in diverse climates, regions, and soils. No matter the region, however, weather and climate characteristics such as temperature, precipitation, carbon dioxide (CO_2), and water availability directly impact the health and well-being of plants and livestock, as well as pasture and rangeland production. The distribution of crops and livestock is also determined by the climatic resources for a given region and U.S. agriculture has benefited from optimizing the adaptive areas of crops and livestock. For any commodity, variation in yield between years is related to growing-season weather effects. These effects also influence how insects, disease, and weeds affect agricultural production.

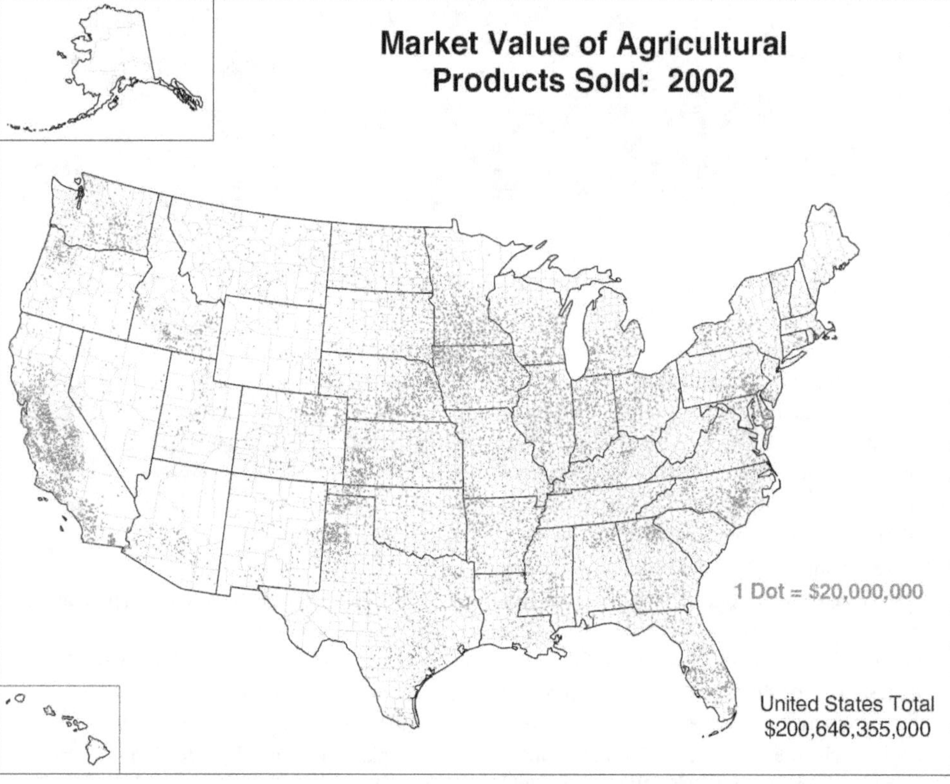

Figure 2.1 The extensive and intensive nature of U.S. agriculture is best represented in the context of the value of the production of crops and livestock. The map above presents the market value of all agricultural products sold in 2002 and their distribution. (USDA National Agricultural Statistics Service.)

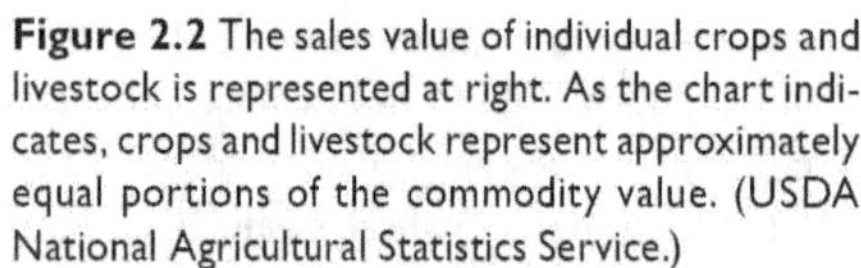

Figure 2.2 The sales value of individual crops and livestock is represented at right. As the chart indicates, crops and livestock represent approximately equal portions of the commodity value. (USDA National Agricultural Statistics Service.)

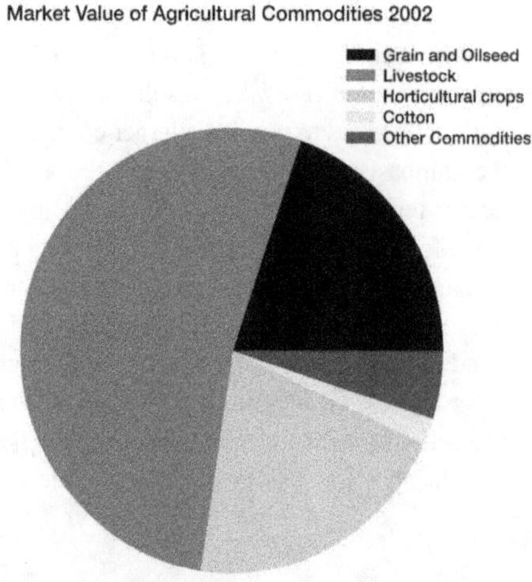

The goal in this chapter is to provide a synthesis of the potential impacts of climate on agriculture that can be used as a baseline to understand the consequences of climate variability. A variety of agricultural crops will be considered in this report. Among them is corn (*Zea mays*), the most widely distributed U.S. crop after pastureland and rangeland; wheat, which is grown in most states, but has a concentration in the upper Great Plains and northwest United States; and orchard crops, which are restricted to regions with moderate winter temperatures. For any of these crops, shifts in climate can affect production through, for instance, variance in temperature during spring (flowering) and fall (fruit maturity).

Additionally, this chapter will look at beef cow production, which is ubiquitous across the United States (Figure 2.3). Because of the regular presence of beef cows across the nation, beef cow vitality provides an effective indicator of the regional impact of climate change. While beef cows are found in every state, the greatest number are raised in regions that have an abundance of native or planted pastures (Figure 2.4), which provide easy access to accessible feed supplies for the grazing animals.

Over the past 25 years, there has been a decline in land classified as rangeland, pastureland, or grazed forest. Many of these shifts relate to changing land use characteristics, such as population growth (Table 2.1); the growing eastern U.S. has experienced the greatest reduction in such land resources (Table 2.2). This chapter will provide an overview of the state of pasturelands and rangelands as defined by the USDA. Pastureland is a land cover/use category of land managed primarily for the production of introduced forage plants for livestock grazing. Pastureland cover may consist of a single species in a pure stand, a grass mixture, or a grass-legume mixture. Management usually consists of cultural treatments: fertilization, weed control, reseeding or renovation, and control of grazing. Rangeland is a land cover/use category on which the climax or potential plant cover is

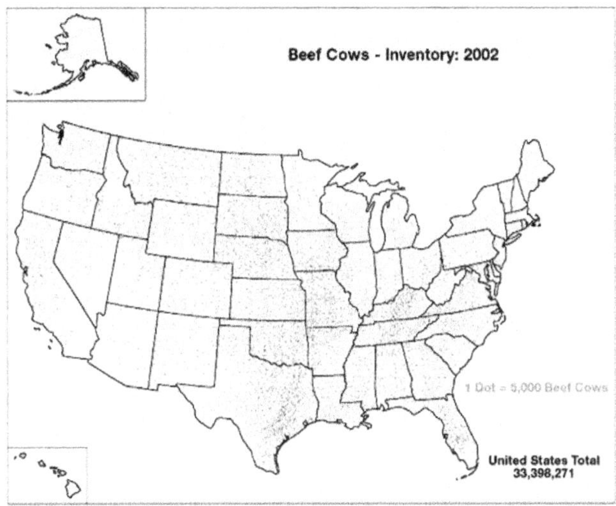

Figure 2.3 Distribution of beef cow inventory across the United States in 2002. (USDA National Agricultural Statistics Service.)

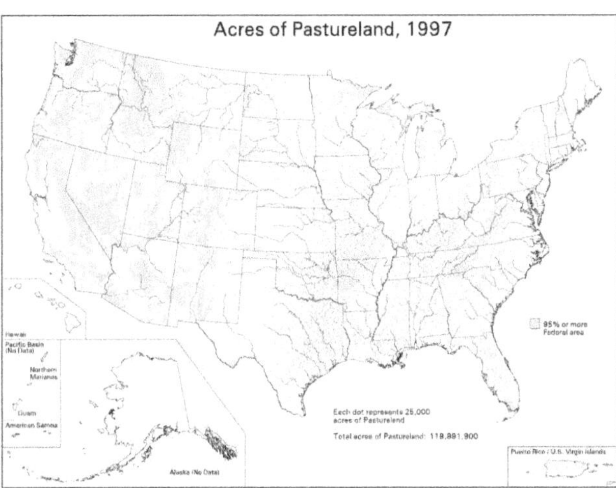

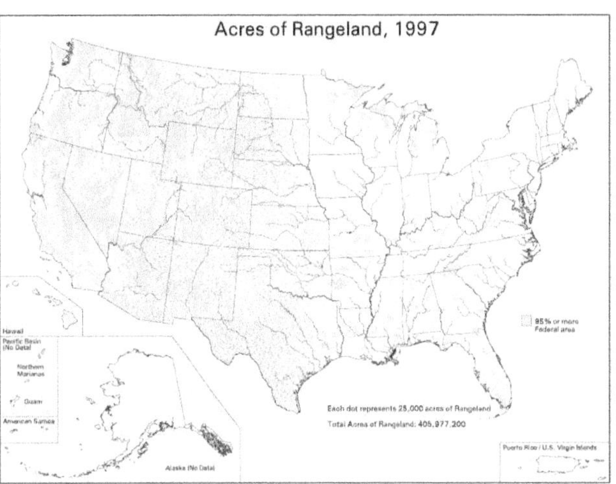

Figures 2.4a and 2.4b Distribution of pastureland and rangeland across the United States in 1997.

composed principally of native grasses, grass-like plants, forbs, or shrubs suitable for grazing and browsing, and introduced forage species that are managed like rangeland. This would include areas where introduced hardy and persistent grasses, such as crested wheatgrass, are planted and such practices as deferred grazing, burning, chaining, and rotational grazing are used, with little or no chemicals or fertilizer applied. This chapter will also consider the effects of climate on these areas.

2.2 OBSERVED CHANGES AND RESPONSES

2.2.1 Crops

2.2.1.1 Scope of the Agricultural Systems

As noted earlier, agriculture is a diverse system that covers a wide range of species and production systems across the United States. However, this chapter's scope includes species covered in the available scientific literature that evaluates observed responses to changing climate

Table 2.1 Non-federal grazing land (in millions of acres). Source: Natural Resources Conservations Service (NRCS).

Year	Rangeland	Pastureland (millions of acres)	Grazed Forest land (millions of acres)	Total (millions of acres)
1982	415.5	131.1	64.3	610.9
1992	406.7	125.2	61.0	592.9
1997	404.9	119.5	58.0	582.4
2001	404.9	119.2	55.2	579.3
2003	405.1	117.0	54.3	576.4

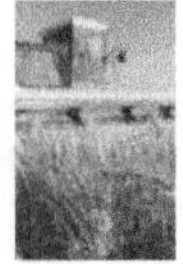

Table 2.2 Changes in pasturelands by major water resource areas (in millions of acres). Source: www.nrcs.usda.gov/technical/land/nri03/national_landuse.html

	1982	1992	2003
Arkansas-White-Red	18.6	19.0	19.8
California / Great Basin	2.3	2.2	2.3
Great Lakes	5.8	4.7	4.4
Lower Colorado / Upper Colorado	0.8	0.9	0.9
Lower Mississippi	5.6	5.4	5.0
Missouri	20.4	19.2	18.0
New England / Mid Atlantic	7.4	6.3	5.6
Ohio / Tennessee River	20.9	19.8	17.7
Pacific Northwest	4.6	4.7	4.3
Souris- Red-Rainy / Upper Mississippi	14.5	12.7	11.7
South Atlantic-Gulf	15.5	15.9	13.9
Texas-Gulf / Rio Grande	14.7	14.4	13.4
Totals	**131.1**	**125.2**	**117.0**

characteristics. In the crops section, the focus is on maize (corn), soybean (*Glycine max*), wheat (*Triticum aestivum*), rice (*Oryza sativa*), sorghum (*Sorghum bicolor*), cotton (*Gossypium hirsutum*), peanut (*Arachis hypogea*), dry kidney bean (*Phaseolus vulgaris*), cowpea (*Vigna unguiculata*), and tomato (*Lycopersicon esculentum*).

Animal production systems cover beef cattle, dairy, swine, and poultry as the primary classes of animals. While climate changes affect all of these animals, the literature predominantly addresses beef, dairy, and swine. Poultry are primarily grown in housed operations, so the effect of climate change more directly affects the energy requirements for building operations compared to a direct effect on the animal. Similar statements can be made for swine production since the vast majority of the animals are housed. Temperature affects animals being moved from buildings to processing plants, but because these animals are moved quickly from production to processing, this is a problem only in extreme conditions.

Both pasture and rangeland are reviewed in this chapter. In the pastureland section, 13 species are considered in the analysis; for rangeland, species include a complex mixture of grasses and forbs, depending on the location.

As much as possible, the conclusions about the effects of global change on agriculture and other ecosystems are based on observed trends as much as possible. However, an immediate obstacle to using this observational approach is that the productivity of most agricultural enterprises has increased dramatically over the past decades due to improvements in technology, and the responses to these changes in technology overwhelm responses to global change that almost certainly are present but are statistically undetectable against the background of large technological improvements. Fortunately, numerous manipulative experiments have been conducted on these managed agricultural systems wherein temperature, CO_2, ozone (O_3), and/or other factors have been varied. From such experiments, the relative responses to the changing climate variables can be deduced. A second challenge, however, is that the details of each experiment have been different – different temperature changes have been explored, different concentrations of CO_2, different crop varieties and so forth. The problem remains as to how to represent such experimental variability in methods in a way that provides a consistent baseline for comparison.

As noted in the Introduction, in about 30 years, CO_2 concentrations are expected to have increased about 60 ppm (from today's 380 ppm to about 440 ppm), and temperatures over the contiguous United States are expected to have increased by an average of about 1.2°C. We have therefore used these increments as baseline comparison points compared to current CO_2 and temperatures to estimate the likely responses of crops to global change for the 30-year time horizon of this report. We have done this by constructing mathematical response functions for crops and experiments that use the experimental data available.

The goal in this chapter is to provide a synthesis of the potential impacts of climate on agriculture that can be used as a baseline to understand the consequences of climate variability.

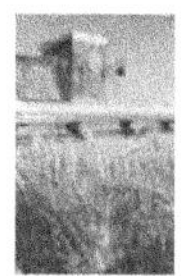

2.2.1.2 Plant Response to Temperature

2.2.1.2.1 General Response

Crop species differ in their cardinal temperatures (critical temperature range) for life cycle development. There is a base temperature for vegetative development, at which growth commences, and an optimum temperature, at which the plant develops as fast as possible. Increasing temperature generally accelerates progression of a crop through its life cycle (phenological) phases, up to a species-dependent optimum temperature. Beyond this optimum temperature, development (node and leaf appearance rate) slows. Cardinal temperature values are presented below, in Tables 2.3 and 2.4, for selected annual (non-perennial) crops under conditions in which temperature is the only limiting variable.

One caveat is that the various scenarios for global change predict increasing air temperatures, but plants often are not growing at air temperature. For example, under arid conditions, amply irrigated crops can easily be 10°C cooler than air temperature due to transpirational cooling. Solar and sky radiation, wind speed, air humidity, and plant stomatal conductance are all variables that affect the difference in temperature between plants and air. While recognizing this problem, it is important to understand that published cardinal temperatures such as those in Tables

2.3 and 2.4 are based on air temperature, rather than vegetation temperature. That is because air temperatures are much easier to measure than plant temperatures, and usually only air temperatures are reported from experiments; also many crop growth models assume that plants are growing at air temperature rather than at their own vegetation temperature. Nevertheless, crop canopy temperatures are sufficiently coupled to air temperatures that for a first approximation, we expect future crop canopy temperatures to increase by about the same amount as air temperatures with global warming.

Faster development of non-perennial crops is not necessarily ideal. A shorter life cycle results in smaller plants, shorter reproductive phase duration, and lower yield potential. Because of this, the optimum temperature for yield is nearly always lower than the optimum temperature for leaf appearance rate, vegetative growth, or reproductive progression. In addition, temperatures that fall below or above specific thresholds at critical times during development can also have significant impact on yield. Temperature affects crop life cycle duration and the fit of given cultivars to production zones. Day-length sensitivity also plays a major role in life cycle progression in many crops, but especially for soybean. Higher temperatures during the reproductive stage of development affect pollen viability, fertilization, and grain or fruit formation. Chronic as well as short-term exposure to high temperatures during the pollination stage of initial grain or fruit set will reduce yield potential. This phase of development is one of the most critical stages of growth in response to temperatures extremes. Each crop has a specific temperature range at which vegetative and reproductive growth will proceed at the optimal rate and exposures to extremely high temperatures during these phases can impact growth and yield; however, acute exposure from extreme events may be most detrimental during the reproductive stages of development.

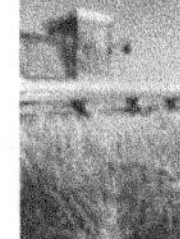

Table 2.3. For several economically significant crops, information is provided regarding cardinal, base, and optimum temperatures (°C) for vegetative development and reproductive development, optimum temperature for vegetative biomass, optimum temperature for maximum grain yield, and failure (ceiling) temperature at which grain yield fails to zero yield. The optimum temperatures for vegetative production, reproductive (grain) yield, and failure point temperatures represent means from studies where diurnal temperature range was up to 10°C.

Crop	Base Temp Veg	Opt Temp Veg	Base Temp Repro	Opt Temp Repro	Opt Temp Range Veg Prod	Opt Temp Range Reprod Yield	Failure Temp Reprod Yield
Maize	8[1]	34[1]	8[1]	34[1]		18-22[2]	35[3]
Soybean	7[4]	30[4]	6[5]	26[5]	25-37[6]	22-24[6]	39[7]
Wheat	0[8]	26[8]	1[8]	26[8]	20-30[9]	15[10]	34[11]
Rice	8[12]	36[13]	8[12]	33[12]	33[14]	23-26[13,15]	35-36[13]
Sorghum	8[16]	34[16]	8[16]	31[17]	26-34[18]	25[17,19]	35[17]
Cotton	14[20]	37[20]	14[20]	28-30[20]	34[21]	25-26[22]	35[23]
Peanut	10[24]	>30[24]	11[24]	29-33[25]	31-35[26]	20-26[26,27]	39[26]
Bean					23[28]	23-24[28,29]	32[28]
Tomato	7[30]	22[30]	7[30]	22[30]		22-25[30]	30[31]

[1]Kiniry and Bonhomme (1991):, [2]Muchow et al. (1990); [3]Herrero and Johnson (1980); [4]Hesketh et al. (1973); [5]Boote et al. (1998); [6]Boote et al. (1997); [7]Boote et al. (2005); [8]Hodges and Ritchie (1991); [9]Kobza and Edwards (1987); [10]Chowdury and Wardlaw (1978); [11]Tashiro and Wardlaw (1990); [12]Alocilja and Ritchie (1991); [13]Baker et al. (1995); [14]Matsushima et al. (1964); [15]Horie et al. (2000); [16]Alagarswamy and Ritchie 1991); [17]Prasad et al. (2006a); [18]Maiti (1996); [19]Downs (1972); [20]K.R. Reddy et al. (1999, 2005); [21]V.R. Reddy et al. (1995); [22]K.R. Reddy et al. (2005); [23]K.R. Reddy et al. (1992a, 1992b); [24]Ong (1986); [25]Bolhuis and deGroot (1959); [26]Prasad et al. (2003); [27]Williams et al. (1975); [28]Prasad et al. (2002); [29]Laing et al. (1984); [30]Adams et al. (2001); [31]Peat et al. (1998).

For most perennial, temperate fruit and nut crops, winter temperatures play a significant role in productivity (Westwood 1993). There is considerable genotypic variation among fruit and nut crops in their winter hardiness (that is, the ability to survive specific low temperature extremes), and variation in their "winter chilling" requirement for optimum flowering and fruit set in the spring and summer (Table 2.5). Placement of fruit and nut crops within specific areas are related to the synchrony of phenological stages to the climate and the climatic resources of the region. Marketable yield of horticultural crops is highly sensitive to minor environmental stresses related to temperatures outside the optimal range, which negatively affect visual and flavor quality (Peet and Wolfe 2000).

2.2.1.2.2 Temperature effects on crop yield

Yield responses to temperature vary among species based on the crop's cardinal temperature requirements. Plants that have an optimum range at cooler temperatures will exhibit significant decreases in yield as temperature increases above this range. However, reductions in yield with increasing temperature in field conditions may not be due to temperature alone, as high temperatures are often associated with lack of rainfall in many climates. The changes in temperature do not produce linear responses with increasing temperature because the biological response to temperature is nonlinear, therefore, as the temperature increases these effects will be larger. The interactions of temperature and water deficits negatively affect crop yield.

Table 2.4 Temperature thresholds for selected vegetable crops. Values are approximate, and for relative comparisons among groups only. For frost sensitivity: "+" = sensitive to weak frost; "-" = relatively insensitive; "( )" = uncertain or dependent on variety or growth stage. Adapted from Krug (1997) and Rubatzky and Yamaguchi (1997).

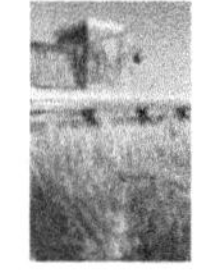

Climatic Classification	Crop	Acceptable Temp (C) For Germination	Opt Temp (C) For Yield	Acceptable Temp (C) Growth Range	Frost Sensitivity
Hot	Watermelon	21-35	25-27	18-35	+
	Okra	21-35	25-27	18-35	+
	Melon	21-32	25-27	18-35	+
	Sweet Potato	21-32	25-27	18-35	+
Warm	Cucumber	16-35	20-25	12-30(35)	+
	Pepper	16-35	20-25	12-30(35)	+
	Sweet corn	16-35	20-25	12-30(35)	+
	Snap bean	16-30	20-25	12-30(35)	+
	Tomato	16-30	20-25	12-30(35)	+
Cool-Warm	Onion	10-30	20-25	7-30	-
	Garlic	7-25	20-25	7-30	-
	Turnip	10-35	18-25	5-25	-
	Pea	10-30	18-25	5-25	( )
Cool	Potato	7-26	16-25	5-25(30)	+
	Lettuce	5-26	16-25	5-25(30)	(+)
	Cabbage	10-30	16-18(25)	5-25	-
	Broccoli	10-30	16-18(25)	5-25	-
	Spinach	4-16	16-18(25)	5-25	-

2.2.1.2.2.1 Maize

Increasing temperature causes the maize life cycle and duration of the reproductive phase to be shortened, resulting in decreased grain yield (Badu-Apraku et al. 1983; Muchow et al. 1990). In the analyses of Muchow et al. (1990), the highest observed (and simulated) grain yields occurred at locations with relatively cool temperatures (growing season mean of 18.0 to 19.8°C in Grand Junction, Colo.), which allowed long maize life cycles, compared to warmer sites (e.g., 21.5 to 24.0°C in Champaign, Ill.), or compared to warm tropical sites (26.3 to 28.9°C). For the Illinois location, simulated yield decreased 5 to 8 percent per 2°C temperature increase. Using this relationship, a temperature rise of 1.2°C over the next 30 years in the Midwest may decrease yield by about 4 percent (Table 2.6) under irrigated or water-sufficient management.

Lobell and Asner (2003) evaluated maize and soybean production relative to climatic variation in the United States, reporting a 17 percent reduction in yield for every 1°C rise in temperature, but this response is unlikely because the confounding effect of rainfall was

Marketable yield of horticultural crops is highly sensitive to minor environmental stresses related to temperatures outside the optimal range, which negatively affect visual and flavor quality.

Table 2.5 Winter chill requirement, winter hardiness (minimum winter temperature), and minimum frost-free period (growing season requirements) for selected woody perennial fruit and nut crops. Not shown in this table is the fact that flowers and developing fruit of all crops are sensitive to damage from mild to moderate frosts (e.g., 0 to -5°C), and high temperature stress (e.g., >35°C), specific damaging temperatures varying with crop and variety. Values are approximate and for relative comparisons only. Adapted from Westwood (1993).

Winter Chill Requirement (hours)[1]

Crop	Common Varieties	Other	Minimum Winter Temp (C)	Minimum Frost-Free Period (days)
Almond	100-500		-10	>180
Apple	1000-1600	400-1800	-46 to -4	<100 (+)
Blueberry	400-1200 (northern highbush)	0-200	-35 to -12	<100 (+)
Cherry	900-1200	600-1400	-29 to -1	<100 (+)
Citrus	0		-7 to 4	>280
Grape (European)	100-500		-25 to 4	>120
Grape (American)	400-2000 (+)		-46 to -12	<100 (+)
Peach	400-800	200-1200	-29 to 4	>120
Pear	500-1500		-35 to -1	>100
Pecan	600-1400		-10	>180
Pistachio	600-1500	400-600 (Asian)	-10	>180
Plum	800-1200	500-600 (Japanese)	-29 to 4	>140
Raspberry	800-1700	100-1800	-46 (+)	<100 (+)
Strawberry	300-400		-12	<100 (+)
Walnut	400-1500		-29	>100

[1]Winter chilling for most fruit and nut crops occurs within a narrow temperature range of 0 to 15°C, with maximum chill-hour accumulation at about 7.2°C. Temperatures below or above this range do not contribute to the chilling requirement, and temperatures above 15°C may even negate previously accumulated chill.

not considered. In a recent evaluation of global maize production response to both temperature and rainfall over the period 1961-2002, Lobell and Field (2007) reported an 8.3 percent yield reduction per 1°C rise in temperature. Runge (1968) documented maize yield responses to the interaction of daily maximum temperature and rainfall during the period 25 days prior to, and 15 days after, anthesis of maize. If rainfall was low (0-44 mm per 8 days), yield was reduced by 1.2 to 3.2 percent per 1°C rise. Alternately, if temperature was warm (maximum temperature (Tmax) of 35°C), yield was reduced 9 percent per 25.4 mm rainfall decline. The Muchow et al. (1990) model, also used to project temperature effects on crops, may underestimate yield reduction with rising temperature because it had no temperature modification on assimilation or respiration, and did not provide for any failures in grain-set with rising temperature. Given the disagreement in literature estimates and lack of real manipulative temperature experiments on maize, the certainty of the estimate in Table 2.6 is only possible to likely.

Yield decreases caused by elevated temperatures are related to temperature effects on pollination and kernel set. Temperatures above 35°C are lethal to pollen viability (Herrero and Johnson 1980; Schoper et al. 1987; Dupuis and Dumas 1990). In addition, the critical duration of pollen viability (prior to silk reception) is a function of pollen moisture content, which is strongly dependent on vapor pressure deficit (Fonseca and Westgate 2005). There is limited data on sensitivity of kernel set in maize to elevated temperature, although in-vitro evidence suggests that the thermal environment during endosperm cell division phase (eight to 10 days post-anthesis) is critical (Jones et al. 1984). A temperature of 35°C, compared to 30°C during the endosperm division phase, dramatically reduced subsequent kernel growth rate (potential) and final kernel size, even if ambient temperature returns to 30°C (Jones et al. 1984). Temperatures above 30°C increasingly impaired cell division and amyloplast replication in maize kernels, and thus reduced grain sink strength and yield (Commuri and Jones 2001). Leaf photosynthesis rate of maize has a high temperature optimum of 33°C to 38°C. There is a minimal sensitivity of light use (quantum) efficiency to these elevated temperatures (Oberhuber and Edwards 1993; Edwards and Baker 1993); however, photosynthesis rate is reduced above 38°C (Crafts-Brandner and Salvucci 2002).

2.2.1.2.2.2 Soybean

Reproductive development (time to anthesis) in soybean has cardinal temperatures that are somewhat lower than those of maize. A base temperature of 6°C and optimum temperature of 26°C are commonly used (Boote et al. 1998), having been derived, in part, from values of 2.5°C and 25.3°C developed from field data by Grimm et al. (1993). The post-anthesis phase for soybean has a surprisingly low optimum temperature of about 23°C, and life cycle is slower and longer if mean daily temperature is above 23°C (Pan 1996; Grimm et al. 1994). This 23°C optimum cardinal temperature for post-anthesis period closely matches the optimum temperature for single seed growth rate (23.5°C), as reported by Egli and Wardlaw (1980), and the 23°C optimum temperature for seed size (Egli and Wardlaw 1980; Baker et al. 1989; Pan 1996; Thomas 2001; Boote et al. 2005). As mean temperature increases above 23°C, seed growth rate, seed size, and intensity of partitioning to grain (seed harvest index) in soybean decrease until reaching zero at 39°C mean (Pan 1996; Thomas 2001).

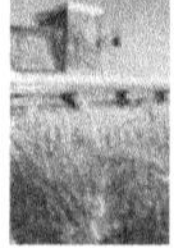

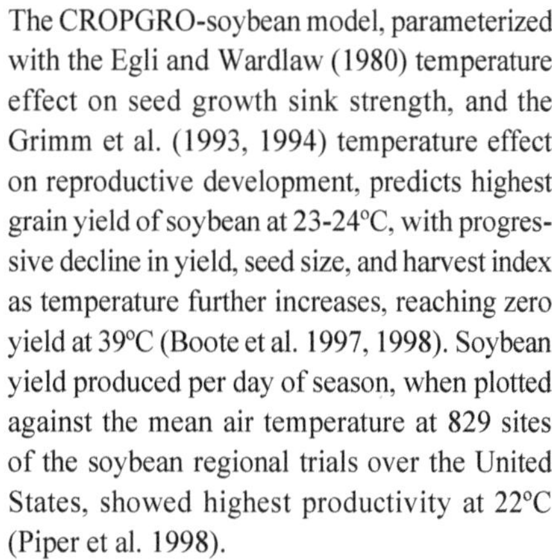

The CROPGRO-soybean model, parameterized with the Egli and Wardlaw (1980) temperature effect on seed growth sink strength, and the Grimm et al. (1993, 1994) temperature effect on reproductive development, predicts highest grain yield of soybean at 23-24°C, with progressive decline in yield, seed size, and harvest index as temperature further increases, reaching zero yield at 39°C (Boote et al. 1997, 1998). Soybean yield produced per day of season, when plotted against the mean air temperature at 829 sites of the soybean regional trials over the United States, showed highest productivity at 22°C (Piper et al. 1998).

Pollen viability of soybean is reduced if temperatures exceed 30°C (optimum temperature), but has a long decline slope to failure at 47°C (Salem et al. 2007). Averaged over many cultivars, the cardinal temperatures (base temperature (Tb), optimum temperature (Topt), and Tmax) were 13.2°C, 30.2°C, and 47.2°C, respectively, for pollen germination, and 12.1°C, 36.1°C, and

47.0°C, respectively, for pollen tube growth. Minor cultivar differences in cardinal temperatures and tolerance of elevated temperature were present, but differences were not very large or meaningful. Salem et al. (2007) evaluated soybean grown at 38/30°C versus 30/22°C (day/night) temperatures. The elevated temperature reduced pollen production by 34 percent, pollen germination by 56 percent, and pollen tube elongation by 33 percent. The progressive reduction in seed size (single seed growth rate) above 23°C, along with reduction in fertility (i.e., percent seed set) above 30°C, results in reduction in seed harvest index at temperatures above 23-27°C (Baker et al.1989; Boote et al. 2005). Zero seed harvest index occurs at 39°C (Pan 1996; Thomas 2001; Boote et al. 2005).

The implication of a temperature change on soybean yield is thus strongly dependent on the prevailing mean temperature during the post-anthesis phase of soybean in different regions. For the upper Midwest, where mean soybean growing season temperatures are about 22.5°C, soybean yield may actually increase 2.5 percent with a 1.2°C rise (Table 2.6). By contrast, soybean production in the southern United States, where mean growing season temperatures are 25°C to 27°C, soybean yield would be progressively reduced – 3.5 percent for 1.2°C increase from the current 26.7°C mean (Table 2.7) (Boote et al. 1996, 1997). Lobell and Field (2007) reported a 1.3 percent decline in soybean yield per 1°C increase in temperature, taken from global production against global average

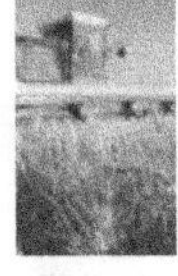

Table 2.6 Percent grain yield and evapotranspiration responses to increased temperature (1.2°C), increased CO_2 (380 to 440 ppm), and the net effects of temperature plus increased CO_2 assuming additivity. Current mean air temperature during reproductive growth is shown in parentheses for each crop/region to give starting references, although yield of all the cereal crops declines with a temperature slope that originates below current mean air temperatures during grain filling.

	Grain Yield			Evapotranspiration	
Crop	Temperature (1.2°C) [1]	CO_2 (380 to 440 ppm) [2]	Temp/CO_2 Combined Irrigated	Temp (1.2°C) [3]	CO_2 (380 to 440 ppm) [4]
	% change				
Corn – Midwest (22.5°C)	-4.0	+1.0	-3.0	+1.8	
Corn – South (26.7°C)	-4.0	+1.0	-3.0	+1.8	
Soybean – Midwest (22.5°C)	+2.5	+7.4	+9.9	+1.8	-2.1
Soybean – South (26.7°C)	-3.5	+7.4	+3.9	+1.8	-2.1
Wheat – Plains (19.5°C)	-6.7	+6.8	+0.1	+1.8	-1.4
Rice – South (26.7°C)	-12.0	+6.4	-5.6	+1.8	-1.7
Sorghum (full range)	-9.4	+1.0	-8.4	+1.8	-3.9
Cotton – South (26.7°C)	-5.7	+9.2	+3.5	+1.8	-1.4
Peanut – South (26.7°C)	-5.4	+6.7	+1.3	+1.8	
Bean – relative to 23°C	-8.6	+6.1	-2.5	+1.8	

[1]Response to temperature summarized from literature cited in the text. [2]Response to CO_2 with Michaelis-Menten rectangular hyperbola interpolation of literature values shown in Table 2.7. [3]From Table 2.8 the sensitivity of a standard alfalfa crop to warming at constant relative humidity on clear summer day would be 1.489% per °C, so assuming the crop ET will respond similarly with warming by 1.2°C, the expected change in ET would be 1.8%. [4]From Table 2.7 assuming linear ET response to 60 ppm increase in CO_2 interpolated from the range, 350 to 700 ppm or 370 to 570 ppm for sorghum.

temperature during July-August, weighted by production area. These two estimates are in agreement and likely, considering that Lobell and Field (2007) averaged over cool and warm production areas.

2.2.1.2.2.3 Wheat

Grain-filling period of wheat and other small grains shortens dramatically with rising temperature (Sofield et al. 1974, 1977; Chowdhury and Wardlaw 1978; Goudrian and Unsworth 1990). Assuming no difference in daily photosynthesis, which can be inferred from the sink removal studies of Sofield et al. (1974, 1977), yield will decrease in direct proportion to the shortening of grain filling period as temperature increases. This temperature effect is already a major reason for the much lower wheat yield potential in the Midwest than in northern Europe, even with the water limitation removed.

The optimum temperature for photosynthesis in wheat is 20-30°C (Kobza and Edwards 1987). This is 10°C higher than the optimum (15°C) for grain yield and single grain growth rate (Chowdhury and Wardlaw 1978). Any increase in temperature beyond the 25-35°C range that is common during grain filling of wheat will reduce the grain filling period and, ultimately, yields. Applying the nonlinear slope of reduction in grain filling period from Chowdury and Wardlaw (1978), relative to the mean temperatures during grain fill in the wheat growing regions of the Great Plains, reduction in yield is about 7 percent per 1°C increase in air temperature between 18 and 21°C, and about 4 percent per 1°C increase in air temperature above 21°C, not considering any reduction in photosynthesis or grain-set. Similarly, Lawlor and Mitchell (2000) stated that a 1°C rise would shorten the reproductive phase by 6 percent, grain filling duration by 5 percent, and would reduce grain yield and harvest index proportionately. Bender et al. (1999) analyzed spring wheat grown at nine sites in Europe and found a 6 percent decrease in yield per 1°C temperature rise. Lobell and Field (2007) reported a 5.4 percent decrease in global mean wheat yield per 1°C increase in temperature. Grain size will also be reduced slightly. These four references are very much in agreement, so the projected temperature effect on yield in Table 2.6 is considered very likely. Effects of rising temperature on photosynthesis should be viewed as an additional reduction factor on wheat yield, primarily influenced via water deficit effects (Paulsen 1994). Temperatures of 36/31°C (maximum/minimum) for two to three days prior to anthesis causes small unfertilized kernels with symptoms of parthenocarpy – that is, small shrunken kernels with notching and chalking of kernels (Tashiro and Wardlaw 1990). Increased temperature also reduces starch synthesis in wheat endosperm (Caley et al. 1990).

2.2.1.2.2.4 Rice

The response of rice to temperature has been well studied (Baker and Allen 1993a, 1993b; Baker et al. 1995; Horie et al. 2000). Leaf-appearance rate of rice increases with temperature from a base of 8°C, until reaching 36-40°C, the thermal threshold of survival (Alocilja and Ritchie 1991; Baker et al. 1995), with biomass increasing up to 33°C (Matsushima et al. 1964); however, the optimum temperature for grain formation and yield of rice is lower (25°C) (Baker et al. 1995). Baker et al. (1995) summarized many of their experiments from sunlit controlled-environment chambers and concluded that the optimum mean temperature for grain formation and grain yield of rice is 25°C. They found that grain yield is reduced about 10 percent per 1°C temperature increase above 25°C, until reaching zero yield at 35-36°C mean temperature, using a 7°C day/night temperature differential (Baker and Allen 1993a; Peng et al. 2004).

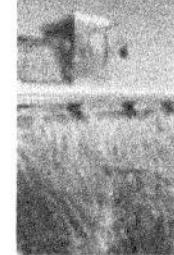

Grain number, percent filled grains, and grain harvest index followed nearly the same optimum and failure curve points. Declining yield above 25°C is initially attributed to shorter grain filling duration (Chowdhury and Wardlaw 1978; Snyder 2000), and then to progressive failure to produce filled grains – the latter is caused by reduced pollen viability and reduced production of pollen (Kim et al. 1996; Matsui et al. 1997; Prasad et al. 2006b). Pollen viability and production begins to decline as daytime maximum temperature exceeds 33°C, and reaches zero at Tmax of 40°C (Kim et al. 1996). Because flowering occurs at mid-day in rice, Tmax is the best indicator of heat stress on spikelet sterility. Grain size of rice tends to hold mostly constant, declining only slowly across increasing temperature, until the pollination failure point (Baker and Allen 1993a). Rice ecotypes, *japonica* and *indica*, mostly do not differ in the upper temperature threshold (Snyder 2000; Prasad et al. 2006b), although the *indica* types are more sensitive to cool temperature (night temperature less than 19°C) (Snyder 2000).

Screening of rice genotypes and ecotypes for heat tolerance (33.1/27.3°C versus 28.3/21.3°C mean day/night temperatures) by Prasad et al. (2006b) demonstrated significant genotypic variation in heat tolerance for percent filled grains, pollen production, pollen shed, and pollen viability. The most tolerant cultivar had the smallest decreases in spikelet fertility, grain yield and harvest index at elevated temperature. This increment of temperature caused, for the range of 14 cultivars, 9-86 percent reduction in spikelet fertility, 0-93 percent reduction in grain weight per panicle, and 16-86 percent reduction in harvest index. Mean air temperature during the rice grain filling phase in summer in the southern United States and many tropical regions is about 26-27°C. These are above the 25°C optimum, which illustrates that elevated temperature above current will likely reduce U.S. and tropical region rice yield by about 10 percent per 1°C rise, or about 12 percent for a 1.2°C rise.

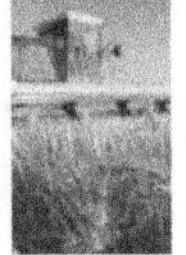

2.2.1.2.2.5 Sorghum

In general, the base and optimum temperatures for vegetative development are 8°C and 34°C, respectively (Alagarswamy and Ritchie 1991), while the optimum temperature for reproductive development is 31°C (Prasad et al. 2006a). Optimum temperature for sorghum vegetative growth is between 26°C and 34°C, and for reproductive growth 25°C and 28°C (Maiti 1996). Maximum dry matter production and grain yield occur at 27/22°C (Downs 1972). Grain filling duration is reduced as temperature increases over a wide range (Chowdury and Wardlaw 1978; Prasad et al. 2006a). Nevertheless, as temperature increased above 36/26°C to 40/30°C (diurnal maximum/minimum), panicle emergence was delayed by 20 days, and no panicles were formed at 44/34°C (Prasad et al. 2006a). Prasad et al. (2006a) found that grain yield, harvest index, pollen viability, and percent seed-set were highest at 32/22°C, and progressively reduced as temperature increased, falling to zero at 40/30°C. Vegetative biomass was highest at 40/30°C and photosynthesis was high up to 44/34°C. Seed size was reduced above 36/26°C. Rice and sorghum have exactly the same sensitivity of grain yield, seed harvest index, pollen viability, and success in grain formation (Prasad et al. 2006a). In addition, maize, a related warm-season cereal, may have the same temperature sensitivity. Basing the yield response of sorghum only on shortening of filling period (Chowdury and Wardlaw 1978), yield would decline 7.8 percent per 1°C temperature rise from 18.5-27.5°C (a 9.4 percent yield reduction for a 1.2°C increase). However, if site temperature is cooler than optimum for biomass/photosynthesis (27/22°C), then yield loss from shorter filling period would be offset by photosynthesis increase. The response from Chowdury and Wardlaw (1978) is supported by the 8.4 percent decrease in global mean sorghum yield per 1°C increase in temperature reported for sorghum by Lobell and Field (2007); therefore, the reported responses are likely.

2.2.1.2.2.6 Cotton

Cotton is an important crop in the southern United States, and is considered to have adapted to high-temperature environments. Despite this perception, reproductive processes of cotton have been shown to be adversely affected by elevated temperature (Reddy et al. 2000, 2005). Being a tropical crop, cotton's rate of leaf appearance has a relatively high base temperature of 14°C, and a relatively high optimum temperature of 37°C, thus leaf and vegetative growth appear to tolerate elevated temperature (Reddy et al. 1999, 2005). On the other hand, reproductive progression (emergence to first flower) has a temperature optimum of 28-30°C, along with a high base temperature of about 14°C (Reddy et al. 1997, 1999). Maximum growth rate per boll occurred at 25-26°C, declining at higher temperatures, while boll harvest index was highest at 28°C, declining at higher temperatures, reaching zero boll harvest index at 33-34°C (Reddy et al. 2005).

Boll size was largest at temperatures less than 20°C, declining progressively as temperature increased. Initially there was compensation with increased boll number set as temperature increased up to 35/27°C day/night temperature, but above 30°C mean temperature, percent boll set, boll number, boll filling period, rate of boll growth, boll size, and yield all decreased (Reddy et al. 2005). Instantaneous air temperature above 32°C reduces pollen viability, and temperature above 29°C reduces pollen tube elongation (Kakani et al. 2005), thus acting to progressively reduce successful boll formation to the point of zero boll yield at 40/32°C day/night (35°C mean) temperature (Reddy et al. 1992a, 1992b). Pettigrew (2008) evaluated two cotton genotypes under a temperature regime 1°C warmer than current temperatures and found lint yield was 10 percent lower in the warm regime. The reduced yields were caused by a 6 percent reduction in boll mass and 7 percent less seed in the bolls. These failure point temperatures show that cotton is more sensitive to elevated temperature than soybean and peanut, but similar in sensitivity to rice and sorghum. There is no well-defined cotton-yield response to temperature in the literature, but if cotton yield is projected with a quadratic equation from its optimum at 25°C to its failure temperature of 35°C, then a 1.2°C increase from 26.7°C to 27.9°C would give a possible yield decrease of 5.7 percent.

2.2.1.2.2.7 Peanut

Peanut is another important crop in the southern United States. The base temperature for peanut-leaf-appearance rate and onset of anthesis are 10°C and 11°C, respectively (Ong 1986). The optimum temperature for leaf appearance rate is above 30°C, while the optimum for rate of vegetative development to anthesis is 29-33°C (Bolhuis and deGroot 1959). Leaf photosynthesis has a fairly high optimum temperature of about 36°C. Cox (1979) observed that 24°C was the optimum temperature for single pod growth rate and pod size, with slower growth rate and smaller pod size occurring at higher temperatures. Williams et al. (1975) evaluated temperature effects on peanut by varying elevation, and found that peanut yield was highest at a mean temperature of 20°C (27/15°C max/min), a temperature that contributed to a long life cycle and long reproductive period. Prasad et al. (2003) conducted studies in sunlit controlled environment chambers, and reported that the optimum mean temperature for pod yield, seed yield, pod harvest index, and seed size occurred at a temperature lower than 26°C; quadratic projections to peak and minimum suggest that the optimum temperature was 23-24°C, with a failure point temperature of 40°C for zero yield and zero harvest index. Pollen viability and percent seed-set in that study began to fail at about 31°C, reaching zero at about 39-40°C (44/34°C treatment) (Prasad et al. 2003). For each individual flower, the period sensitive to elevated temperature starts six days prior to opening of a given flower and ends one day after, with greatest sensitivity on the day of flower opening (Prasad et al. 1999; Prasad et al. 2001). Percent fruit-set is first reduced at bud temperature of 33°C, declining linearly to zero fruit-set at 43°C bud temperature (Prasad et al. 2001).

Genotypic differences in heat-tolerance of peanut (pollen viability) have been reported (Craufurd et al. 2003). As air temperature in the southern United States already averages 26.7°C during the peanut growing season, any temperature increase will reduce seed yields (4.5 percent per 1°C, or 5.4 percent for a 1.2°C rise in range of 26-28°C) using the relationship of Prasad et al. (2003). At higher temperatures, 27.5-31°C, peanut yield declines more rapidly (6.9 percent per 1°C) based on unpublished data of Boote. A recent trend in peanut production has been the move of production from south Texas to west Texas, a cooler location with higher yield potential.

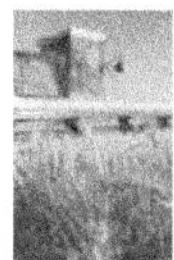

2.2.1.2.2.8 Dry Bean and Cowpea

Dry bean is typical of many vegetable crops and is grown in relatively cool regions of the United States. Prasad et al. (2002) found that red kidney bean, a large-seeded ecotype of dry bean, is quite sensitive to elevated temperature, having highest seed yield at 28/18°C (23°C mean) or lower (lower temperatures were not tested), with linear decline to zero yield as temperature increased to 37/27°C (32°C mean). In that study, pollen production per flower was reduced above 31/21°C, pollen viability was dramatically reduced above 34/24°C, and seed size was decreased above 31/21°C. Laing et al. (1984) found highest bean yield at 24°C, with a steep decline at higher temperatures. Gross and Kigel (1994) reported reduced fruit-set when flower buds were exposed to 32/27°C during the six to 12 days prior to anthesis and at anthesis, caused by non-viable pollen, failure of anther dehiscence, and reduced pollen tube growth. Heat-induced decreases in seed and fruit set in cowpea have been associated with formation of non-viable pollen (Hall 1992). Hall (1992) also reported genetic differences in heat tolerance of cowpea lines. Screening for temperature-tolerance within bean cultivars has not been done explicitly, but the Mesoamerican lines are more tolerant of warm tropical locations than are the Andean lines, which include the red kidney bean type (Sexton et al. 1994). Taking the initial slope of decline from data of Prasad et al. (2002), bean yield will likely decrease 7.2 percent per 1°C temperature rise, or 8.6 percent for 1.2°C above 23°C (Table 2.6).

2.2.1.2.2.9 Tomato

Tomato is an important vegetable crop known to suffer heat stress in mid-summer in southern U.S. locations. The base and optimum temperature is 7° and 22°C for rate of leaf appearance, rate of truss appearance, and rate of progress to anthesis (Adams et al. 2001). Leaf photosynthesis of tomato has a base at 6-8°C (Duchowski and Brazaityte 2001), while its optimum is about 30°C (Bunce 2000). The rate of fruit development and maturation has a base temperature of 5.7°C and optimum of 26°C, and rate of individual fruit growth has its optimum at 22-25°C (Adams et al. 2001). Largest fruit size occurs at 17-18°C, and declines at progressively higher temperature (Adams et al. 2001; De Koning 1996). Rate of fruit addition (fruit-set, from pollination) has an optimum at or lower than 26°C and progressively fails as temperature reaches 32°C (Adams et al. 2001). Peat et al. (1998) observed that the number of fruits per plant (or percent fruit-set) at 32/26°C day/night (29°C mean) was only 10 percent of that at 28/22°C (25°C mean). The projected failure temperature was about 30°C. Sato et al. (2000) found that only one of five cultivars of tomato successfully set any fruit at chronic exposures to 32/26°C, although fruit-set recovered if the stressful temperature was relieved.

Sato et al. (2000) also noted that pollen release and pollen germination were critical factors affected by heat stress. The anticipated temperature effect on tomato production will depend on the region of production and time of sowing (in the southern United States); however, at optima of 22°C for leaf/truss development, 22-26°C for fruit addition, 22-25°C for fruit growth, and fruit-set failures above 26°C, temperatures exceeding 25°C will likely reduce tomato production. Depending on region of production, tomato yield is projected to decrease 12.6 percent for 1.2°C rise above 25°C, assuming a non-linear yield response and assuming optimum temperature and failure temperatures for yield of 23.5°C and 30°C, respectively.

2.2.1.3 CROP RESPONSES TO CO_2

2.2.1.3.1 Overview of Individual Crop Responses to CO_2

Reviews of the early enclosure CO_2 studies indicate a 33 percent increase in average yield for many C_3 crops under a doubling CO_2 scenario (Kimball 1983) at a time when doubling meant increase from 330 to 660 parts per million (ppm) CO_2. The general phenomenon was expressed as increased numbers of tillers-branches, panicles-pods, and numbers of seeds, with minimal effect on seed size. The C_4 species response to doubling of CO_2 was reported by Kimball (1983) to be 10 percent. High temperature stress during reproductive development can negate CO_2's beneficial effects on yield, even though total biomass accumulation maintains a CO_2 benefit (e.g., for Phaseolus bean, Jifon and Wolfe 2000). Unrestricted root growth, optimum fertility, and excellent control of weeds, insects, and disease are also required to maximize CO_2 benefits (Wolfe 1994). Most C_3 weeds benefit more than C_3 crop species from elevated CO_2 (Ziska 2003).

In recent years, new field "free-air CO_2 enrichment" (FACE) technology has allowed evaluation of a few select crops to better understand their response under field conditions without enclosure-confounding effects. Generally, the FACE results corroborate previous enclosure studies (Ziska and Bunce 2007), although some FACE results suggest yield responses are less than previously reported (Long et al. 2006). Although the continuously increasing "ambient" reference concentration is a cause for lesser response, the smaller increment of CO_2 enrichment requires even better replication and sampling in FACE to evaluate the response. Enclosures are not the only concern; single-spaced plants, or unbordered plants may respond too much, and potted plants that are root bound may not respond well. Additional research, data analysis, and evaluation of a broader range of crops using FACE techniques will be required to sort discrepancies where they exist.

Table 2.7 Percent response of leaf photosynthesis, total biomass, grain yield, stomatal conductance, and canopy temperature or evapotranspiration, to a doubling in CO_2 concentration (usually 350 to 700 ppm, but sometimes 330 to 660 ppm). *Responses to increase from ambient to 550 or 570 ppm (FACE) are separately noted.

Crop	Leaf Photosynthesis	Total Biomass	Grain Yield	Stomatal Conductance	Canopy T, ET
	% change				
Corn	3[1*]	4[1,2,3,4]	4[1,2]	-34[5]	
Soybean	39[6]	37[6]	38[6], 34[7]	-40[6]	-9[8],-12[9,10*]
Wheat	35[11]	15-27[12]	31[13]	-33 to -43[14*]	-8[15,16*]
Rice	36[17]	30[17]	30[17,18]		-10[19,27]
Sorghum	9[20,21*]	3[22*]	8[20], 0[22*]	-37[21*]	-13[23*]
Cotton	33[24]	36[24]	44[24]	-36[24]	-8[25]
Peanut	27[26]	36[26]	30[26]		
Bean	50[26]	30[26]	27[26]		

References: [1]Leakey et al. (2006)*; [2]King and Greer (1986); [3]Ziska and Bunce (1997); [4]Maroco et al. (1999); [5]Leakey et al. (2006)*; [6]Ainsworth et al. (2002); [7]Allen and Boote (2000); [8]Allen et al. (2003); [9]Jones et al. (1985); [10]Bernacchi et al. (2007)*; [11]Long (1991); [12]Lawlor and Mitchell (2000); [13]Amthor (2001); [14]Wall et al. (2006)*; [15]Andre and duCloux (1993); [16]Kimball et al. (1999)*; [17]Horie et al. (2000); [18]Baker and Allen (1993a); [19]Baker et al. (1997a); [20]Prasad et al. (2006a); [21]Wall et al. (2001); [22]Ottman et al. (2001)*; [23]Triggs et al. (2004)*; [24]K.R. Reddy et al. (1995,1997); [25]Reddy et al. (2000); [26]Prasad et al. (2003); [27]Yoshimoto et al. (2005).

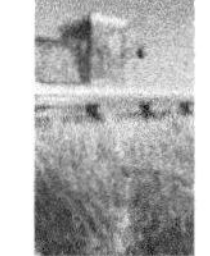

Effects of doubling of CO_2 on leaf photosynthesis, total biomass, grain or fruit yield, conductance, and canopy temperature or evapotranspiration (ET) of important non-water-stressed crops are shown in Table 2.7. (In addition to the specific references cited below, Kimball et al. (2002) provide CO_2 responses of several more crop and soil parameters for a variety of species.)

Maize, being a C_4 species, is less responsive to increased atmospheric CO_2. Single leaf photosynthesis of maize shows no effect of CO_2 on quantum efficiency, but there is a minor increase in leaf rate at light saturation (3 percent for 376 to 542 ppm; Leakey et al. 2006). There is a paucity of data for maize grown to maturity under elevated CO_2 conditions. Until 2006, there was only one data set for maize grown to maturity under CO_2 treatments: King and Greer (1986) observed 6.2 percent and 2.6 percent responses to increasing CO_2 from 355 to 625 and 875 ppm, respectively, in a 111-day study. The mean of the two levels gives about 4.4 percent increase to doubling or more of CO_2.

Leakey et al. (2006) conducted a full-season FACE study of maize grown to maturity, and reported no significant response of maize to a 50 percent increase in CO_2 (376 to 542 ppm (target: 370 to 550 ppm)). However, they used a very small biomass sample size in their FACE study (four random plant samples per replicate). This small sample size coupled with the small increment of CO_2 increase raises concern about whether these experimental measurements were sufficient to detect a statistically significant response. Ziska and Bunce (1997) reported a 2.9 percent increase in biomass when CO_2 was increased from 371 to 674 ppm during a 33-day, glasshouse study. Maroco et al. (1999) reported a 19.4 percent biomass increase when CO_2 was increased from 350 to 1,100 ppm during a 30-day growth period at very high light (supplemented above outdoor ambient) for a short duration on young plants. Thus, 4 percent increases in both biomass and grain yield of maize are possible, with increase in CO_2 from 350 to 700 ppm. This is less than the simulated 10 percent increase for C_4 species to incremental CO_2 increases (330 to 660 ppm) as parameterized in the CERES-Maize (Crop Environment Resource Synthesis) or EPIC (Environmental Policy Integrated Climate) models based on sparse data (Tubiello et al. 2007).

In summary, the evidence for maize response to CO_2 is sparse and questionable, resulting in only a possible degree of certainty. The expected increment of CO_2 increase over the next 30 years is anticipated to have a negligible effect (i.e., 1 percent) on maize production, unless there is a water-savings effect in drought years (Table 2.6). Sorghum, another important C_4 crop, gave 9, 34, and 8 percent increases in leaf photosynthesis, biomass, and grain yield, respectively, with doubling of CO_2 when grown in 1-by-2-meter, sunlit controlled-environment chambers (Prasad et al. 2005a). Over an entire season, with a CO_2 increase from 368 to 561 ppm, sorghum grown as part of a FACE study in Arizona gave 3 and 15 percent increases in biomass, and -4 percent and +20 percent change in grain yield, under irrigated versus water-limited conditions, respectively (Ottman et al. 2001).

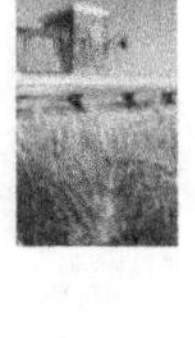

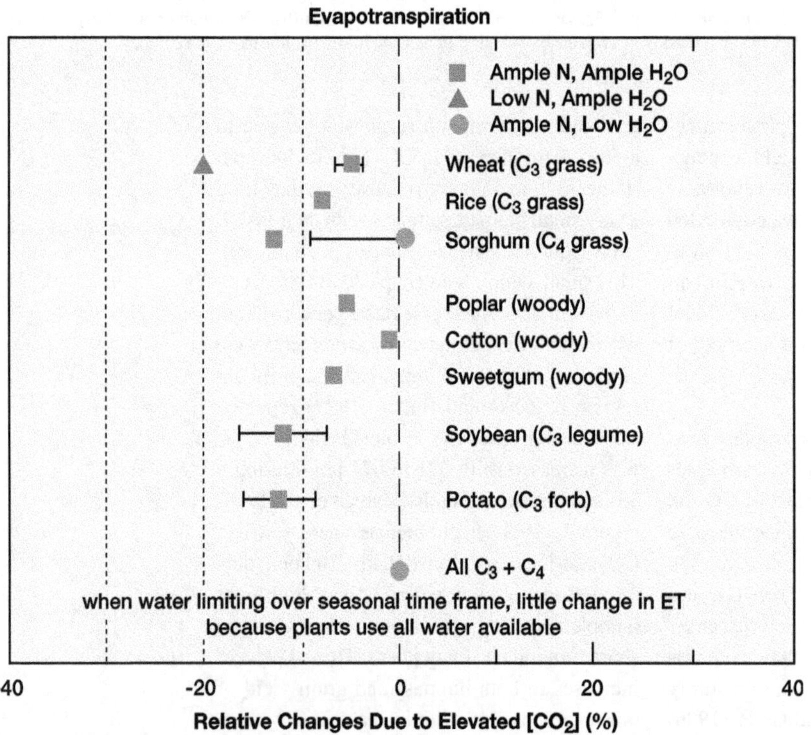

Figure 2.5 Relative changes in evapotranspiration due to elevated CO_2 concentrations in FACE experiments at about 550 ppm. [Wheat and cotton data from Table 2 of Kimball et al. (2002); rice datum from Yoshimoto et al. (2005); sorghum datum from Triggs et al. (2004); poplar datum from Tommasi et al. (2002); sweetgum from Wullschleger and Norby (2001); soybean datum from Bernacchi et al. (2007); and potato datum from Magliulo et al. (2003)].

Soybean is a C_3 legume that is quite responsive to CO_2. Based on the metadata summarized by Ainsworth et al. (2002), soybean response to a doubling of CO_2 is about 39 percent for light-saturated leaf photosynthesis, 37 percent for biomass accumulation, and 38 percent for grain yield. (These values are only from soybean raised in large, ≥1-square-meter crop stands grown in soil because yield response to CO_2 potted plants was shown to be affected by pot size). Allen and Boote (2000) reported a response of 34 percent in sunlit controlled-environment chambers to increases in CO_2 from 330 to 660 ppm. Ainsworth et al. (2002) found that under similar conditions, leaf conductance was reduced by 40 percent, which is consistent with other C_3 and C_4 species (Morison 1987), and seed harvest index was reduced by 9 percent. The C_3 photosynthetic response to CO_2 enrichment is well documented, and generally easy to predict using either the Farquhar and von Cammerer (1982) equations, or simplifications based on those equations. The CROPGRO-soybean model (Boote et al. 1998), parameterized with Farquhar kinetics equations (Boote and Pickering 1994; Alagarswamy et al. 2006), was used to simulate soybean yield to CO_2 rises from 350 to 700 ppm. The CROPGRO-soybean model predicted 29-41 percent increase in biomass, and 29 to 34 percent increase in grain yield (Boote et al. 1997), values that are comparable to metadata summarized by Ainsworth et al. (2002) and Allen and Boote (2000). Crop models can be used to project yield responses to CO_2 increase from past to present and future levels. Simulations by Boote et al. (2003) suggested that soybean yield in Iowa would have increased 9.1 percent between 1958 and 2000, during which time the CO_2 increased from 315 to 370 ppm; thus some of the past yield trend of soybean was associated with global change rather than technological innovation.

Using the same type of Michaelis-Menten rectangular hyperbola projection for soybean as used for all other crops, a CO_2 increase from 380 to 440 ppm is projected to increase yield by 7.4 percent (Table 2.7) in the dominant soybean-growing regions in the Midwest. For this region, expected temperatures are so close to the optimum for soybean yield, and the temperature increment so small (1.2°C) that the net effect of climate change on soybean yield is dominated by the CO_2 increment. To the extent that water-use efficiency increases with CO_2 enrichment and conserves soil water, yield response for rainfed regions will be enhanced by a net 0.9 percent increase in ET.

Other C_3 field crop species exhibit similar responses to increasing CO_2. For wheat, a cool-season cereal, doubling of CO_2 (350 to 700 ppm) increased light-saturated leaf photosynthesis by 30-40 percent (Long 1991), and grain yield by about 31 percent, averaged over many data sets (Amthor 2001). For rice, doubling CO_2 (330 to 660 ppm) increased canopy assimilation, biomass, and grain yield by about 36, 30, and 30 percent, respectively (Horie et al. 2000). Baker and Allen (1993a) reported a 31 percent increase in grain yield, averaged over five experiments, with increase of CO_2 from 330 to 660 ppm. Rice shows photosynthetic acclimation associated with decline in leaf nitrogen (N) concentration, and a 6-22 percent reduction in leaf rubisco content per unit leaf area (Vu et al. 1998).

For peanut, a warm-season grain legume, doubling CO_2 increased light-saturated leaf photosynthesis, total biomass and pod yield of peanut by 27, 36, and 30 percent, respectively (Prasad et al. 2003). Doubling CO_2 (350 to 700 ppm) increased light-saturated leaf photosynthesis, biomass, and seed yield of dry bean by 50, 30, and 27 percent (Prasad et al. 2002).

For cotton, a warm-season non-legume, doubling CO_2 (350 to 700 ppm) increased light-saturated leaf photosynthesis, total biomass, and boll yield by 33 percent, 36 percent, and 44 percent (K. R. Reddy et al. 1995, 1997), respectively, and decreased stomatal conductance by 36 percent (V. R. Reddy et al. 1995). Under well-watered conditions, leaf and canopy photosynthesis of cotton increased about 27 percent with CO_2 enrichment, to 550 ppm CO_2 in a FACE experiment in Arizona (Hileman et al. 1994). Mauney et al. (1994) reported 37 percent and 40 percent increases in biomass and boll yield of cotton with CO_2 enrichment to 550 ppm. Even larger increases in yield and biomass of cotton were obtained under the same enrichment for cotton under water-deficit situations (Kimball and Mauney 1993). An important consideration relative to cotton responses in Arizona is that the large vapor pressure deficit may have given more benefit to elevated CO_2 via water conservation effects. So, the degree of responsiveness in arid region studies may differ from that in humid regions. There were no reported effects of doubled CO_2 on vegetative or reproductive growth stage progression in cotton (Reddy et al. 2005), soybean (Allen and Boote 2000; Pan 1996), dry bean (Prasad et al. 2002), and peanut (Prasad et al. 2003).

In recent years, new field "free-air CO_2 enrichment" (FACE) technology has allowed evaluation of a few select crops to better understand their response under field conditions without enclosure-confounding effects.

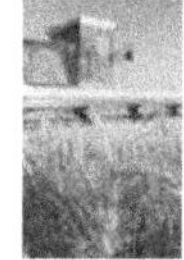

The certainty level of biomass and yield response of these C_3 crops to CO_2 is likely to very likely, given the large number of experiments and the general agreement in response across the different C_3 crops.

2.2.1.3.2 *Effects of CO_2 Increase in Combination with Temperature Increase*

There could be beneficial interaction of CO_2 enrichment and temperature on dry matter production (greater response to CO_2 as temperature rises) for the vegetative phase of non-competitive plants, as highlighted by Idso et al. (1987). This effect may be beneficial to production of radish (*Raphanus sativus*), lettuce (*Lactuca sativa*), or spinach (*Spinacea olervicea*), mainly because any factor that speeds leaf area growth (whether CO_2 or temperature) speeds the exponential phase of early growth. However, this "beta" factor effect does not appear to apply to closed canopies or to reproductive grain yield processes.

There are no reported beneficial interactions in grain yield caused by the combined effects of CO_2 and temperature increase for rice (Baker and Allen 1993a, 1993b; Baker et al. 1995; Snyder 2000), wheat (Mitchell et al. 1993), soybean (Baker et al. 1989; Pan 1994), dry bean (Prasad et al. 2002), peanut (Prasad et al. 2003), or sorghum (Prasad et al. 2005a). In other words, the separate main effects of CO_2 and temperature were present, but yield response to CO_2 was not enhanced as temperature increased. By contrast, there are three reported negative effects caused by elevated CO_2 and temperature in terms of fertility. Elevated CO_2 causes greater sensitivity of fertility to temperature in rice (Kim et al. 1996; Matsui et al. 1997), sorghum (Prasad et al. 2006a), and dry bean (Prasad et al. 2002). For rice, the relative enhancement in grain yield with doubled CO_2 decreases, and actually goes negative as Tmax increases in the range 32-40°C (Kim et al. 1996). Likewise, the relative CO_2 enhancement of grain yield of soybean (Baker et al. 1989) lessened as temperature increased from optimum to super-optimum. In the case of rice, sorghum, and dry bean, failure point temperature (i.e., the point at which reproduction fails) is about 1-2°C lower at elevated CO_2 than at ambient CO_2. This likely occurs because elevated CO_2 causes warming of the foliage (doubled CO_2 canopies of dry bean were 1.5°C warmer) (Prasad et al. 2002); doubled CO_2 canopies of soybean were 1-2°C warmer (Allen et al. 2003); doubled CO_2 canopies of sorghum averaged 2°C warmer during daytime period (Prasad et al. 2006a). The higher canopy temperature of rice, sorghum, and dry bean adversely affected fertility and grain-set. Increases in canopy temperature for wheat, rice, sorghum, cotton, poplar, potato, and soybean have been reported in FACE experiments (Kimball and Bernacchi 2006).

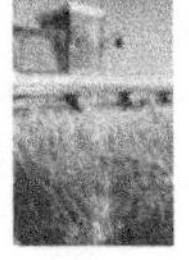

In cotton, there was progressively greater photosynthesis and vegetative growth response to CO_2 as temperature increased up to 34°C (Reddy 1995), but this response did not carry over to reproductive growth (Reddy et al. 1995). The reproductive enhancement from doubled CO_2 was largest (45 percent) at the 27°C optimum temperature for boll yield, and there was no beneficial interaction of increased CO_2 on reproductive growth at elevated temperature, reaching zero boll yield at 35°C (Reddy et al. 1995).
Mitchell et al. (1993) conducted field studies of wheat grown at ambient and +4°C temperature differential, and at elevated versus ambient CO_2 in England. While interactions of CO_2 and temperature did not affect yield, higher temperatures reduced grain yield at both CO_2 levels such that yields were significantly greater at ambient CO_2 and ambient temperature compared to elevated CO_2 and high temperature. Batts et al. (1997) similarly reported no beneficial interactions of CO_2 and temperature on wheat yield.

In studies with bean (Jifon and Wolfe 2005) and potato (Peet and Wolfe 2000), there were no significant beneficial effects of CO_2 on yield in high temperature treatments that negatively affected reproductive development, although the beneficial effects on vegetative biomass were maintained. These results suggest that in those regions and for those crops where climate change impairs crop reproductive development because of an increase in the frequency of high temperature stress events, the potential beneficial effects of elevated CO_2 on yield may not be fully realized.

For peanut, there was no interaction of elevated temperature with CO_2 increase, as the extent of temperature-induced decrease in pollination, seed-set, pod yield, seed yield, and seed harvest index was the same at ambient and elevated CO_2 levels (Prasad et al. 2003). For dry bean, Prasad et al. (2002) found no beneficial interaction of elevated temperature with CO_2 increase, as the temperature-induced decrease in pollination, seed-set, pod yield, seed yield, and seed harvest index were the same or even greater at elevated than at ambient CO_2 levels. The temperature-sensitivity of fertility (grain-set) and yield for sorghum was significantly greater at elevated CO_2 than at ambient CO_2 (Prasad et al. 2006a), thus showing a negative interaction with temperature associated with fertility and grain-set, but not photosynthesis.

2.2.1.3.3 Interactions of Elevated CO_2 with Nitrogen Fertility

For non-legumes like rice, there is clear evidence of an interaction of CO_2 enrichment with nitrogen (N) fertility regime. For *japonica* rice, Nakagawa et al. (1994) reported 17, 26, and 30 percent responses of biomass to CO_2 enrichment, at N applications of 40, 120, and 200 kg N ha-1, respectively. For *indica* rice, 0, 29, and 39 percent responses of biomass to CO_2 enrichment were reported at N applications of 0, 90, and 200 kg N per hectare, respectively (Ziska et al. 1996). For C_4 bahiagrass (*Paspalum notatum*), Newman et al. (2006) observed no biomass

Table 2.8 Sensitivity of evapotranspiration (ET; percent change in ET per °C change in temperature or percent change in ET per percent change in variable other than temperature) to changes in weather and plant variables as calculated by Kimball (2007) from the ASCE standardized hourly reference equation for alfalfa (Allen et al. 2005). The weather data were from the AZMET network (Brown 1987) for Maricopa, AZ, on a clear summer day (21 June 2000), and for the whole 2000 year. Calculations were made hourly then summed for the clear summer day and whole year.

Weather or Plant Variable	ET Sensitivity (°C or % change)	
	Summer Day	Whole Year
T_{ah}, air temperature with absolute humidity constant, EC	2.394	3.435
T_{rh}, air temperature with relative humidity constant, EC	1.489	2.052
R_s, solar radiation, %	0.585	0.399
e_a, absolute vapor pressure, %	-0.160	-0.223
u, wind speed, %	0.293	0.381
g_s, surface or canopy conductance, %	0.085	0.160
LAI, leaf area index, %	0.085	0.160

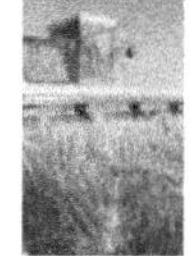

response to doubled CO_2 at low N fertilization rate, but observed 7-17 percent increases with doubled CO_2 when fertilized with 320 kg N per hectare. Biomass production in that study was determined over four harvests in each of two years (the 7 percent response in year one was non-significant, but 17 percent response in year two was significant).

2.2.1.3.4 *Effects of CO_2 Increase on Water Use and Water Use Efficiency*

2.2.1.3.4.1 *Changes in Crop Water Use due to Increasing Temperature, CO_2, and O_3*

Water use (i.e., ET) of crop plants is a physical process but is mediated by crop physiological and morphological characteristics (e.g., Kimball 2007). It can be described by the Penman-Monteith equation, whose form was recently standardized (Allen et al. 2005) (Table 2.8). The equation reveals several mechanisms by which the climate change parameters – temperature, CO2, and O3 – can affect water use. These include: (1) direct effects on crop growth and leaf area, (2) alterations in leaf stomatal aperture and consequently their conductance for water vapor loss, and (3) physical changes in the vapor pressure inside leaves.

When plants are young and widely spaced, increases in leaf area are approximately proportional to the increases in growth, and transpiration increases accordingly. More importantly, duration of leaf area will affect total seasonal crop water requirements. Thus, the lengthening of growing seasons due to global warming likely will increase crop water requirements. On the other hand, for some determinate cereal crops, increasing temperature can hasten plant maturity, thereby shortening the leaf area duration with the possibility of reducing the total season water requirement for such crops.

Elevated CO_2 causes partial stomatal closure, which decreases conductance, and reduces loss of water vapor from leaves to the atmosphere. Reviews of the effects of elevated CO_2 on stomatal conductance from chamber-based studies have reported that, on average, a doubling of CO_2 (from about 340 to 680 ppm) reduces stomatal conductance about 34 percent (e.g., Kimball and Idso 1983). Morison (1987) calculated an average reduction of about 40 percent, with no difference between C_3 and C_4 species. More recently, Wand et al. (1999) performed a meta-analysis on observations reported for wild C_3 and C_4 grass species, and found that with no stresses, elevated CO_2 reduced stomatal conductance by 39 and 29 percent for C_3 and C_4 species, respectively. The stomatal conductance of woody plants appears to decrease less than that of herbaceous plants in elevated CO_2, as

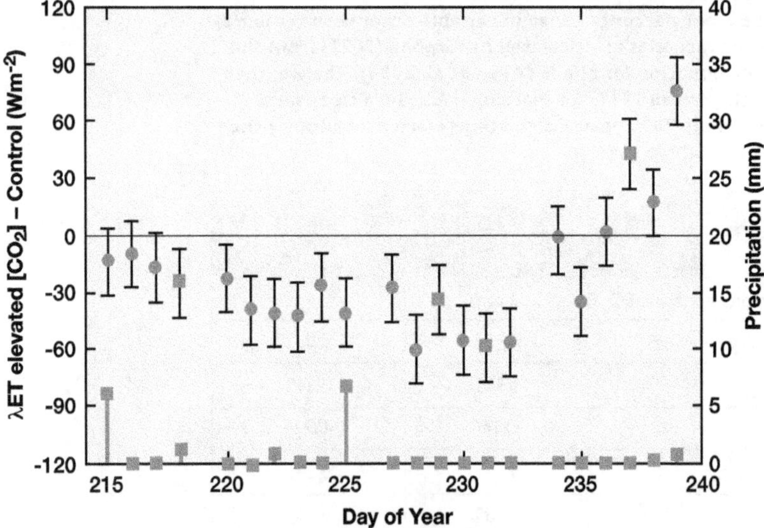

Figure 2.6 Differences in evapotranspiration rate (latent energy, W m-2) between soybean plots enriched to 550 ppm from free-air CO_2 enrichment (FACE) and plots at today's ambient CO_2 levels at Urbana, IL, versus day of year (circles, left axis). Corresponding precipitation is also shown (squares, right axis). Adapted from Bernacchi et al. 2007.

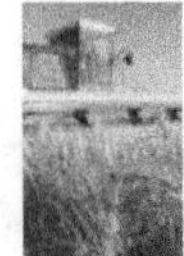

indicated by an 11 percent reduction in the meta-analysis of woody plant data by Curtis and Wang (1998). Ainsworth et al. (2002) found an average reduction of about 40 percent in conductance of soybean for a wide range of CO_2 concentrations, with the reduction for a doubling being about 30 percent. Meta-analysis by Ainsworth and Long (2005) and Ainsworth and Rogers (2007) of data generated by free-air CO_2 enrichment experiments, for which the daytime concentrations were 550-600 ppm, versus ambient concentrations of about 360 ppm, produced an average reduction in stomatal conductance of 20 and 22 percent, respectively. They did not detect any significant difference between C_3 and C_4 species. Projecting out 30 years, the atmospheric CO_2 concentration likely will be about 440 ppm (see Introduction). Interpolating from these reviews, it appears very likely that an increase in CO_2 concentration from 380 to 440 ppm will cause reductions in stomatal conductance on the order of 10 percent compared to today's values.

However, as plants shift from vegetative to reproductive growth during their life cycles, proportionately more of the accumulating biomass is partitioned to other organs, such as developing grain. At this point, leaf area and biomass accumulation are no longer proportional. Also, as plants grow and leaf area index (LAI) increases, the mutual shading and interference among the leaves within a plant canopy cause plant transpiration to plateau (Ritchie 1972; Villalobos and Fereres 1990; Sau et al. 2004). Further, considering that a doubling of CO_2 from present-day levels is likely to increase average C_3 species growth on the order of 30 percent (e.g., Kimball 1983, 2007; Kimball et al. 2002; Table 2.7), so projecting out 30 years to a CO_2 concentration of about 440 ppm suggests increases in C_3 plant growth only on the order of 10 percent. Therefore, because changes in growth affect ET mostly while plants are small (i.e., after planting), and progressively less after canopy closure, changes in ET rates over the next 30 years due to leaf area index effects are likely to be minor (Figure 2.5).

Elevated CO_2 concentrations – approximately 550 ppm or about 180 ppm above ambient – in FACE experiments have reduced water use in experimental plots by about 2-13 percent, depending on species (Figure 2.6). Interpolating linearly to 440 ppm of CO_2, the corresponding reductions likely would be about one-third of those observed in the FACE experiments (i.e., 1-4 percent). Because there are fetch considerations in extrapolating FACE plot data to larger areas (see discussion in Triggs et al. 2004), reductions in crop water requirements due to elevated CO_2 likely will be significant, but smaller yet.

Less research has been done on the effects of elevated O_3 on stomatal conductance compared to elevated CO_2, but some pertinent work has been published. Barnes et al. (1995) and Balaguer et al. (1995) measured stomatal conductance of wheat exposed to elevated CO_2 (700 ppm), elevated O_3 (about 75 ppb), and combined elevated CO_2 plus O_3 in controlled environment chambers. The ozone treatment reduced conductance by about 20 percent, while both CO_2 and CO_2+O_3 reduced conductance by 40 percent. Wheat was exposed by Donnelly et al. (2000) to elevated CO_2 (680 ppm) and O_3 (50 or 90 ppb) and CO_2+O_3 in open-top chambers, and they found that all three treatments produced reductions in stomatal conductance of approximately 50 percent, with relative order changing with days after sowing and year. Using open-top chambers with potato, both Lawson et al. (2002) and Finnan et al. (2002) report 50 percent reduction of stomatal conductance with elevated CO_2 (680 ppm) and a similar amount in combination with elevated O_3, but their results are variable and mutually inconsistent among treatments. In a FACE project that included both CO_2 and O_3 treatments, Noormets et al. (2001) measured stomatal conductance of aspen leaves. Results varied with leaf age and aspen clone, but generally it appears that conductance had the following treatment rank: Control>O_3>CO_2+O_3>CO_2. Morgan et al. (2003) performed a meta-analysis of 53 prior chamber studies in which O_3 was elevated by 70 ppm above clean air, and found that stomatal conductance was reduced by 17 percent on average. However, in a recent FACE soybean experiment in which O_3 was elevated by 50 percent above ambient conditions, Bernacchi et al. (2007) detected no significant effect of O_3 on stomatal conductance. Thus, while chamber studies comparing the effects of O_3 on stomatal conductance showed that reductions can occur, in the case of field-grown plants exposed to present-day ambient levels of O_3 that are considerably above zero, the effects on conductance of the additional increases in O_3 levels that are likely to occur in the next 30 years are likely to be rather small.

Water vapor pressure (e) inside leaves is tightly coupled to leaf temperature (T) and increases exponentially (e.g., as described by the Teten's equation, e=0.61078*exp(17.269*T/(T+237.3)). Therefore, anything that affects the energy balance and temperature of a crop's leaf canopy will affect leaf water vapor pressure, and ultimately water consumption. Consequently, so long as there are no significant concomitant compensatory changes in other factors such as humidity, it is virtually certain that air temperature increases will also increase crop canopy temperature, leaf water vapor pressure, and ET (Figure 2.5). Based on the sensitivity analysis of Kimball (2007; Table 2.8), an increase of about 1.2°C with constant relative humidity, such as expected in 30 years (see Introduction), is likely to cause a small increase of about 1.8% in summer-day ET of a standard alfalfa reference crop if CO_2 concentrations were to remain at today's level. As already dicussed, CO_2 concentrations of about 440 ppm are likely to cause small decreases in ET, so therefore, the net effect of increased temperature plus CO_2 likely will result in insignificant changes in ET within the next 30 years.

Another aspect to consider is the dynamics of crop water use and the timing of rain/irrigation events. The latent energy associated with ET from soybean was 10 to 60 W/m2 less in the FACE plots compared to the control plots at ambient CO_2 when the crop had ample water (Figure 2.6).

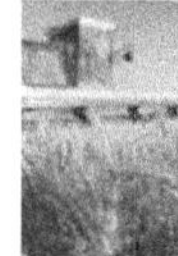

However, on about Day-of-Year (DOY) 233, the control plots had exhausted the water supply, and their water use declined (Bernacchi et al. 2006) (Figure 2.6). In contrast, the water conservation in the elevated-CO_2 plots enabled plants to keep their stomata open and transpiring, and for DOYs 237-239, the FACE plots transpired more water than the controls. During this latter period, the FACE plants had their stomata open, while those of the control plots were closed. As a result, the FACE plots were able to continue photosynthesizing and growing while the controls were not. In other words, elevated concentrations of CO_2 can enable some conservation of soil water for rain-fed agriculture, which often experiences periods of drought, and can sustain crop productivity over more days than is true at today's CO_2 levels.

The net irrigation requirement is the difference between seasonal ET for a well-watered crop and the amounts of precipitation and soil water storage available during a growing season. A few researchers have attempted to estimate future changes in irrigation water requirements based on projected climate changes (including rainfall changes) from general circulation models (GCMs), and estimates of decreased stomatal conductance due to elevated CO_2 (e.g., Allen et al. 1991; Izaurralde et al. 2003). Izaurralde et al. (2003) used EPIC, a crop growth model, to calculate growth and yield, as well as future irrigation requirements of corn and alfalfa. Following Stockle et al. (1992a, b), EPIC was modified to allow stomatal conductance to be reduced with increased CO_2 concentration (28 percent reduction corresponding to 560 μmol CO_2 mol-1), as well as increasing photosynthesis via improved radiation use efficiency. For climate change projections, they used scenarios generated for 2030 by the Hadley Centre's (HadCM2J) GCM, which was selected because its climate sensitivity is in the midrange of most of the GCMs. For corn, Izaurralde et al. (2003) calculated that by 2030 irrigation requirements will change from -1 (Lower Colorado Basin) to +451 percent (Lower Mississippi Basin), because of rainfall variation. Given the variation in the sizes and baseline irrigation requirements of U.S. basins, a representative figure for the overall U.S. increase in irrigation requirements is 64 percent if stomatal effects are ignored, or 35 percent if they are included. Similar calculations were made for alfalfa, for which overall irrigation requirements are predicted to increase 50 and 29 percent in the next 30 years in the cases of ignoring and including stomatal effects, respectively. These increases are more likely due to the decrease in rainfall during the growing season and the reduction in soil water availability.

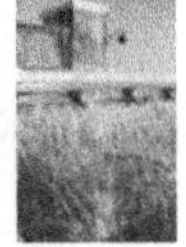

2.2.1.3.4.2 Implications for Irrigation and Water Deficit

As mentioned above, stomatal conductance is reduced about 40 percent for doubling of CO_2 for both C_3 and C_4 species (Morison 1987), thus causing water conservation effects, and potentially less water deficit. However, actual reduction in crop transpiration and ET will not be as great as the reduction in stomatal conductance because warming of the foliage to solve the energy balance will increase both latent heat loss (transpiration) and sensible heat loss. Allen et al. (2003) concluded that both increased foliage temperature, and increased LAI associated with CO_2 enrichment were responsible for the compensatory effects on ET (small to non-existent reductions). Jones et al. (1985) reported 12 percent reduction in season-long transpiration and 51 percent increase in water use efficiency (WUE) measured for canopies of soybean crops grown in ambient and doubled CO_2 in sunlit, controlled environment chambers. In experimental studies in the same chambers, foliage temperatures measured by infrared sensors have typically been increased 1-2°C for soybean, 1.5°C for dry bean, and 2°C for sorghum in response to doubled CO_2 (Pan 1996; Prasad et al. 2002; Prasad et al. 2006a). Similarly, in FACE experiments at about 550 ppm CO_2 foliage temperatures increased by an average 0.6°C for wheat (Kimball et al. 2002), 0.4°C for rice (Yoshimoto et al. 2005), 1.7°C for sorghum (Triggs et al. 2004), 0.8°C for cotton (Kimball et al. 2002), 0.8°C for potato (Magliuo et al. 2003), and 0.2 to 0.5°C for soybean (Bernacchi et al. 2007).

Allen et al. (2003) reported that soybean foliage at doubled CO_2 was, on average, 1.3°C warmer at mid-day. Andre and du Cloux (1993) reported an 8 percent decrease in transpiration of wheat in response to doubled CO_2, which compares well to a 5 percent reduction in ET of wheat for a 200 ppm CO_2 increase in FACE studies (Hunsaker et al. 1997; Kimball et al. 1999) (Figure 2.5). Reddy et al. (2000), using similar chambers, found an 8 percent reduction in transpiration of cotton canopies at doubled CO_2, averaged over five temperature treatments, while Kimball et al. (1983) found a 4 percent reduction in seasonal water use of cotton at ambient versus 650 ppm CO_2 in lysimeter experiments in Arizona. Soybean canopies grown at 550 compared to 375 ppm in FACE experiments in Illinois had 9-16 percent decreases in ET depending on season. Their data show an average 12 percent reduction over three years (Bernacchi et al. 2007). Allen et al. (2003) observed 9 percent reduction in ET of soybean with doubling of CO_2 in the sunlit, controlled environment chambers for a 28/18°C treatment (about the same mean temperature as the Illinois site), but they observed no reduction in ET for a high temperature treatment 40/30°C. The extent of CO_2-related reduction in ET appears to be dependent on temperature. In their review, Horie et al. (2000) reported the same phenomenon in rice, where doubling CO_2 caused 15 percent reduction in ET at 26°C, but resulted in increased ET at higher temperatures (29.5°C). At 24-26°C, rice's WUE increased 50 percent with doubled CO_2, but the CO_2 enrichment effect declined as temperature increased. At higher temperature, CO_2-induced reduction in conductance lessened.

Using observed sensitivity of soybean stomatal conductance to CO_2 in a crop climate model,

Allen (1990) showed that CO_2 enrichment from 330 to 800 ppm should cause an increase in foliage temperature of about 1°C when air vapor pressure deficit is low, but an increase of about 2.5 and 4°C with air vapor pressure deficit of 1.5, and 3 kPa, respectively. At the higher vapor pressure deficit values, the foliage temperatures simulated with this crop climate model (Allen 1990) exceeded the differential observed under larger vapor pressure deficit in the sunlit, controlled-environment chambers (Prasad et al. 2002; Allen et al. 2003; Prasad et al. 2006a). Allen et al. (2003) found that soybean canopies increased their conductance (lower resistance) at progressively larger vapor pressure deficit (associated with higher temperature), such that foliage temperature did not increase as much as supposed by the crop-climate model. Concurrently, the anticipated degree of reduction in ET with doubling of CO_2, while being 9 percent less at cool temperatures (28/18°C), became progressively less and was non-existent (no difference) at very high temperatures (40/30°C and 44/34°C). In other words, the CO_2-induced reduction in conductance became less as temperature increased.

Boote et al. (1997) used a version of the CROPGRO-Soybean model with hourly energy balance and feedback of stomatal conductance on transpiration and leaf temperature (Pickering et al.1995), to study simulated effects of 350 versus 700 ppm CO_2 for field weather from Ohio and Florida. The simulated transpiration was reduced 11-16 percent for irrigated sites and 7 percent for a rainfed site in Florida, while the ET was reduced 6-8 percent for irrigated sites and 4 percent for the rainfed site. Simulated water use efficiency was increased 53-61 percent, which matches the 50-60 percent increase in soybean WUE reported by Allen et al. (2003) for doubling of CO_2. The smaller reduction in transpiration and ET for the rainfed site was associated with more effective prolonged use of the soil water, also giving a larger yield response (44 percent) for rainfed crop than for irrigated (32 percent). The model simulated reductions in transpiration were close (11-16 percent) to those measured (12 percent) by Jones et al. (1985), and the reduction was much less than the reduction in leaf conductance. The model simulations also produced a 1°C higher foliage temperature at mid-day under doubled CO_2.

Interactions of CO_2 enrichment with climatic factors of water supply and evaporative demand will be especially evident under water deficit conditions. The reduction in stomatal conductance with elevated CO_2 will cause soil water conservation and potentially less water stress, especially for crops grown with periodic soil water deficit, or under high evaporative demand. This reduction in water stress effects on photosynthesis, growth, and yield has been documented for both C_3 wheat (Wall et al. 2006) and C_4 sorghum (Ottman et al. 2001; Wall et al. 2001; Triggs et al. 2004). Sorghum grown in the Arizona FACE site showed significant CO_2-induced enhancement of biomass and grain yield for water deficit treatments, but no significant enhancement for sorghum grown with full irrigation (Ottman et al. 2001). In the sorghum FACE studies, the stomatal conductance was reduced 32-37 percent (Wall et al. 2001), while ET was reduced 13 percent (Triggs et al. 2004).

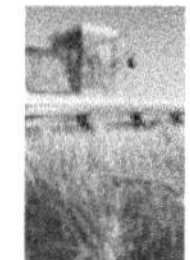

2.2.1.4 Crop Response to Tropospheric Ozone

Ozone at the land surface has risen in rural areas of the United States, particularly over the past 50 years, and is forecast to continue increasing during the next 50 years. The Midwest and eastern U.S. have some of the highest rural ozone levels on the globe. Average ozone concentrations rise toward the east and south, such that average levels in Illinois are higher than in Nebraska, Minnesota, and Iowa. Only western Europe and eastern China have similarly high levels. Argentina and Brazil, like most areas of the Southern Hemisphere, have much lower levels of ozone, and are forecast to see little increase over the next 50 years. Increasing ozone tolerance will therefore be important to the competitiveness of U.S. growers. Numerous models for future changes in global ozone concentrations have emerged that are linked to IPCC scenarios, so the impacts of ozone can be considered in the context of wider global change. For example, a model that incorporates expected economic development and planned emission controls in individual countries projects increases in annual mean surface ozone concentrations in all major agricultural areas of the Northern Hemisphere (Dentener et al. 2005).

Ozone at the land surface has risen in rural areas of the United States, particularly over the past 50 years, and is forecast to continue increasing during the next 50 years.

Ozone is a secondary pollutant resulting from the interaction of nitrogen oxides with sunlight and hydrocarbons. Nitrogen oxides are produced in the high-temperature combustion of any fuel. They are stable and can be transported thousands of miles in the atmosphere. In the presence of sunlight, ozone is formed from these nitrogen oxides and, in contrast to most pollutants, higher levels are observed in rural than urban areas. This occurs because rural areas have more hours of sunshine and less haze, and city air includes short-lived pollutants that react with, and remove, ozone. These short-lived pollutants are largely absent from rural areas. Levels of ozone during the day in much of the Midwest now reach an average of 60 parts of ozone per billion parts of air (ppb), compared to less than 10 ppb 100 years ago. While control measures on emissions of NOx and volatile organic carbons (VOCs) in North America and western Europe are reducing peak ozone levels, global background tropospheric ozone concentrations are on the rise (Ashmore 2005). Ozone is toxic to many plants, but studies in greenhouses and small chambers have shown soybean, wheat, peanut, and cotton are the most sensitive of our major crops (Ashmore 2002).

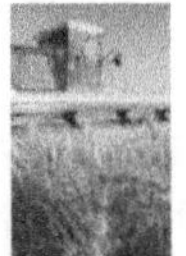

Ozone effects on soybean crops have been most extensively studied and best analyzed. This is because soybean is the most widely planted dicotyledonous crop, and is our best model of C_3 annual crops. The response of soybean to ozone can be influenced by the ozone profile and dynamics, nutrient and moisture conditions, atmospheric CO_2 concentration, and even the cultivar investigated, which creates a very complex literature to interpret. Meta-analytic methods are useful to quantitatively summarize treatment effects across multiple studies, and thereby identify commonalities. A meta-analysis of more than 50 studies of soybean, grown in controlled environment chambers at chronic levels of ozone, show convincingly that ozone exposure results in decreased photosynthesis, dry matter, and yield (Morgan et al. 2003). Even mild chronic exposure (40-60 ppb) produces such losses, and these losses increase linearly with ozone concentration (Morgan et al. 2003) as anticipated from the exposure/response relationship shown by Mills et al. (2000).

The meta-analytic summary further reveals that chronic ozone lowers the capacity of carbon uptake in soybean by reducing photosynthetic capacity and leaf area. Soybean plants exposed to chronic ozone levels were shorter with less dry mass and fewer set pods, which contained fewer, smaller seeds. Averaged across all studies, biomass decreased 34 percent, and seed yield was 24 percent lower, but photosynthesis was depressed by only 20 percent. Ozone damage increased with the age of the soybean, consistent with the suggestion that ozone effects accumulate over time (Adams et al. 1996; Miller et al. 1998), and may additionally reflect greater sensitivity of reproductive developmental stages, particularly seed filling (Tingey et al. 2002). The meta-analysis did not reveal any interactions with other stresses, even stresses expected to lower stomatal conductance and therefore ozone entry into the leaf (Medlyn et al. 2001). However, all of the ozone effects on soybean mentioned above were less under elevated CO_2, a response generally attributed to lower stomatal conductance (Heagle et al. 1989).

Plant growth in chambers can be different compared to the open field (Long et al. 2006), and therefore the outcomes of chamber experiments have been questioned as a sole basis for projecting yield losses due to ozone (Elagoz and Manning 2005). FACE experiments in which soybeans were exposed to a 20 percent elevation above ambient ozone levels indicate that ozone-induced yield losses were at least as large under open air treatment. In 2003, the background ozone level in central Illinois was unusually low over the growing season, averaging 45 parts per billion (ppb). Elevation of ozone by 20 percent in this year raised the ozone concentration to the average of the previous 10 years. In the plots with elevated ozone in 2003, yields were reduced approximately 25 percent (Morgan et al. 2006). This suggests that, in a typical year under open-air field conditions, yield loss due to ozone is even greater than predictions from greenhouse experiments (Ashmore 2002).

Analysis in the soybean FACE results showed a significant decrease in leaf area (Dermody et al. 2006), a loss of photosynthetic capacity during grain filling, and earlier senescence of leaves (Morgan et al. 2004). This may explain why yield loss is largely due to decreased seed size rather than decreased seed number (Morgan et al. 2006). On average, yield losses in Illinois soybean FACE experiments between 2002 and 2005 were 0.5 percent per ppb ozone increase over the 30 ppb threshold, which is twice the ozone sensitivity as determined in growth chamber studies (Ashmore 2002). These results suggest that during an average year, ozone is currently causing soybean yield losses of 10-25 percent in the Midwest, with even greater losses in some years. The IPCC forecasts that ozone levels will continue to rise in the rural Midwest by about 0.5 ppb per year, suggesting that soybean yields may continue to decline by 1 percent every two to four years. The IPCC also forecasts that ozone, which is low in South America, will remain low in that region over the next 50 years.

Meta-analysis has not been conducted for the effects of ozone on any crops other than soybean, or across different crops. Nevertheless, there is little doubt that current tropospheric ozone levels are limiting yield in many crops (e.g., Heagle 1989) and further increases in ozone will reduce yield in sensitive species further. The effect of exposure to ozone on yield and yield parameters from studies conducted prior to 2000 are compiled in Table 4 of Black et al. (2000), which reveals that, in addition to soybean, the yield of C_3 crops, such as wheat, oats, French and snap bean, pepper, rape, and various cucurbits, are highly sensitive to chronic ozone exposure. Yield of woody perennial cotton is also highly sensitive to ozone (e.g., Temple 1990; Heagle et al. 1996). While there are isolated reports that maize yield is reduced by ozone (e.g., Rudorff et al. 1996), C_4 crops are generally much less sensitive to ozone. Recent studies by Booker et al. (2007) and Burkey et al. (2007) on peanuts that evaluated the effect of ozone under CO_2 levels from 375 to 730 ppm, and ozone levels of 22-75 ppb, showed that CO_2 increases offset the effects of ozone. Increasing CO_2 levels overcame the effect of ozone on peanut yield; however, in none of the treatments was there a change in seed quality, or protein or oil content of the seed (Burkey et al. 2007).

2.2.2 Pastureland

In general, grassland species have received less attention than cropland species for their response to projected changes in temperature, precipitation, and atmospheric CO_2 concentration associated with climate change (Newman et al. 2001). Pastureland response to climate change is complex because, in addition to the major climatic drivers (CO_2 concentration, temperature, and precipitation), other plant and management factors affect this response (e.g., plant competition, perennial growth habits, seasonal productivity, etc.). Many of the studies in our review of published materials that report on temperate-climate pasture responses to changes in temperature, precipitation, and CO_2 concentrations originate from regions outside the United States.

An early comprehensive greenhouse study examined the photosynthetic response of 13 pasture species (Table 2.9) to elevated CO_2 (350 and 700 ppm) and temperature (12/7°C, 18/13°C, and 28/23°C for daytime/nighttime temperatures) (Greer et al. 1995). On average, photosynthetic rates increased by 40 percent under elevated CO_2 in C_3 species, while those for C_4 species remained largely unaffected. The response of C_3 species to elevated CO_2 decreased as temperatures increased from 12-28°C. However, the temperatures at which the maximum rates of photosynthesis occurred varied with species and level of CO_2 exposure. At 350 ppm, four species (*L. multiflorum*, *A. capillaris*, *C. intybus*, and *P. dilatatum*) showed maximum rates of photosynthesis at 18°C while, for the rest, the maximum occurred at 28°C. At 700 ppm, rates shifted upwards from 18-28°C in *A. capillaries*, and downward from 28-18°C in *L. perenne*, *F. arundinacea*, *B. wildenowii*, and *T. subterraneum*. However, little if any correlation existed between the temperature response of photosynthesis and climatic adaptations of the pasture species.

In Florida, a 3-year study examined the effects of elevated atmospheric CO_2 (360 and 700 ppm), and temperature (ambient temperature or baseline (B), B+1.5°C, B+3.0°C, and B+4.5°C) on dry matter yield of rhizoma peanut (a C_3 legume), and bahiagrass (a C_4 grass) (Newman et al. 2001). On average, yields increased by

Table 2.9 Pasture species studied for response to CO_2 and temperature changes. Adapted from Greer et al. (1995).

Species	Common name	Photosynthetic pathway	Growth characteristics
Lolium multiflorum	Italian ryegrass	C_3	Cool season annual grass
Bromus wildenowii		C_3	Cool season perennial grass
Lolium perenne	Ryegrass	C_3	Cool season perennial grass
Phalaris aquatica		C_3	Cool season perennial grass
Trifolium dubium		C_3	Cool season annual broadleaf
Trifolium subterraneum	Subterraneum clover	C_3	Cool season annual broadleaf
Agrostis capillaris		C_3	Warm season perennial grass
Dactylis glomerata	Orchardgrass	C_3	Warm season perennial grass
Festuca arundinacea	Tall fescue	C_3	Warm season perennial grass
Cichorium intybus		C_3	Warm season perennial broadleaf
Trifolium repens	White clover	C_3	Warm season perennial broadleaf
Digitaria sanguinalis	Crabgrass	C_4	Warm season annual grass
Paspalum dilatatum	Dallisgrass	C_4	Warm season perennial grass

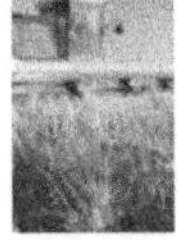

25 percent in rhizoma peanut plots exposed to elevated CO_2, but exhibited only a positive trend in bahiagrass plots under the same conditions. These results are consistent with C_3- and C_4-type plant responses to elevated CO_2.

The response of forage species to elevated CO_2 may be affected by grazing and aboveground/belowground interactions (Wilsey 2001). In a phytotron study, Kentucky bluegrass and timothy (Phleum pratense L.) were grown together in pots during 12 weeks under ambient (360 ppm) and elevated CO_2 (650 ppm), with and without aboveground defoliation, and with and without the presence of *Pratylenchus penetrans*, a root-feeding nematode commonly found in old fields and pastures. Timothy was the only species that responded to elevated CO_2 with an increase in shoot biomass, leading to its predominance in the pots. This suggests that Kentucky bluegrass might be at the lower end of the range in the responsiveness of C_3 grasses to elevated CO_2, especially under low nutrient conditions. Defoliation increased productivity only under ambient CO_2; thus, the largest response to elevated CO_2 was observed in non-defoliated plants. Timothy was the only species that showed an increase in root biomass under elevated CO_2. Defoliation reduced root biomass. Elevated CO_2 interacted with the presence of nematodes in reducing root biomass. In contrast, defoliation alleviated the effect of root biomass reduction caused by the presence of nematodes. This study demonstrates the importance of using aboveground/belowground approaches when investigating the environmental impacts of climate change (Wardle et al. 2004).

Kentucky bluegrass might not be the only species showing low response to elevated CO_2. Perennial ryegrass (Lolium perenne) has been reported to have low or even negative yield response to elevated CO_2 under field conditions but, contradictorily, often shows a strong response in photosynthetic rates (Suter et al. 2001). An experiment at the Swiss FACE examined the effects of ambient (360 ppm) and elevated (600 ppm) CO_2 on regrowth characteristics of perennial ryegrass (Suter et al. 2001). Elevated CO_2 increased root mass by 68 percent, pseudostems by 38 percent, and shoot necromass below cutting height by 45 percent during the entire regrowth period. Many of the variables measured (e.g., yield, dry matter, and leaf area index) showed a strong response to elevated CO_2 during the first regrowth period but not during the second, suggesting a lack of a strong sink for the extra carbon fixed during the latter period.

When combined, rising CO_2 and projected changes in temperature and precipitation may significantly change the growth and chemical composition of plant species. However, it is not clear how the various forage species that harbor mutualistic relationships with other organisms would respond to elevated CO_2. Newman et al. (2003) studied the effects of endophyte infection, N fertilization, and elevated CO_2 on growth parameters and chemical composition of tall fescue. Fescue plants, with and without endophyte infection (*Neotyphodium coenophialum*), were transplanted to open chambers and exposed to ambient (350 ppm) and elevated (700 ppm) levels of CO_2. All chambers were fertilized with uniform rates of phosphorus (P) and potassium (K). Nitrogen fertilizer was applied at rates of 6.7 and 67.3 g m^{-2}. The results revealed complex interactions of the effects of elevated CO_2 on the mutualistic relationship between a fungus and its host, tall fescue. After 12 weeks of growth, plants grown under elevated CO_2 exhibited apparent photosynthetic rates 15 percent higher than those grown under ambient conditions. The presence of the endophyte fungus in combination with N fertilization enhanced the CO_2 fertilization effect. Elevated CO_2 accelerated the rate of tiller appearance and increased dry matter production by at least 53 percent (under the low N treatment). Contrary to previous findings, Newman et al. (2003) found that elevated CO_2 decreased lignin concentrations by 14 percent. Reduced lignin concentration would favor the diet of grazing animals, but hinder stabilization of carbon in soil organic matter (Six et al. 2002).

Climate change may cause reduction in precipitation and, in turn, induce soil moisture limitations in pasturelands. An experiment in New Zealand examined the interaction of elevated CO_2 and soil moisture limitations on the growth of temperate pastures (Newton et al. 1996). Intact turves (plural of turf) composed primarily of perennial ryegrass and dallisgrass (*Paspalum dilatatum*) were grown for 324 days under two levels of CO_2 (350 and 700 ppm), with air temperatures and photoperiod designed to emulate the monthly climate of the region. After this equilibration period, half the turves in each CO_2 treatment underwent soil moisture deficit for 42 days. Turves under elevated CO_2 continued to exchange CO_2 with the atmosphere, while turves under ambient CO_2 did not. Root density measurements indicated that roots acted as sinks for the carbon fixed during the soil moisture deficit period. Upon rewatering, turves under ambient CO_2 had a vigorous rebound in growth while those under elevated CO_2 did not exhibit additional growth, suggesting that plants may exhibit a different strategy in response to soil moisture deficit depending on the CO_2 concentration.

Climate change may cause reduction in precipitation and, in turn, induce soil moisture limitations in pasturelands.

2.2.2.1 Predictions of Pastureland Forage Yields and Nutrient Cycling under Climate Change

To evaluate the effect of climate scenarios on a forage crop, alfalfa production was simulated with the EPIC agroecosystem model (Williams 1995), using various climate change projections from the HadCM2 (Izaurralde et al. 2003), and GCMs from Australia's Bureau of Meteorology Research Centre (BMRC), and the University of Illinois, Urbana-Champaigne (UIUC) (Thomson et al. 2005). All model runs were driven with CO_2 levels of 365 and 560 ppm without irrigation.

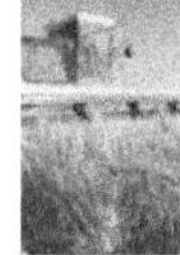

The results give an indication of pastureland crop response to changes in temperature, precipitation, and CO_2 for major regions of the United States (Table 2.10). Of these three factors, variation in precipitation had the greatest impact on regional alfalfa yield. Under the HadCM2 projected climate, alfalfa yields increase substantially in eastern regions, with declines through the central part of the country where temperature increases are greater and precipitation is lower. Slight alfalfa yield increases are predicted for western regions. The BMRC model projects substantially higher temperatures and consistent declines in precipitation over the next several decades, leading to a nationwide decline in alfalfa yields. In contrast, the UIUC model projects more moderate temperature increases along with higher precipitation, leading to modest increases in alfalfa yields throughout the central and western regions. While these results illustrate the uncertainty of model projections of crop yields due to the variation in global climate model projections of the future, they also underscore the primary importance of future precipitation changes on crop yield. Analysis of the results shown in Table 2.10 reveals that precipitation was the explanatory variable in yield changes followed by CO_2 and temperature change. Comparing the BMRC, HadCM2, and UIUC models showed that future changes in precipitation will be extremely important in alfalfa yields with a 1 percent decrease in alfalfa yields for every 4 mm decrease in annual precipitation.

Thornley and Cannell (1997) argued that experiments on elevated CO_2, and temperature effects on photosynthesis and other ecosystem processes may have limited usefulness for at least two reasons. First, laboratory or field experiments incorporating sudden changes in temperature or elevated CO_2 are short term and thus rarely produce quantitative changes in net primary productivity (NPP), ecosystem C, or other ecosystem properties connected to long-term responses to gradual climate change.

Table 2.10 Change in alfalfa yields in major U.S. regions as a percentage of baseline yield with average temperature and precipitation change under the selected climate model for early century (2030) climate change projections. Data in table from the simulations provided in Izaurralde et al. (2003).

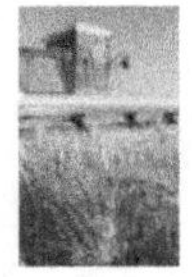

Region	CO_2	HadCM2			BMRC			UIUC		
		ΔT (°C)	ΔP (mm)	Yield % change	ΔT (°C)	ΔP (mm)	Yield % change	ΔT (°C)	ΔP (mm)	Yield % change
Great Lakes	365	1.13	74	17.0	1.79	-6	-0.4	0.96	19	-1.3
	560			20.6			0.0			-1.0
Ohio	365	0.70	80	12.5	1.66	-16	-5.2	0.86	25	-3.7
	560			13.9			-5.0			-3.8
Upper Mississippi	365	1.24	74	10.9	1.71	-14	-3.4	0.89	29	-2.2
	560			14.8			-2.5			-2.1
Souris-Red-Rainy	365	1.40	-30	-30.7	1.73	-3	-1.9	0.96	12	-0.4
	560			-25.4			2.1			2.6
Missouri	365	1.42	34	-9.2	1.50	-18	-9.4	0.92	41	3.5
	560			-7.1			-9.1			3.1
Arkansas	365	1.77	-2	-18.6	1.53	-32	-9.6	0.76	61	3.8
	560			-14.2			-7.3			5.1
Rio Grande	365	3.11	12	5.0	1.41	-20	-9.3	0.84	25	16.2
	560			5.3			-8.7			17.8
Upper Colorado	365	2.21	76	5.0	1.48	-18	-15.3	0.97	40	16.2
	560			5.4			-14.1			16.7
Lower Colorado	365	1.43	2	7.3	1.31	-23	-16.0	0.97	27	7.8
	560			11.9			-19.4			4.7
Great Basin	365	0.62	21	-4.7	1.36	-15	-6.3	1.07	45	24.2
	560			-4.5			-7.1			23.7
Pacific Northwest	365	0.45	3	0.4	1.24	-6	2.0	1.11	54	8.4
	560			1.7			1.9			8.1
California	365	0.95	58	8.7	1.13	-45	-5.5	1.08	17	6.3
	560			9.3			-3.5			4.6

Second, the difficulty of incorporating grazing in these experiments prevents a full analysis of its effects on ecosystem properties such as NPP, LAI, belowground process, and ecosystem C.

Thornley and Cannell (1997) used their Hurley Pasture Model to simulate ecosystem responses of ungrazed and grazed pastures to increasing trends in CO_2 concentrations and temperature. The simulations revealed three important results: 1) rising CO_2 induces a carbon sink, 2) rising temperatures alone produce a carbon source, and 3) a combination of the two effects is likely to generate a carbon sink for several decades (5-15 g C m^{-2} yr^{-1}). Modeling the dynamics of mineral N availability in grazed pastures under elevated CO_2, Thornley and Cannell (2000) ascertained the role of the mineral N pool and its turnover rate in slowly increasing C content in plants and soils.

2.2.2.2 Implications of Altered Productivity, Nitrogen Cycle (Forage Quality), Phenology, and Growing Season on Species Mixes, Fertilizer, and Stocking

In general, the response of pasture species to elevated CO_2 deduced from these studies is consistent with the general response of C_3 and C_4 type vegetation to elevated CO_2, although significant exceptions exist. Pasture species with C_3-type metabolism increased their photosynthetic rates by up to 40 percent, but not those with a C_4 pathway (Greer et al. 1995). The study of Greer et al. (1995) suggests shifts in optimal temperatures for photosynthesis under elevated CO_2, with perennial ryegrass and tall fescue showing a downward shift in their optimal temperature from 28-18°C. Unlike croplands, the literature for pasturelands is sparse in providing quantitative information to predict the yield change of pastureland species under a temperature increase of 1.2°C. The projected increases in temperature and the lengthening of the growing season should be, in principle, beneficial for livestock produced by increasing pasture productivity and reducing the need for forage storage during the winter period.

Naturally, changes in CO_2 and temperature will be accompanied by changes in precipitation, with the possibility of more extreme weather causing floods and droughts. Precipitation changes will likely play a major role in determining NPP of pasture species as suggested by the simulated 1 percent change in yields of dryland alfalfa for every 4-mm change in annual precipitation (Izaurralde et al. 2003; Thomson et al. 2005).

Another aspect that emerges from this review is the need for comprehensive studies of the impacts of climate change on the pasture ecosystem including grazing regimes, mutualistic relationships (e.g., plant roots-nematodes; N-fixing organisms), as well as C, nutrient, and water balances. Despite their complexities, the studies by Newton et al. (1996) and Wilsey (2001) underscore the importance, difficulties, and benefits of conducting multifactor experiments. To augment their value, these studies should include the use of simulation modeling (Thornley and Cannell 1997) in order to test hypotheses regarding ecosystem processes.

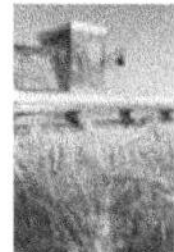

2.2.3 Rangelands

The overall ecology of rangelands is determined primarily by the spatial and temporal distribution of precipitation and consequences of precipitation patterns for soil water availability (Campbell et al. 1997; Knapp, Briggs and Koelliker 2001; Morgan 2005). Rising CO_2 in the atmosphere, warming and altered precipitation patterns all impact strongly on soil water content and plant water relations (Alley et al. 2007; Morgan et al. 2004b), so an understanding of their combined effects on the functioning of rangeland ecosystems is essential.

2.2.3.1 Ecosystem Responses to CO_2 and Climate Drivers

2.2.3.1.1 Growing Season Length and Plant Phenology

Although responses vary considerably among species, in general warming should accelerate plant metabolism and developmental processes, leading to earlier onset of spring green-up, and lengthening of the growing season in rangelands (Badeck et al. 2004). The effects of warming are also likely to be seen as changes in the timing of phenological events such as flowering and fruiting. For instance, experimental soil warming of approximately 2°C in a tallgrass prairie (Wan et al. 2005) extended the growing season by three weeks, and shifted the timing and duration of reproductive events variably among species; spring blooming species flowered earlier, late blooming species flowered later (Sherry et al. 2007). Extensions and contractions in lengths of the reproductive periods were also observed among the species tested (see also Cleland et al. 2006). Different species responses to warming suggest strong selection pressure for altering future rangeland community structure, and for the associated trophic levels that depend on the plants for important stages of their life cycles. Periods of drought stress may suppress warming-induced plant activity (Gielen et al. 2005), thereby effectively decreasing plant development time. CO_2 may also impact phenology of herbaceous plant species, although species can differ widely in their developmental responses to CO_2 (Huxman and Smith; 2001 Rae et al. 2006), and the implications for these changes in rangelands are not well understood. Thus, temperature is the primary climate driver that will determine growing season length and plant phenology, but precipitation variability and CO_2 may cause deviations from the overall patterns set by temperature.

2.2.3.1.2 Net Primary Production

Increases in CO_2 concentration and in precipitation and soil water content expected for rangelands generally enhance NPP, whereas increased air temperature may either increase or reduce NPP.

2.2.3.1.2.1 CO_2 Enrichment

Most forage species on rangelands have either the C_3 or C_4 photosynthetic pathway. Photosynthesis of C_3 plants, including most woody species and herbaceous broad-leaf species (forbs), is not CO_2-saturated at the present atmospheric concentration, so carbon gain and productivity usually are very sensitive to CO_2 in these species (Drake et al. 1997). Conversely, photosynthesis of C_4 plants, including many of the warm-season perennial grass species of rangelands, is nearly CO_2-saturated at current atmospheric CO_2 concentrations (approximately 380 ppm) when soil water is plentiful, although the C_4 metabolism does not preclude photosynthetic and growth responses to CO_2 (Polley et al. 2003). In addition, CO_2 effects on rates of water loss (transpiration) and plant WUE are at least as important as photosynthetic response to CO_2 for rangeland productivity. Stomata of most herbaceous plants partially close as CO_2 concentration increases, thus reducing plant transpiration. Reduced water loss improves plant and soil water relations, increases plant production under water limitation, and may lengthen the growing season for water-limited vegetation (Morgan et al. 2004b).

CO_2 enrichment will stimulate NPP on most rangelands, with the amount of increase dependent on precipitation and soil water availability. Indeed, there is evidence that the historical increase in CO_2 of about 35 percent has already enhanced rangeland NPP. Increasing CO_2 from pre-industrial levels to elevated concentrations (250 to 550 ppm) increased aboveground NPP of mesic grassland in central Texas between 42-69 percent (Polley et al. 2003). Biomass increased by similar amounts at pre-industrial to current, and current to elevated concentrations. Comparisons between CO_2-induced production responses of semi-arid Colorado shortgrass steppe with the sub-humid Kansas tall grass prairie suggest that Great Plains rangelands respond more to CO_2 enrichment during dry than wet years, and that the potential for CO_2-induced production enhancements are greater in drier rangelands (Figure 2.7). However, in the still-drier Mojave Desert, CO_2 enrichment-enhanced shrub growth occurred most consistently during relatively wet years (Smith et al. 2000). CO_2 enrichment stimulated total biomass (aboveground + belowground) production in one study on annual grassland in California (Field et al. 1997), but elicited no production response in a second experiment (Shaw et al. 2002).

2.2.3.1.2.2 Temperature

Like CO_2 enrichment, increasing ambient air and soil temperatures may enhance rangeland NPP, although negative effects of higher temperatures also are possible, especially in dry and hot regions. Temperature directly affects plant physiological processes, but rising ambient temperatures may indirectly affect plant production by extending growing season length, increasing soil nitrogen (N) mineralization and availability, altering soil water content, and shifting plant species composition and community structure (Wan et al. 2005). Rates of biological processes for a given species typically peak at plant temperatures that are intermediate in the range over which a species is active, so direct effects of warming likely will vary within and among years, and among plant species. Because of severe cold-temperature restrictions on growth rate and duration, warmer plant temperatures alone should stimulate production in high- and mid-latitude, and high-altitude rangelands. Conversely, increasing plant temperature during summer months may reduce NPP.

Increasing daily minimum air temperature and mean soil temperature (2.5 cm depth) by 2°C increased aboveground NPP of tallgrass prairie in Oklahoma between 0-19 percent during the first three years of study, largely by increasing NPP of C_4 grasses (Wan et al. 2005). Warming stimulated biomass production in spring and autumn, but aboveground biomass in summer declined as soil temperature increased. Positive effects of warming on production may be lessened by an accompanying increase in the rate of water loss. Warming reduced the annual mean of soil water content in tallgrass prairie during one year (Wan et al. 2005), but actually increased soil water content in California annual grassland by accelerating plant senescence (Zavaleta et al. 2003b).

2.2.3.1.2.3 Precipitation

Historic changes in climatic patterns have always been accompanied by changes in grassland vegetation because grasslands have both high production potential and a high degree of variability in precipitation (Knapp and Smith 2001). In contrast, aboveground NPP (ANPP) variability in forest systems appears to be limited by invariant rainfall patterns, while production potential more strongly limits desert and arctic/alpine systems. Projected altered rainfall regimes are likely to elicit important changes in rangeland ecology, including NPP.

On most rangelands where total annual precipitation is sufficiently low that soil water limits productivity more than other soil resources, the timing of precipitation can play an important role in regulating NPP. Increased rainfall variability caused by altered rainfall timing (no change in rainfall amount) led to lower and more variable soil water content (between 0-30 cm depth), an approximate 10 percent reduction in ANPP, which was species-specific, and increased root-to-shoot ratios in a native tallgrass prairie ecosystem in northeastern Kansas (Fay

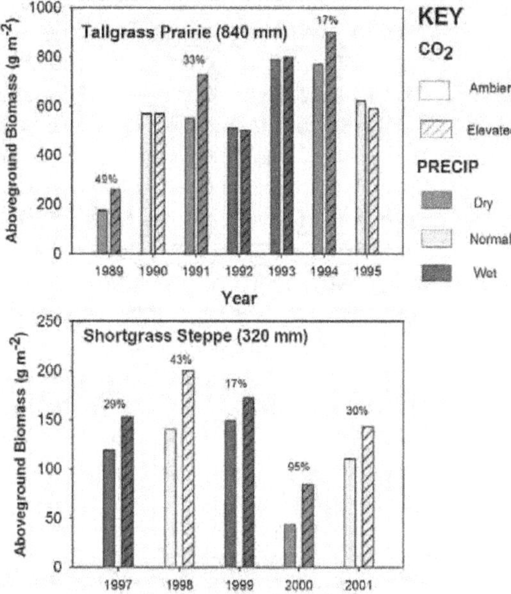

Figure 2.7 Aboveground plant biomass of native Kansas tallgrass prairie (Owensby et al. 1999; 1989-1995) and Colorado Shortgrass steppe (Morgan et al. 2004a; 1997-2001), harvested during summer-time seasonal peak. These grasses were grown in similarly-designed Open Top Chambers maintained at present (ambient, approximately 370 parts per million CO_2 in air; no cross-hatches) and elevated (approximately 720 parts per million CO_2 in air; cross-hatches) atmospheric CO_2 concentrations. Histograms from different years are color-coded (red for dry; yellow for normal; blue for wet) according to the amount of annual precipitation received during that particular year compared to long-term averages for the two sites (840 mm for the tallgrass prairie, and 320 mm for shortgrass steppe). Where production increases due to elevated CO_2 were observed, the percentage-increased production is given within a year above the histograms. The involvement of water in the CO_2 responses is seen in two ways: the relative plant biomass responses occur more commonly and in greater magnitude in the shortgrass steppe than in the tallgrass prairie, and the relative responses in both systems are greater in dry than wet years.

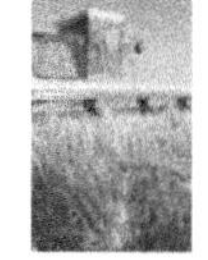

et al. 2003). In general, vegetation responses to rainfall timing (no change in amount) were at least equal to changes caused by rainfall quantity (30 percent reduction, no change in timing). Reduced ANPP most likely resulted from direct effects of soil moisture deficits on root activity, plant water status, and photosynthesis.

The seasonality of precipitation is also an important factor determining NPP through its affects on locally adapted species, which can differ depending on the particular ecosystem. For example, herbaceous plants in the Great Basin are physiologically adapted to winter/early spring precipitation patterns, where reliable soil water recharge occurs prior to the growing season

(Svejcar et al. 2003). Similarly, Northern Great Plains grasslands are dominated by cool-season plant species that complete most of their growth by late spring to early summer, and NPP primarily depends on sufficient soil moisture going into the growing season (Heitschmidt and Haferkamp 2003). Productivity of herbaceous species in both of these rangeland systems is highly dependent on early spring soil moisture, which can be significantly affected by winter precipitation. In contrast, oak savannas of the southwestern United States experience a strongly seasonal pattern of precipitation, with a primary peak in summer and lesser peak in winter (Weltzin and McPherson 2003). Here, herbaceous biomass is more sensitive to summer precipitation than to winter precipitation.

2.2.3.1.3 Environmental Controls on Species Composition

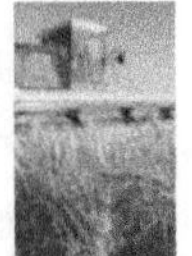

At regional scales, species composition of rangelands is determined mostly by climate and soils, with fire regime, grazing, and other land uses locally important. The primary climatic control on the distribution and abundance of plants is water balance (Stephenson 1990). On rangelands in particular, species composition is highly correlated with both the amount of water plants use and its availability in time and space.

On rangelands in particular, species composition is highly correlated with both the amount of water plants use and its availability in time and space.

Each of the global changes considered here – CO_2 enrichment, altered precipitation regimes, and higher temperatures – may change species composition by altering water balance. Unless stomatal closure is compensated by atmospheric or other feedbacks, CO_2 enrichment should affect water balance by slowing canopy-level ET (Polley et al. 2007) and the rate or extent of soil water depletion (Morgan et al. 2001; Nelson et al. 2004). The resultant higher soil water content has been hypothesized to favor deep-rooted woody plants in future CO_2-enriched atmospheres because of their greater access to stored soil water compared to relatively shallow-rooted grasses (Polley 1997). A warmer climate will likely be characterized by more rapid evaporation and transpiration, and an increase in frequency of extreme events like heavy rains and droughts. Changes in timing and intensity of rainfall may be especially important on arid rangelands where plant community dynamics are 'event-driven' and the seasonality of precipitation determines which plant growth strategies are successful. The timing of precipitation also affects the vertical distribution of soil water, which regulates relative abundances of plants that root at different depths (Ehleringer et al. 1991; Weltzin and McPherson 1997), and influences natural disturbance regimes, which feed back to regulate species composition. For example, grass-dominated rangelands in the eastern Great Plains were historically tree-free due to periodic fire. Fires occurred frequently because the area is subject to summer droughts, which dessicated the grasses and provided abundant fuel for wildfires.

In addition to its indirect effect on water balance, the direct effect of temperature on plant physiology has long been acknowledged as an important determinant of plant species distribution. A good example of this is the distribution of cool-season, C_3 grasses being primarily at northern latitudes and warm-season, C_4 grasses at southern latitudes (Terri and Stowe 1976). Thus, the relative abundances of different plants types (C_3 grasses, C_4 grasses, and shrubs) in grasslands and shrublands of North America are determined in large part by soil water availability and temperatures (Paruelo and Lauenroth 1996).

Observational evidence that global changes are affecting rangelands and other ecosystems is accumulating. During the last century, juniper trees in the arid West grew more than expected from climatic conditions, implying that the historical increase in atmospheric CO_2 concentration stimulated juniper growth (Knapp et al. 2001). The apparent growth response of juniper to CO_2 was proportionally greater during dry than wet years, consistent with the notion that access to deep soil water, which tends to accumulate under elevated CO_2 (Morgan et al. 2004b), gives a growth advantage to deep-rooted woody vegetation (Polley 1997; Morgan et al. 2007). Such observational reports in combination with manipulative experimentation (Morgan et al. 2004b, 2007) suggest that expansion of shrublands over the past couple hundred years has been driven in part by a combination of climate change and increased atmospheric CO_2 concentrations (Polley 1997; Archer et al. 1995).

2.2.3.1.4 Nitrogen Cycle Feedbacks

Plant production on rangelands often is limited by nitrogen (N). Because most terrestrial N occurs in organic forms that are not readily available to plants, rangeland responses to global changes will depend partly on how quickly N cycles between organic and inorganic N compounds. Plant material that falls to the soil surface, or is deposited belowground as the result of root exudation or death, is subject to decomposition by soil fauna and micro flora and enters the soil organic matter (SOM) pool. During decomposition of SOM, mineral and other plant-available forms of N are released. Several of the variables that regulate N-release from SOM may be affected by CO_2 enrichment and climate change, and thus are likely to be important factors determining the long-term responses of rangelands.

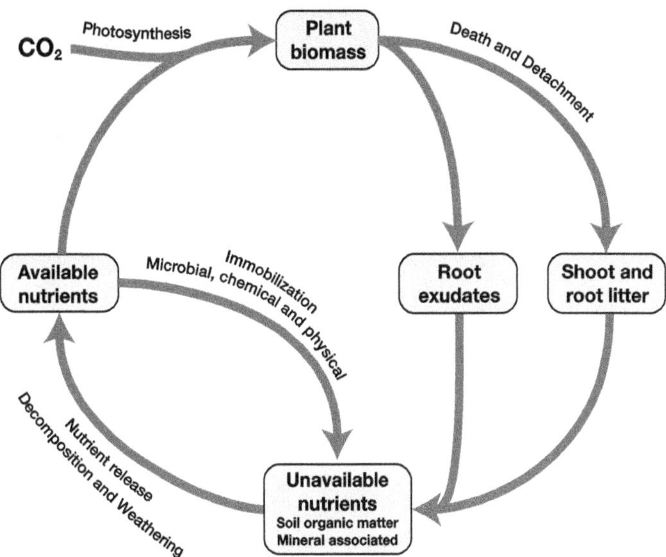

Figure 2.8 Nutrient Cycling Feedbacks. While CO_2 enrichment may lead to increased photosynthesis and enhanced plant growth, the long-term response will depend on nutrient cycling feedbacks. Litter from decaying plants and root exudates enter a large soil nutrient pool that is unavailable to plants until they are broken down and released by microbial activity. Soil microbes may also fix available nutrients into new microbial biomass, thereby temporarily immobilizing them. The balance between these and other nutrient release and immobilization processes determines available nutrients and ultimate plant response. Figure reprinted with permission from *Science* (Morgan 2002).

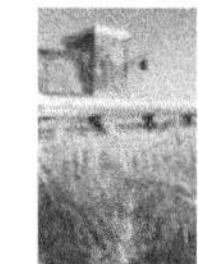

For instance, while CO_2 enrichment above present atmospheric levels is known to increase photosynthesis, particularly in C_3 species, soil feedbacks involving nutrient cycling may constrain the potential CO_2 fertilization response (Figure 2.8). The Progressive Nitrogen Limitation (PNL) hypothesis holds that CO_2 enrichment is reducing plant-available N by increasing plant demand for N, and enhancing sequestration of N in long-lived plant biomass and SOM pools (Luo et al. 2004). The greater plant demand for N is driven by CO_2-enhanced plant growth. Accumulation of N in organic compounds at elevated CO_2 may eventually reduce soil N availability and limit plant growth response to CO_2 or other changes (Reich et al. 2006a, 2006b; van Groenigen et al. 2006; Parton et al. 2007a). Alternatively, greater C input may stimulate N accumulation in soil/plant systems. A number of processes may be involved, including increased biological fixation of N, greater retention of atmospheric N deposition, reduced losses of N in gaseous or liquid forms, and more complete exploration of soil by expanded root systems (Luo et al. 2006). Rangeland plants often compensate for temporary imbalances in C and N availability by maximizing the amount of C retained in the ecosystem per unit of N. Thus, N concentration of leaves or aboveground tissues declined on shortgrass steppe, tallgrass prairie, and mesic grassland at elevated CO_2, and on tallgrass prairie with warming, but total N content of aboveground tissues increased with plant biomass in these ecosystems and on annual grasslands (Owensby et al. 1993; Hungate et al. 1997; King et al. 2004; Wan et al. 2005; Gill et al. 2006). The degree to which N may respond to rising atmospheric CO_2 is presently unknown, but may vary among ecosystems (Luo et al. 2006), and has important consequences for forage quality and soil C storage, as both depend strongly on the available soil N.

Warmer temperatures generally increase SOM decomposition, especially in cold regions (Reich et al. 2006b; Rustad et al. 2001), although warming also may limit microbial activity by drying soil or enhancing plant growth (Wan et al. 2005). Wan et al. (2005) found that warming stimulated N mineralization during the first year of treatment on Oklahoma tallgrass prairie, but in the second year, caused N immobilization by reducing plant N concentration, stimulating plant growth, and increasing allocation of carbon (C) compounds belowground (Wan et al. 2005). Warming can also affect decomposition processes by extending the growing season (Wan et al. 2005). However, as water becomes limiting, decomposition becomes more dependant on soil water content and less on temperature (Epstein et al. 2002; Wan et al. 2005), with lower soil water content leading to reduced decomposition rates. A recent global model of litter decomposition (Parton et al. 2007b) indicates that litter N-concentration, along with temperature and water, are the dominant drivers behind N release and immobilization dynamics, although UV-stimulation of decomposition (Austin and Vivanco 2006) is especially important in controlling surface litter decomposition dynamics in arid systems like rangelands.

Nutrient cycling also is sensitive to changes in plant species composition; this may result because species differ in sensitivity to global changes. Soil microorganisms require organic material with relatively fixed proportions of C and N. The ratio of C to N (C:N) in plant residues thus affects the rate at which N is released during decomposition in soil. Because C:N varies among plant species, shifts in species composition can strongly affect nutrient cycling (Allard et al. 2004; Dijkstra et al. 2006; Gill et al. 2006; King et al. 2004; Schaeffer et al. 2007; Weatherly et al. 2003). CO_2 enrichment may reduce decomposition by reducing the N concentration in leaf litter (Gill et al. 2006), for example, although litter quality may not be the best predictor of tissue decomposition (Norby et al. 2001). Like CO_2, climatic changes may alter litter quality by causing species change (Murphy et al. 2002; Semmartin et al. 2004; Weatherly et al. 2003). Elevated atmospheric CO_2 and/or temperature may also alter the amounts and proportions of micro flora and fauna in the soil microfood web (e.g., Hungate et al. 2000; Sonnemann and Wolters 2005), and/or the activities of soil biota (Billings et al. 2004; Henry et al. 2005; Kandeler et al. 2006). Although changes in microbial communities are bound to have important feedbacks on soil nutrient cycling and C storage, the full impact of global changes on microbes remains unclear (Niklaus et al. 2003; Ayers et al., in press).

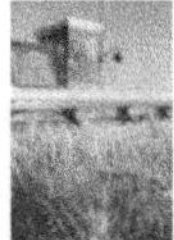

Adverse environmental stress can elicit a panting or shivering response, which increases maintenance requirements of the animal and contributes to decreases in productivity.

Computer simulation models that incorporate decomposition dynamics and can evaluate incremental global changes agree that combined effects of warming and CO_2 enrichment during the next 30 years will stimulate plant production, but disagree on the impact on soil C and N. The Daycent Model predicts a decrease in soil C stocks, whereas the Generic Decomposition And Yield Model (G'Day) predicts an increase in soil C (Pepper et al. 2005). Measurements of N isotopes from herbarium specimens collected over the past hundred years indicate that rising atmospheric CO_2 has been accompanied by increased N fixation and soil N mineralization, decreased soil N losses, and a decline in shoot N concentration (Peñuelas and Estiarte 1997). Collectively, these results indicate that soil N may constrain the responses of some terrestrial ecosystems to CO_2.

2.2.4 Temperature Response of Animals

2.2.4.1 Thermal Stress

The optimal zone (thermoneutral zone) for livestock production is a range of temperatures and other environmental conditions for which the animal does not need to significantly alter behavior or physiological functions to maintain a relatively constant core body temperature. As environmental conditions result in core body temperature approaching and/or moving outside normal diurnal boundaries, the animal must begin to conserve or dissipate heat to maintain homeostasis. This is accomplished through shifts in short-term and long-term behavioral, physiological, and metabolic thermoregulatory processes (Mader et al. 1997b; Davis et al. 2003). The onset of a thermal challenge often results in declines in physical activity and an associated decline in eating and grazing activity (for ruminants and other herbivores). Hormonal changes, triggered by environmental stress, result in shifts in cardiac output, blood flow to extremities, and passage rate of digesta. Adverse environmental stress can elicit a panting or shivering response, which increases maintenance requirements of the animal and contributes to decreases in productivity. Depending on the domestic livestock species, longer term adaptive responses include hair coat gain or loss through growth and shedding processes, respectively. In addition, heat stress is directly related to respiration and sweating rate in most domestic animals (Gaughan et al. 1999, 2000, and 2005).

Production losses in domestic animals are largely attributed to increases in aintenance requirement associated with sustaining a constant body temperature, and altered feed intake (Mader et al. 2002; Davis et al. 2003; Mader and Davis 2004). As a survival mechanism, voluntary feed intake increases (after a one- to two-day decline) under cold stress, and decreases almost immediately under heat stress (NRC 1987). Depending on the intensity and duration of the environmental stress, voluntary feed intake can average as much as 30 percent above normal under cold conditions, to as much as 50 percent below normal in hot conditions.

Domestic livestock are remarkable in their adaptive ability. They can mobilize coping mechanisms when challenged by environmental stressors. However, not all coping capabilities are mobilized at the same time. As a general model for mammals of all species, respiration rate serves as an early warning of increasing thermal stress, and increases markedly above a threshold as animals try to maintain homeothermy by dissipating excess heat. At a higher threshold, body temperature begins to increase as a result of the animal's inability to adequately dissipate the excess heat load by increased respiratory vaporization (Brown-Brandl et al. 2003; Davis et al. 2003; Mader and Kreikemeier 2006). There is a concomitant decrease in voluntary feed intake as body temperature increases, which ultimately results in reduced performance (i.e., production, reproduction), health and well-being if adverse conditions persist (Hahn et al. 1992; Mader 2003).

Thresholds are species dependent, and affected by many factors, as noted in Figure 2.9. For shaded *Bos taurus* feeder cattle, Hahn (1999) reported respiration rate as related to air temperature typically shows increases above a threshold of about 21°C, with the threshold for increasing body temperature and decreasing voluntary feed intake being about 25°C. Recent studies (Brown-Brandl et al. 2006) clearly show the influences of animal condition, genotype, respiratory pneumonia, and temperament on respiration rate of *Bos taurus* heifers.

Even though voluntary feed intake reduction usually occurs on the first day of hot conditions,

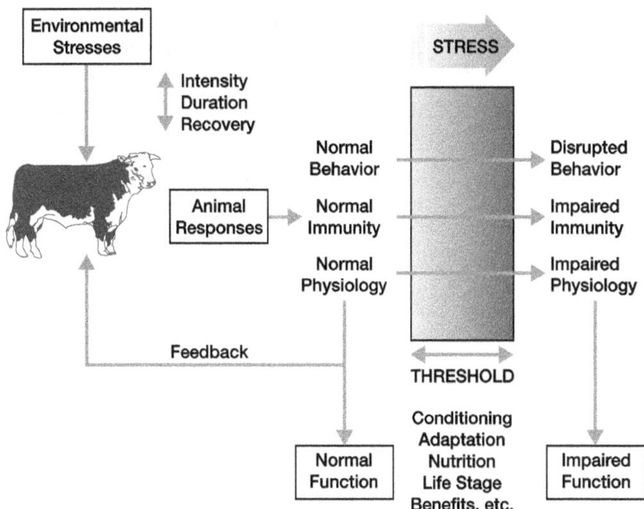

Figure 2.9. Response model for farm animals with thermal environmental challenges (Hahn 1999).

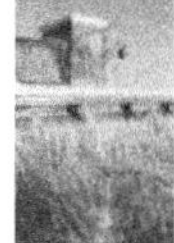

the animals' internal metabolic heat load generated by digesting existing rumen contents adds to the increased external, environmental heat load. Nighttime recovery also has been shown to be an essential element of survival when severe heat challenges occur (Hahn and Mader 1997; Amundson et al. 2006). After about three days, the animal enters the chronic response stage, with mean body temperature declining slightly and voluntary feed intake reduced in line with heat dissipation capabilities. Diurnal body temperature amplitude and phase remain altered. These typical thermoregulatory responses, when left unchecked during a severe heat wave with excessive heat loads, can lead to impaired performance or death (Hahn and Mader 1997; Mader 2003).

2.2.4.1.1 Methods to Identify Environmentally Stressed Animals

Temperature provides a measure of the sensible heat content of air, and represents a major portion of the driving force for heat exchange between the environment and an animal. However, latent heat content of the air, as represented by some measure of the insensible heat content (e.g., dewpoint temperature), thermal radiation (short- and long-wave), and airflow, also impacts the total heat exchange. Because of the limitations of air temperature alone as a measure of the thermal environment, there have been many efforts to combine the effects of two or more thermal measures representing the influence of sensible and latent heat exchanges between the organism and its environment. It is important to recognize that all such efforts produce index values rather than a true temperature (even when expressed on a temperature scale). As such, an index value represents the effect produced by the heat exchange process, which can alter the biological response that might be associated with changes in temperature alone. In the case of humans, the useful effect is the sensation of comfort; for animals, the useful effect is the impact on performance, health, and well-being.

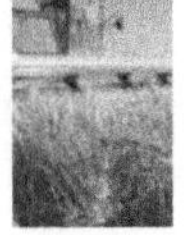

Contrary to the focus of human-oriented thermal indices on comfort, the primary emphasis for domestic animals has been on indices to support rational environmental management decisions related to performance, health, and well-being. Hahn and Mader (1997), Hahn et al. (1999), and Hahn et al. (2001) have used retrospective climatological analyses to evaluate the characteristics of prior heat waves causing extensive livestock losses. Although limited by lack of inclusion of wind speed and thermal radiation effects, the Temperature-Humidity Index (THI) has been a particularly useful tool for profiling and classifying heat wave events (Hahn and Mader 1997; Hahn et al. 1999). In connection with extreme conditions associated with heat waves, the THI has recently been used to evaluate spatial and temporal aspects of their development (Hubbard et al. 1999; Hahn and Mader 1997). For cattle in feedlots, a THI-based classification scheme has also been developed to assess the potential impact of heat waves (Hahn et al. 1999) (Table 2.11). The classifications are based on a retrospective analysis of heat waves that have resulted in extensive feedlot cattle deaths, using a THI-hours approach to assess the magnitude (intensity x duration) of the heat wave events that put the animals at risk. When calculated hourly from records of temperature and humidity, this classification scheme can be used to compute cumulative daily THI-hrs at or above the Livestock Weather Safety Index (LWSI) thresholds for the "Danger" and "Emergency" categories. The THI-hrs provide a measure of the magnitude of daytime heat load (intensity and duration), while the number of hours below THI thresholds of 74 and 72 indicate the opportunity for nighttime recovery from daytime heat.

As applied to *Bos taurus* feedlot cattle during the 1995 Nebraska-Iowa heat wave event, evaluation of records for several weather stations in the region using the THI-hrs approach reinforced the LWSI thresholds for the Danger and Emergency categories of risk and possible death (Hahn and Mader 1997). Based on that event, analysis indicated that over a successive, three-day span, 15 or more THI-hrs per day above a THI base level of 84 could be lethal for vulnerable animals

Table 2.11 Heat wave categories for *Bos taurus* feedlot cattle exposed to single heat wave events (Hahn et al. 1999).

	Descriptive Characteristics			
Category	Duration	THI*-hrs ≥79	THI-hrs ≥84*	Nighttime recovery (hrs # 72 THI*)
1. Slight	Limited: 3-4 days	10-25/day	None	Good: 5-10 hr/night
2. Mild	Limited: 3-4 days	18-40/day	#5/day	Some: 3-8 hr/night
3. Moderate	More persistent (4-6 days usual)	25-50/day	#6/day	Reduced: 1-6 hr/night
4. Strong	Increased persistence (5-7 days)	33-65/day	#6/day	Limited: 0-4 hr/night
5. Severe	Very persistent (usually 6-8 days)	40-80/day	3-15/day on 3 or more successive days	Very limited: 0-2 hr per night
6. Extreme	Very persistent (usually 6-10+ days)	50-100/day	15-30/day on 3 or more successive days	Nil: #1 for 3 or more successive days

***Temperature Humidity Index (THI). Daily THI-hrs are the summation of the differences between the THI and the base level at each hr of the day. For example, if the THI value at 1300 is 86.5 and the base level selected is 84, THI-hr = 2.5. The accumulation for the day is obtained by summing all THI-hr ≥ 84, and can exceed 24.**

(especially those that were ill, recently placed in the feedlot, or nearing market weight). The extreme daytime heat in 1995 was exacerbated by limited nighttime relief (only a few hrs with THI ≤ 74), high solar radiation loads (clear to mostly clear skies), and low to moderate wind speeds in the area of highest risk. During this same period, for cattle in other locations enduring 20 or more daily THI-hrs in the Emergency category (THI ≥ 84) over one or two days, the heat load was apparently dissipated with minimal or no mortality, although these environmental conditions can markedly depress voluntary feed intake (Hahn 1999; NRC 1981) with resultant reduced performance.

Similar analysis of a single heat wave in August 1992 further confirmed that 15 or more THI-hrs above a base level of 84 can cause deaths of vulnerable animals (Hahn et al. 1999). A contributing factor to losses during that event was lack of acclimation to hot weather, as the summer had been relatively cool. In the region under study, only four years between 1887-1998 had fewer days during the summer when air temperature was ≥ 32.2°C (High Plains Regional Climate Center 2000).

There are limitations to the THI caused by airflow and changing solar radiation loads. Modifications to the THI have been proposed to overcome shortcomings related to airflow and radiation heat loads. Based on recent research, Mader et al. (2006) and Eigenberg et al. (2005) have proposed corrections to the THI for use with feedlot cattle, based on measures of windspeed and solar radiation. While the proposed adjustment-factor differences are substantial, there were marked differences in the types and number of animals used in the two studies. Nevertheless, the approach appears to merit further research to establish acceptable THI corrections, perhaps for a variety of animal parameters.

By using body temperatures, a similar approach was developed to derive an Apparent Equivalent Temperature (AET) from air temperature and vapor pressure to develop "thermal comfort zones" for transport of broiler chickens (Mitchell et al. 2001). Experimental studies to link the AET with increased body temperature during exposure to hot conditions indicated potential for improved transport practices.

Gaughan et al. (2002) developed a Heat Load Index (HLI) as a guide to management of unshaded *Bos taurus* feedlot cattle during hot weather (>28°C). The HLI was developed following observation of behavioral responses (respiration rate and panting score) and changes in dry-matter intake during prevailing thermal conditions. The HLI is based on humidity, windspeed, and predicted black globe temperature. As a result of its demonstrated broad success, the THI is currently the most widely accepted thermal index used for guidance of strategic and tactical decisions in animal management during moderate to hot conditions. Biologic response functions, when combined with likelihood of occurrence of the THI for specific locations, provide the basis for economic evaluation to make cost-benefit comparisons for rational strategic decisions among alternatives (Hahn 1981). Developing a climatology of summer weather extremes (in particular, heat waves) for specific locations also provides the livestock manager with information about how often those extremes (with possible associated death losses) might occur (Hahn et al. 2001). The THI has also served well for making tactical decisions about when to apply available practices and techniques (e.g., sprinkling) during either normal weather variability or weather extremes, such as heat waves. Other approaches, such as the AET proposed by Mitchell et al. (2001) for use in poultry transport, also may be appropriate. An enthalpy-based alternative thermal index has been suggested by Moura et al. (1997) for swine and poultry.

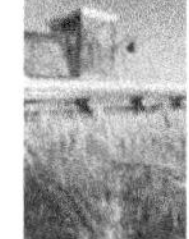

Panting score is one observation method used to monitor heat stress in cattle (Table 2.12). As the temperature increases, cattle pant more to increase evaporative cooling. Respiration dynamics change as ambient conditions change, and surroundings surfaces warm. This is a relatively easy method for assessing genotype differences and determining breed acclimatization rates to higher temperatures. In addition, shivering score or indices also have potential for use as thermal indicators of cold stress. However, recent data were not found regarding cold stress indicators for domestic livestock.

Table 2.12 Panting scores assigned to steers (Mader et al. 2006).

Score	Description
0	Normal respiration
1	Elevated respiration
2	Moderate panting and/or presence of drool or a small amount of saliva
3	Heavy open-mouthed panting, saliva usually present
4	Severe open-mouthed panting accompanied by protruding tongue and excess salivation; usually with neck extended forward

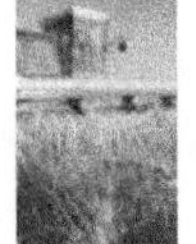

2.2.5 Episodes of Extreme Events

2.2.5.1 Elevated Temperature or Rain fall Deficit

Episodic increases in temperature would have greatest effect when occurring just prior to, or during, critical crop pollination phases. Crop sensitivity and ability to compensate during later, improved weather will depend on the synchrony of anthesis in each crop; for example, maize has a highly compressed phase of anthesis, while spikelets on rice and sorghum may achieve anthesis over a period of a week or more. Soybean, peanut, and cotton will have several weeks over which to spread the success of reproductive structures. For peanut, the sensitivity to elevated temperature for a given flower extends from six days prior to opening (pollen cell division and formation) up through the day of anthesis (Prasad et al. 2001). Therefore, several days of elevated temperature may affect fertility of many flowers, whether still in their formative 6-day phase or just achieving anthesis today. In addition, the first six hours of the day were more critical during pollen dehiscence, pollen tube growth, and fertilization.

For rice, the reproductive processes that occur within one to three hours after anthesis (dehiscence of the anther, shedding of pollen, germination of pollen grains on stigma, and elongation of pollen tubes) are disrupted by daytime air temperatures above 33°C (Satake and Yoshida 1978). Since anthesis occurs between about 9 a.m. and 11 a.m. in rice (Prasad et al. 2006b), damage from temperatures exceeding 33°C may already be common, and may become more prevalent in the future. Pollination processes in other cereals, maize, and sorghum may have a similar sensitivity to elevated daytime temperature as rice. Rice and sorghum have the same sensitivity of grain yield, seed harvest index, pollen viability, and success in grain formation in which pollen viability and percent fertility is first reduced at instantaneous hourly air temperature above 33°C, and reaches zero at 40°C (Kim et al. 1996; Prasad et al. 2006a, 2006b). Diurnal max/min, day/night temperatures of 40/30°C (35°C mean) can cause zero yield for those two species, and the same response would likely apply to maize.

2.2.5.2 Intense Rainfall Events

Historical data for many parts of the United States indicate an increase in the frequency of high-precipitation events (e.g., >5 cm in 48 hours), and this trend is projected to continue for many regions. One economic consequence of excessive rainfall is delayed spring planting, which jeopardizes profits for farmers paid a premium for early season production of high value horticultural crops such as melon, sweet corn, and tomatoes. Field flooding during the growing season causes crop losses associated with anoxia, increases susceptibility to root diseases, increases soil compaction (due to use of heavy farm equipment on wet soils), and causes more runoff and leaching of nutrients and agricultural chemicals into groundwater and surface water. More rainfall concentrated into high precipitation events will increase the likelihood of water deficiencies at other times because of the changes in rainfall frequency (Hatfield and Prueger 2004). Heavy rainfall is often accompanied by wind gusts in storm events, which increases the potential for lodging of crops. Wetter conditions at harvest time could increase the potential for decreasing quality of many crops.

2.3 POSSIBLE FUTURE CHANGES AND IMPACTS

2.3.1 Projections Based on Increment of Temperature and CO_2 for Crops

Using the representative grain crops – maize, soybean, etc. – some expected effects resulting from the projected rise in CO_2 of 380 to 440 ppm along with a 1.2°C rise in temperature over the next 30 years are explored.

The responsiveness of grain yield to temperature is dependent on current mean temperatures during the reproductive phase in different regions (crops like soybean and maize are dominant in both the Midwest and southern regions, while others, like cotton, sorghum, and peanut, are only grown in southern regions). Grain yield response to CO_2 increase of 380 to 440 ppm was 1.0 percent for C_4, and 6.1 to 7.4 percent for C_3 species, except for cotton, which had 9.2 percent response.

For maize, under water sufficiency conditions in the Midwest, the net yield response is -3.0 percent, assuming additivity of the -4.0 percent from 1.2°C rise, and +1.0 percent from CO_2 of 380 to 440 ppm (Table 2.7). The response of maize in the South is possibly more negative. For soybean under water sufficiency in the Midwest, net yield response is +9.9 percent, assuming additivity of the +2.5 percent from 1.2°C rise above current 22.5°C mean, and +7.4 percent from CO_2 increase.

For soybean under water sufficiency in the South, the temperature effect will be detrimental, -3.5 percent, with 1.2°C temperature increment above 26.7°C, with the same CO_2 effect, giving a net yield response of +3.9 percent. For wheat (with no change in water availability), the net yield response would be +0.1 percent coming from -6.7 percent with 1.2°C rise, and +6.8 percent increase from CO_2 increase. For rice in the South, net yield response is -5.6 percent, assuming additivity of the -12.0 percent from 1.2°C rise and +6.4 percent from CO_2 increase. For peanut in the South, the net yield response is +1.3 percent, assuming additivity of the -5.4 percent from 1.2°C rise and +6.7 percent from CO_2 increase. For cotton in the South, the net yield response is +3.5 percent, assuming additivity of the -5.7 percent from 1.2°C rise and +9.2 percent from CO_2 increase. The sorghum response is less certain, although yield reduction caused by shortening filling period is dominant, giving a net yield decrease of 8.4 percent in the South. Dry bean yield response in all regions is less certain, with net effect of -2.5 percent, coming from -8.6 percent response to 1.2°C rise and +6.1 percent from CO_2 increase. The confidence in CO_2 responses is likely to very likely, while the confidence in temperature responses is generally likely, except for less knowledge concerning maize and cotton sensitivity to temperature when these responses are possible.

Projections of crop yield under water deficit should start with the responses to temperature and CO_2 for the water-sufficient cases. However, yield will likely be slightly increased to the same extent (percentage) that increased CO_2 causes reduction in ET but decreased to the extent that rainfall is decreased (but that requires climate scenarios and simulations not presented in Table 2.7). Model simulations with CROPGRO-Soybean with energy balance option and stomatal feedback from CO_2 enrichment (350 to 700 ppm, without temperature increase) resulted in a 44 percent yield increase for water-stressed crops compared to fully-irrigated crops (32 percent). The yield increment was nearly proportional to the decrease in simulated transpiration (11-16 percent). Based on this assumption, the 380 to 440 ppm CO_2 increment would likely further increase yield of C_3 crops (soybean, rice, wheat, and cotton) by an additional 1.4 to 2.1 percent (incremental reduction in ET from CO_2 in Table 2.7). However, the projected 1.2°C would increase ET by 1.8 percent, thereby partially negating this water-savings effect of CO_2.

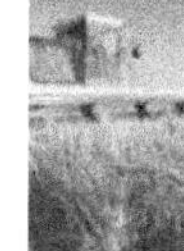

2.3.2 Projections for Weeds

Many weeds respond more positively to increasing CO_2 than most cash crops, particularly C_3 "invasive" weeds that reproduce by vegetative means (roots, stolons, etc.) (Ziska and George 2004; Ziska 2003). Recent research also suggests that glyphosate, the most widely used herbicide in the United States, loses its efficacy on weeds grown at CO_2 levels that likely will occur in the coming decades (Ziska et al. 1999). While many weed species have the C_4 photosynthetic pathway, and therefore show a smaller response to atmospheric CO_2 relative to C_3 crops, in most agronomic situations crops are in competition with a mix of both C_3 and C_4 weeds. In addition, the worst weeds for a given crop are often similar in growth habit or photosynthetic pathway. To date, for all weed/crop competition studies where the photosynthetic pathway is the same, weed growth is favored as CO_2 increases (Ziska and Runion 2006).

To date, for all weed/crop competition studies where the photosynthetic pathway is the same, weed growth is favored as CO_2 increases.

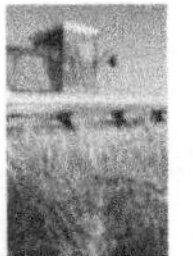

The habitable zone of many weed species is largely determined by temperature, and weed scientists have long recognized the potential for northward expansion of weed species' ranges as the climate changes (Patterson et al. 1999). More than 15 years ago, Sasek and Strain (1990) utilized climate model projections of the -20°C minimum winter temperature zone to forecast the northward expansion of kudzu (*Pueraria lobata, var. montana*), an aggressive invasive weed that currently infests more than one million hectares in the southeastern U.S. While temperature is not the only factor that could constrain spread of kudzu and other invasive weeds, a more comprehensive assessment of potential weed species migration based on the latest climate projections for the United States seems warranted.

2.3.3 Projections for Insects and Pathogens

Plants do not grow in isolation in agroecosystems. Beneficial and harmful insects, microbes, and other organisms in the environment will also be responding to changes in CO_2 and climate. Studies conducted in Western Europe and other regions have already documented changes in spring arrival and/or geographic range of many insect and animal species due to climate change (Montaigne 2004; Goho 2004; Walther et al. 2002). Temperature is the single most important factor affecting insect ecology, epidemiology, and distribution, while plant pathogens will be highly responsive to humidity and rainfall, as well as temperature (Coakley et al. 1999).

There is currently a clear trend for increased insecticide use in warmer, more southern regions of the United States, compared to cooler, higher latitude regions. For example, the frequency of pesticide sprays for control of lepidopteran insect pests in sweet corn currently ranges from 15 to 32 applications per year in Florida (Aerts et al. 1999), to four to eight applications in Delaware (Whitney et al. 2000), and zero to five applications per year in New York (Stivers 1999). Warmer winters will likely increase populations of insect species that are currently marginally over-wintering in high latitude regions, such as flea beetles (*Chaetocnema pulicaria*), which act as a vector for bacterial Stewart's Wilt (*Erwinia sterwartii*), an economically important corn pathogen (Harrington et al. 2001).

An overall increase in humidity and frequency of heavy rainfall events projected for many parts of the United States will tend to favor some leaf and root pathogens (Coakley et al. 1999). However, an increase in short- to medium-term drought will tend to decrease the duration of leaf wetness and reduce some forms of pathogen attack on leaves.

The increasing atmospheric concentration of CO_2 alone may affect plant-insect interactions. The frequently observed higher carbon-to-nitrogen ratio of leaves of plants grown at high CO_2 (Wolfe 1994) can require increased insect feeding to meet nitrogen (protein) requirements (Coviella and Trumble 1999). However, slowed insect development on high CO_2-grown plants can lengthen the insect life stages vulnerable to attack by parasitoids (Coviella and Trumble 1999). In a recent FACE study, Hamilton et al. (2005) found that early season soybeans grown at elevated CO_2 had 57 percent more damage from insects, presumably due in this case to measured increases in simple sugars in leaves of high CO_2-grown plants.

2.3.4 Projections for Rangelands

2.3.4.1 Net Primary Production and Plant Species Changes

By stimulating both photosynthesis and water use efficiency, rising CO_2 has likely enhanced plant productivity on most rangelands over the past 150 years, and will likely continue to do so over the next 30 years. The magnitude of this response will depend on how CO_2 enrichment affects the composition of plant communities and on whether nutrient limitations to plant growth develop as the result of increased carbon input to rangelands. Increasing temperature will likely have both positive and negative effects on plant productivity, depending on the prevailing climate and the extent to which warmer temperature leads to desiccation. Like CO_2 enrichment, warming will induce species shifts because of differing species sensitivities and adaptabilities to temperatures. Modeling exercises suggest generally positive NPP responses of Great Plains native grasslands to increases in CO_2 and temperature projected for the next 30 years (Pepper et al. 2005; Parton et al. 2007a), a response which is supported by experimental results from shortgrass steppe (Morgan et al. 2004a). An important exception to these findings is California annual grasslands, where production appears only minimally responsive to CO_2 or temperature (Dukes et al. 2005). Alterations in precipitation patterns will interact with rising CO_2 and temperature, although uncertainties about the nature of precipitation shifts, especially at regional levels, and the lack of multiple global change experiments that incorporate CO_2, temperature and precipitation severely limit our ability to predict consequences for rangelands. However, if annual precipitation changes little or declines in the southwestern United States as currently predicted (Christensen et al. 2007), plant production in rangelands of that region may respond little to combined warming and rising CO_2, and may even decline due to increased drought.

Plants with the C_3 photosynthetic pathway, forbs and possibly legumes will be favored by rising CO_2, although rising temperature and changes in precipitation patterns may affect these functional group responses to CO_2 (Morgan 2005; Polley 1997). In general, plants that are less tolerant of water stress than current dominants may also be favored in future CO_2-enriched atmospheres where CO_2 significantly enhances plant water use efficiency and seasonal available soil water (Polley et al. 2000). Deep-rooted forbs and shrubs may be particularly favored because of their strong carbon-allocation and nitrogen-use strategies (Polley et al. 2000; Bond and Midgley 2000; Morgan et al. 2007), including the ability of their roots to access deep soil water, which is predicted to be enhanced in future CO_2-enriched environments. Shifts in precipitation patterns toward wetter winters and drier summers, which are predicted to favor woody shrubs over herbaceous vegetation in the desert southwest (Neilson 1986), may reinforce some of the predicted CO_2-induced changes in plant community dynamics. In grasslands of the Northern Great Plains, enhanced winter precipitation may benefit the dominant cool-season, C_3 grass species that rely on early-season soil moisture to complete most of their growth by late spring to early summer (Heitschmidt and Haferkamp 2003). Greater winter precipitation, in addition to rising CO_2, may also benefit woody plants that are invading many grasslands of the central and northern Great Plains (Briggs et al. 2005; Samson and Knopf 1994). However, by itself, warmer temperature will tend to favor C_4 species (Epstein et al. 2002), which may cancel out the CO_2-advantage of C_3 plants in some rangelands.

There is already some evidence that climate change-induced species changes are underway in rangelands. The worldwide encroachment of woody plants into grasslands remains one of the best examples of the combined effects of climate change and management in driving a species change that has had a tremendous negative impact on the range livestock industry. In the southwestern arid and semi-arid grasslands of the United States, mesquite (*Prosopis glandulosa*) and creosote (*Larrea tridentate*) bushes have replaced most of the former warm season, perennial grasses (Figure 2.10), whereas in more mesic grasslands of the Central Great Plains, trees and large shrubs are supplanting C_3 grasslands (Figure 2.11). While both of these changes are due to complex combinations of management (grazing and fire) and a host of environmental factors (Briggs et al. 2005; Peters et al 2006), evidence is accumulating that rising CO_2 and climate change are very likely important factors influencing these transitions (Briggs et al. 2005; Knapp et al. 2001; Polley et al. 2002; Morgan et al. 2007; Peters et al. 2006; Polley 1997). In contrast, the observed loss of woody species and spread of the annual grass *Bromus tectorum* (cheatgrass) throughout the Intermountain region of western North America also appears driven at least in part by the species sensitivity to rising atmospheric CO_2 (Smith et al. 2000; Ziska et al. 2005), and has altered the frequency and timing of wildfires, reducing establishment of perennial herbaceous species by pre-empting soil water early in the growing season (Young 1991).

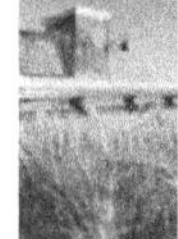

Figure 2.10 Today, in most areas of the Chihuahan desert, mesquite bushes have largely replaced perennial, warm-season grasses that dominated this ecosystem two centuries ago (photograph courtesy of Jornada Experimental Range photo gallery).

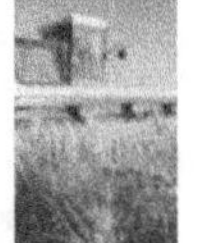

2.3.4.2 LOCAL AND SHORT-TERM CHANGES

Our ability to predict vegetation changes at local scales and over shorter time periods is more limited because at these scales the response of vegetation to global changes depends on a variety of local processes, including soils and disturbance regimes, and how quickly various species can disperse seeds across sometimes fragmented landscapes. Nevertheless, patterns of vegetation response are beginning to emerge.

- Directional shifts in the composition of vegetation occur most consistently when global change treatments alter water availability (Polley et al. 2000; Morgan et al. 2004b).

Figure 2.11 *Gleditsia triacanthos* tree islands in Kansas tallgrass prairie (photograph courtesy of Alan K. Knapp).

- Effects of CO_2 enrichment on species composition and the rate of species change will very likely be greatest in disturbed or early-successional communities where nutrient and light availability are high and species change is influenced largely by growth-related parameters (e.g., Polley et al. 2003).

- Weedy and invasive plant species likely will be favored by CO_2 enrichment (Smith et al. 2000; Morgan et al. 2007) and perhaps by other global changes because these species possess traits (rapid growth rate, prolific seed production) that permit a large growth response to CO_2.

- CO_2 enrichment will likely accelerate the rate of successional change in species composition following overgrazing or other severe disturbances (Polley et al. 2003).

- Plants do not respond as predictably to temperature or CO_2 as to changes in water, N, and other soil resources (Chapin et al. 1995). Progress in predicting the response of vegetation to temperature and CO_2 thus may require a better understanding of indirect effects of global change factors on soil resources. At larger scales, effects of atmospheric and climatic change on fire frequency and intensity and on soil water and N availability will likely influence botanical composition to a much greater extent than global change effects on production. (See Chapter 3, Arid Lands Section for a more complete discussion on the interactions and implications of fire ecology, invasive weeds, and global change for rangelands.)

- Rangeland vegetation will very likely be influenced more by management practices (land use) than by atmospheric and climatic change. Global change effects will be superimposed on and modify those resulting from land use patterns in ways that are as of yet uncertain.

2.3.4.3 Forage Quality

2.3.4.3.1 Plant-animal Interface

Animal production on rangelands, as in other grazing systems, depends on the quality as well as the quantity of forage. Key quality parameters for rangeland forage include fiber content and concentrations of crude protein, non-structural carbohydrates, minerals, and secondary toxic compounds. Ruminants require forage with at least 7 percent crude protein (as a percentage of dietary dry matter) for maintenance, 10-14 percent protein for growth, and 15 percent protein for lactation. Optimal rumen fermentation also requires a balance between ruminally-available protein and energy. The rate at which digesta pass through the rumen decreases with increasing fiber content, which depends on the fiber content of forage. High fiber content slows passage and reduces animal intake.

2.3.4.3.2 Climate Change Effects on Forage Quality

Based on expected vegetation changes and known environmental effects on forage protein, carbohydrate, and fiber contents, both positive and negative changes in forage quality are possible as a result of atmospheric and climatic change (Table 2.13). Non-structural carbohydrates can increase under elevated CO_2 (Read et al. 1987), thereby potentially enhancing forage quality. However, plant N and crude protein concentrations often decline in CO_2-enriched atmospheres, especially when plant production is enhanced by CO_2. This reduction in crude protein reduces forage quality and counters the positive effects of CO_2 enrichment on plant production and carbohydrates (Cotrufo et al. 1998; Milchunas et al. 2005). Limited evidence suggests that the decline is greater when soil nitrogen availability is low than high (Bowler and Press 1996; Wilsey 1996), implying that rising CO_2 possibly reduces the digestibility of forages that are already of poor quality for ruminants. Experimental warming also reduces tissue N concentrations (Wan et al. 2005), but reduced precipitation typically has the opposite effect. Such reductions in forage quality could possibly have pronounced negative effects on animal growth, reproduction, and mortality (Milchunas et al. 2005; Owensby et al. 1996), and could render livestock production unsustainable unless animal diets are supplemented with N (e.g., urea, soybean meal). On shortgrass steppe, for example, CO_2 enrichment reduced the crude protein concentration of autumn forage below

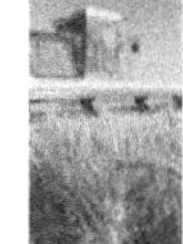

Table 2.13 Potential changes in forage quality arising from atmospheric and climatic change.

Change	Examples of positive effects on forage quality	Examples of negative effects on forage quality
Life-form distributions	Decrease in proportion of woody shrubs and increase in grasses in areas with increased fire frequency.	Increase in the proportion of woody species because of elevated CO_2, increases in rainfall event sizes and longer intervals between rainfall events.
Species or functional group distributions	Possible increase in C_3 grasses relative to C_4 grasses at elevated CO_2.	Increase in the proportion of C_4 grasses relative to C_3 grasses at higher temperatures. Increase in abundance of perennial forb species or perennial grasses of low digestibility at elevated CO_2. Increase in poisonous or weedy plants.
Plant biochemical properties	Increase in non-structural carbohydrates at elevated CO_2. Increase in crude protein content of forage with reduced rainfall.	Decrease in crude protein content and digestibility of forage at elevated CO_2 or higher temperatures. No change or decrease in crude protein in regions with more summer rainfall.

critical maintenance levels for livestock in three out of four years and reduced the digestibility of forage by 14 percent in mid-season and by 10 percent in autumn (Milchunas et al. 2005). Significantly, the grass most favored by CO_2 enrichment also had the lowest crude protein concentration. Plant tissues that re-grow following defoliation generally are of higher quality than older tissue, so defoliation could ameliorate negative effects of CO_2 on forage quality. This however did not occur on shortgrass steppe (Milchunas et al. 2005). Changes in life forms, species, or functional groups resulting from differential responses to global changes will very likely vary among rangelands depending on the present climate and species composition, with mixed consequences for domestic livestock (Table 2.13).

2.3.5 Climatic Influences on Livestock

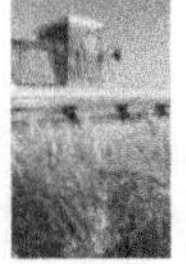

Climate changes, as suggested by some GCMs, could impact the economic viability of livestock production systems worldwide. Surrounding environmental conditions directly affect mechanisms and rates of heat gain or loss by all animals (NRC 1981). Lack of prior conditioning to weather events most often results in catastrophic losses in the domestic livestock industry. In the central U.S. in 1992, 1995, 1997, 1999, 2005, and 2006, some feedlots (intensive cattle feeding operations) lost in excess of 100 head each during severe heat episodes. The heat waves of 1995 and 1999 were particularly severe with documented cattle losses in individual states approaching 5,000 head each year (Hahn and Mader 1997; Hahn et al. 2001). The intensity and/or duration of the 2005 and 2006 heat waves were just as severe as the 1995 and 1999 heat waves, although the extent of losses could not be adequately documented.

The winter of 1996-97 also caused hardship for cattle producers because of greater than normal snowfall and wind velocity, with some feedlots reporting losses in excess of 1,000 head. During that winter, up to 50 percent of the newborn calves were lost, and more than 100,000 head of cattle died in the Northern Plains of the United States.

Additional snowstorm losses were incurred with the collapse of and/or loss of power to buildings that housed confined domestic livestock. Early snowstorms in 1992 and 1997 resulted in the loss of more than 30,000 head of feedlot cattle each year in the southern plains of the United States (Mader 2003).

Economic losses from reduced cattle performance (morbidity) likely exceed those associated with cattle death losses by several-fold (Mader 2003). In addition to losses in the 1990s, conditions during the winter of 2000-2001 resulted in decreased efficiencies of feedlot cattle in terms of overall gain and daily gain of approximately 5 and 10 percent, respectively, from previous years as a result of late autumn and early winter moisture, combined with prolonged cold stress conditions (Mader 2003). In addition, the 2006 snowstorms, which occurred in the southern plains around year end, appear to be as devastating as the 1992 and 1997 storms. These documented examples of how climate can impact livestock production illustrate the potential for more drastic consequences of increased variability in weather patterns, and extreme events that may be associated with climate change.

2.3.5.1 Potential Impact of Climate Change on Livestock

The risk potential associated with livestock production systems due to global warming can be characterized by levels of vulnerability, as influenced by animal performance and environmental parameters (Hahn 1995). When combined performance level and environmental influences create a low level of vulnerability, there is little risk. As performance levels (e.g., rate of gain, milk production per day, eggs/day) increase, the vulnerability of the animal increases and, when coupled with an adverse environment, the animal is at greater risk. Combining an adverse environment with high performance pushes the level of vulnerability and consequent risk to even higher levels. Inherent genetic characteristics or management scenarios that limit the animal's ability to adapt to or cope with environmental factors also puts the animal at risk. At very high performance levels, any environment other than near-optimal may increase animal vulnerability and risk.

The potential impacts of climatic change on overall performance of domestic animals can be determined using defined relationships between climatic conditions and voluntary feed intake, climatological data, and GCM output. Because ingestion of feed is directly related to heat production, any change in voluntary feed intake and/or energy density of the diet will change the amount of heat produced by the animal (Mader et al. 1999b). Ambient temperature has the greatest influence on voluntary feed intake. However, individual animals exposed to the same ambient temperature will not exhibit the same reduction in voluntary feed intake. Body weight, body condition, and level of production affect the magnitude of voluntary feed intake and ambient temperature at which changes in voluntary feed intake begin to be observed. Intake of digestible nutrients is most often the limiting factor in animal production. Animals generally prioritize available nutrients to support maintenance needs first, followed by growth or milk production, and then reproduction.

Based on predicted climate outputs from GCM scenarios, production and response models for growing confined swine and beef cattle, and milk-producing dairy cattle have been developed (Frank et al. 2001). The goal in the development of these models was to utilize climate projections – primarily average daily temperature – to generate an estimate of direct climate-induced changes in daily voluntary feed intake and subsequent performance during summer in the central portion of the United States (the dominant livestock producing region of the country), and across the entire country. The production response models were run for one current (pre-1986 as baseline) and two future climate scenarios: doubled CO_2 (~2040) and a triple of CO_2 (~2090) levels. This data base employed the output from two GCMs – the Canadian Global Coupled (CGC) Model, Version I, and the United Kingdom Meteorological Office/Hadley Center for Climate Prediction and Research model – for input to the livestock production/response models. Changes in production of swine and beef cattle data were represented by the number of days to reach the target weight under each climate scenario and time period. Dairy production is reported in kilograms of milk produced per cow per season. Details of this analysis are reported by Frank (2001) and Frank et al. (2001).

In the central U.S. (MINK region = Missouri, Iowa, Nebraska, and Kansas), days-to-slaughter weight for swine associated with the CGC 2040 scenario increased an average of 3.7 days from the baseline of 61.2 days. Potential losses under this scenario averaged 6 percent and would cost swine producers in the region $12.4 million annually. Losses associated with the Hadley scenario are less severe. Increased time-to-slaughter weight averaged 1.5 days, or 2.5 percent, and would cost producers $5 million annually. For confined beef cattle reared in the central U.S., time-to-slaughter weight associated with the CGC 2040 scenario increased 4.8 days (above the 127-day baseline value) or 3.8 percent, costing producers $43.9 million annually. Climate changes predicted by the Hadley model resulted in loss of 2.8 days of production, or 2.2 percent. For dairy, the projected CGC 2040 climate scenario would result in a 2.2 percent (105.7 kg/cow) reduction in milk output, and cost producers $28 million annually. Production losses associated with the Hadley scenarios would average 2.9 percent and cost producers $37 million annually. Figures 2.12, 2.13, and 2.14 indicate predicted changes in productivity in swine, beef and dairy, respectively, for the various regions of the United States.

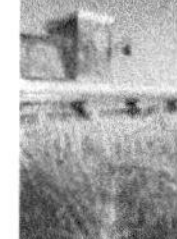

Across the entire United States, percent increase in days to market for swine and beef, and the percent decrease in dairy milk production for the 2040 scenario, averaged 1.2 percent, 2.0 percent, and 2.2 percent, respectively, using the CGC model, and 0.9 percent, 0.7 percent, and 2.1 percent, respectively, using the Hadley model. For the 2090 scenario, respective changes averaged 13.1 percent, 6.9 percent, and 6.0 percent using the CGC model, and 4.3 percent, 3.4 percent, and 3.9 percent using the Hadley model. In general, greater declines in productivity are found with the CGC model than with the Hadley model. Swine and beef production were affected most in the south-central and southeastern United States. Dairy production was affected the most in the U.S. Midwest and Northeast regions.

In earlier research, Hahn et al. (1992) also derived estimates of the effects of climate change of swine growth rate and dairy milk production during summer, as well as other periods during the year. In the east-central United States, per animal milk production was found to decline

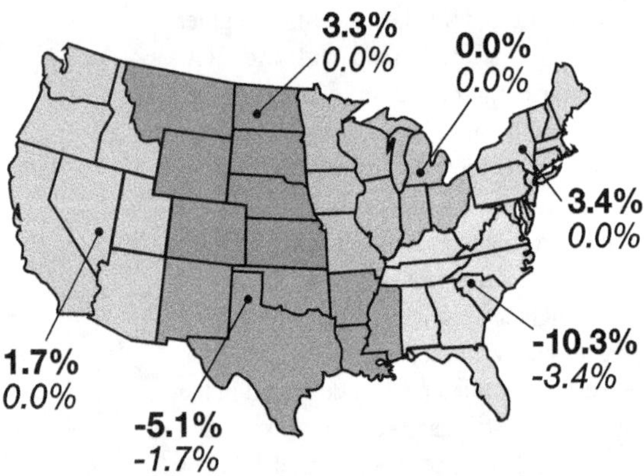

Figure 2.12 *Percent change from baseline to 2040 of days for swine to grow from 50 to 110 kg, beginning June 1 under CGC (bold text) and Hadley (italicized text) modeled climate (Frank 2001; Frank et al. 2001).*

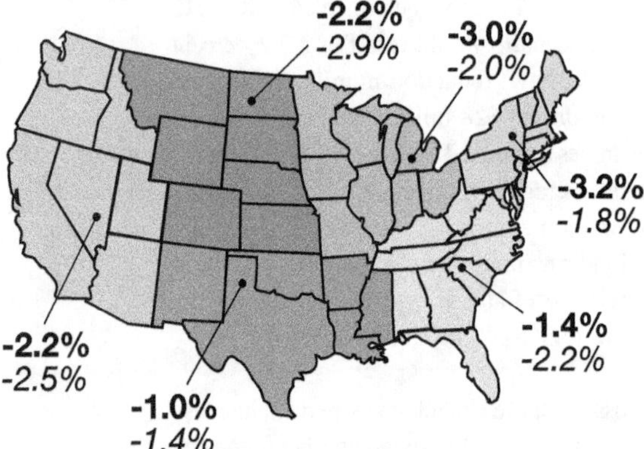

Figure 2.13 *Numerical values represent changes in beef productivity based on the number of days required to reach finish weights from baseline to 2040, beginning June 1 under CGC (bold text) and Hadley (italicized text) modeled climate (Frank 2001; Frank et al. 2001).*

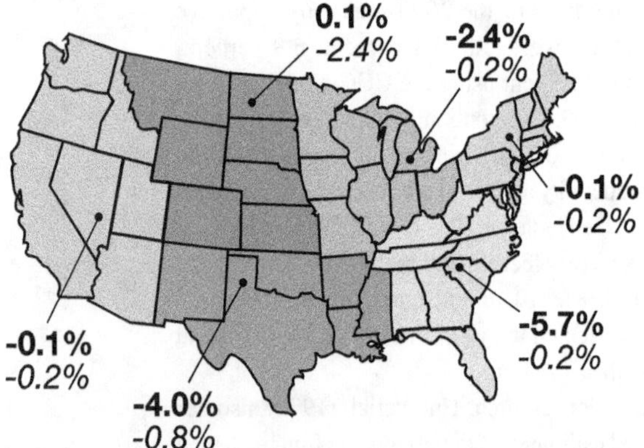

Figure 2.14 *Percent change of kg fat-corrected milk (FCM) yield/cow/season (June 1 to October 31) from baseline to 2040, under CGC (bold text) and Hadley (italicized text) modeled climate (Frank 2001; Frank et al. 2001).*

388 kg (~4 percent) for a July through April production cycle, and 219 kg (~2.2 percent) for an October through July production cycle as a result of global warming. Swine growth rate in this same region was found to decline 26 percent during the summer months, but increased nearly 12 percent during the winter months as a result of global warming. Approximately one-half of these summer domestic livestock production declines are offset by improvements in productivity during the winter. In addition, high producing animals will most likely be affected to a greater extent by global climate change than animals with lower production levels.

A production area in which global climate change may have negative effects that are not offset by positive winter effects is conception rates, particularly in instances when the breeding season primarily occurs in the spring and summer months. (This will particularly affect cattle.)

Hahn (1995) reported that conception rates in dairy cows were reduced 4.6 percent for each unit change when the THI reaches above 70. Amundson et al. (2005) reported a decrease in pregnancy rates of *Bos taurus* cattle of 3.2 percent for each increase in average THI above 70, and a decrease of 3.5 percent for each increase in average temperature above 23.4°C. These data were obtained from beef cows in a range or pasture management system. Amundson et al. (2006) also reported that of the environmental variables studied, minimum temperature had the greatest influence on the percent of cows getting pregnant. Clearly, increases in temperature and/or humidity have the potential to affect conception rates of domestic animals not adapted to those conditions. Summertime conception rates are considerably lower in the Gulf States compared with conception rates in the Northern Plains (Sprott et al. 2001).

In an effort to maintain optimum levels of production, climate change will likely result in livestock producers selecting breeds and breed types that have genetically adapted to conditions that are similar to those associated with the climate change. However, in warmer climates, breeds that are found to be more heat tolerant are generally those that have lower productivity levels, which is likely the mechanism by which they were able to survive as a dominant regional breed. In addition, climate change and associated variation in weather patterns will likely result in more livestock being managed in or near facilities that have capabilities for imposing microclimate modifications (Mader et al. 1997a, 1999a; Gaughan et al. 2002). Domestic livestock, in general, can cope with or adapt to gradual changes in environmental conditions; however, rapid changes in environmental conditions or extended periods of exposure to extreme conditions drastically reduce productivity and are potentially life threatening.

Estimates of livestock production efficiency suggest that negative effects of hotter weather in summer outweigh positive effects of warmer winters (Adams et al. 1999). The largest change occurred under a 5°C increase in temperature, when livestock yields fell by 10 percent in cow-calf and dairy operations in Appalachia, the Southeast, Mississippi Delta, and southern Plains regions of the United States. The smallest change was one percent under 1.5°C warming in the same regions.

Another area of concern is the influence of climate change on diseases and parasites that affect domestic animals. Incidences of disease, such as bovine respiratory disease, are known to be increasing (Duff and Gaylean 2007). However, causes for this increase can be attributed to a number of non-environmentally related factors. As for parasites, similar insect migration and over-wintering scenarios observed in cropping systems may be found for some parasites that affect livestock.

Baylis and Githeko (2006) describe the potential of how climate change could affect parasites and pathogens, disease hosts, and disease vectors for domestic livestock. The potential clearly exists for increased rate of development of pathogens and parasites due to spring arriving earlier and warmer winters that allow greater proliferation and survivability of these organisms. For example, bluetongue was recently reported in Europe for the first time in 20 years (Baylis and Githeko 2006). Warming and changes in rainfall distribution may lead to changes in spatial or temporal distributions of those diseases sensitive to moisture such as anthrax, blackleg, haemorrhagic septicaemia, and vector-borne diseases. However, these diseases, as shown by climate-driven models designed for Africa, may decline in some areas and spread to others (Baylis and Githeko 2006).

2.4 OBSERVING/MONITORING SYSTEMS

2.4.1 Monitoring Relevant to Crops

2.4.1.1 Environmental Stress on Crop Production

Stress symptoms on crop production include warmer canopies associated with increased CO_2 (but the increment may be too small to detect over 30 years), smaller grain size or lower test weight from heat stress, more failures of pollination associated with heat stress, and greater variability in crop production. However, elevated CO_2 will have a helpful effect via reduced water consumption.

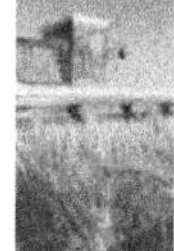

Heat stress could potentially be monitored by satellite image processing over the 30-year span, but causal factors for crop foliage temperature need to be properly considered (temporary water deficit from periodic low rainfall periods, effects of elevated CO_2 to increase foliage temperature, direct effects of elevated air temperature, offset by opposite effect from prolonged water extraction associated with CO_2-induced water conservation). Increased variability in crop yield and lower test weight associated with greater weather variability relative to thresholds for increased temperature can be evaluated both at the buying point, and by using annual USDA crop statistics for rainfed crops. Assessments of irrigated crops can be done in the same way, but with less expectation of water deficit as a causal factor for yield loss. The extent of water requirement for irrigated crops could be monitored by water management district records and pumping permits, but the same issue is present for understanding the confounding effects of temperature, radiation, vapor pressure deficit, rainfall, and CO_2 effects.

2.4.1.2 Phenological Responses to Climate Change

A recent analysis of more than 40 years of spring bloom data from the northeastern United States, the "lilac phenology network," which was established by the USDA in the 1960s, provided robust evidence of a significant biological response to climate change in the region during the latter half of the 20th century (Wolfe et al. 2005).

2.4.1.3 Crop Pest Range Shifts in Collaboration with Integrated Pest Management (IPM) Programs

IPM specialists, and the weather-based weed, insect, and pathogen models they currently utilize, will provide an important link between climate science and the agricultural community. The preponderance of evidence indicates an overall increase in the number of outbreaks and northward migration of a wide variety of weeds, insects, and pathogens. The existing IPM infrastructure for monitoring insect and disease populations could be particularly valuable for tracking shifts in habitable zone of potential weed, insect, and disease pests, and for forecasting outbreaks.

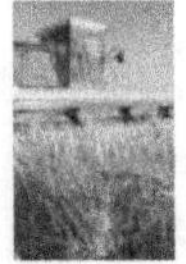

2.4.2 Monitoring Relevant to Pasturelands

Efforts geared toward monitoring the long-term response of pasturelands to climate change should be as comprehensive as possible. When possible, monitoring efforts should include observation of vegetation dynamics, grazing regimes, animal behavior (e.g., indicators of animal stress to heat), mutualistic relationships (e.g., plant-root nematodes; N-fixing organisms), and belowground processes, such as development and changes in root mass, carbon inputs and turnover, nutrient cycling, and water balance. To augment their value, these studies should include use of simulation modeling in order to test hypotheses regarding ecosystem processes as affected by climate change. The development of protocols for monitoring the response of pasturelands to climate change should be coordinated with the development of protocols for rangelands and livestock.

2.4.3 Monitoring Relevant to Rangelands

Soil processes are closely linked to rangeland productivity and vegetation dynamics. As a result, future efforts to track long-term rangeland-vegetation responses to climate change and CO_2 should also involve monitoring efforts directed toward tracking changes in soils. While considerable progress has been made in the application of remote sensing for monitoring plant phenology and productivity, there remains a need for tracking critical soil attributes, which will be important in driving ecological responses of rangelands to climate change.

Nationwide, rangelands cover a broad expanse and are often in regions with limited accessibility. Consequently, ranchers and public land managers need to periodically evaluate range resources (Sustainable Rangeland Roundtable Members 2006). Monitoring of rangelands via remote sensing is already an important research activity, albeit with limited rancher acceptance (Butterfield and Malmstrom 2006). A variety of platforms are currently being evaluated, from low-flying aerial photography (Booth and Cox 2006) to satellite imagery (Afinowicz et al. 2005; Everitt et al. 2006; Phillips et al. 2006; Weber 2006), plus hybrid approaches (Afinowicz et al. 2005) for use in evaluating a variety of attributes considered important indicators of rangeland health – plant cover and bare ground, presence of important plant functional groups or species – documenting changes in plant communities including weed invasion, primary productivity, and forage N concentration.

Although not explicitly developed for global change applications, the goal of many of these methodologies to document changing range conditions suggests tools that could be employed for tracking vegetation change in rangelands, and correlated to climatic or CO_2 data, as done by Knapp et al. (2001). For example, state-and-transition models (Bestelmeyer et al. 2004; Briske et al. 2005) could be expanded to incorporate knowledge of rangeland responses to global change. Integration of those models with existing monitoring efforts and plant developmental data bases, such as the National Phenology Network, could provide a cost-effective monitoring strategy for enhancing knowledge of how rangelands are being impacted by global change, as well as offering management options.

Fundamental soil processes related to nutrient cycling – which may ultimately determine how rangeland vegetation responds to global change – are more difficult to assess. At present, there are no easy and reliable means by which to accurately ascertain the mineral and carbon state of rangelands, particularly over large land areas. The Natural Resources Conservation Service (NRCS) National Soil Characterization Data Base is an especially important baseline of soils information that can be useful for understanding how soils might respond to climate change. However, this data base does not provide a dynamic record of responses through time. Until such information is easily accessible, or reliable methodologies are developed for monitoring rangeland soil properties, predictions of rangeland responses to future environments will be limited. However, much can be ascertained about N cycling responses to global change from relatively easily determined measures of leaf-N chemistry (Peñuelas and Estiarte 1997). As a result, sampling of ecologically important target species in different rangeland ecosystems would be a comparatively low-cost measure to monitor biogeochemical response to global change.

2.5 INTERACTIONS AMONG SYSTEMS

2.5.1 Climate Change and Sustainability of Pasturelands

The current land use system in the United States requires high resource inputs, from the use of synthetic fertilizer on crops to the transport of crops to animal feeding operations. In addition to being inefficient with regard to fuel use, this system creates environmental problems from erosion to high nutrient degradation of water supplies. Recently, scientists have been examining the potential for improved profitability and improved sustainability with a conversion to integrated crop-livestock farming systems (Russelle et al. 2007). This could take many forms. One possible scenario involves grain crops grown in rotation with perennial pasture that also integrates small livestock operations into the farming system. Planting of perennial pastures decreases nitrate leaching and soil erosion, and planting of perennial legumes also reduces the need for synthetic N fertilizer. Diversifying crops also reduces incidence of pests, diseases, and weeds, imparting resilience to the agro-ecosystem. This resilience will become increasingly important as a component of farm adaptation to climate change.

2.6 FINDINGS AND CONCLUSIONS

2.6.1 Crops

2.6.1.1 Grain and Oilseed Crops

Crop yield response to temperature and CO_2 for maize, soybean, wheat, rice, sorghum, cotton, peanut, and dry bean in the United States was assembled from the scientific literature. Cardinal base, optimum, and upper failure-point temperatures for crop development, vegetative, and reproductive growth and slopes-of-yield decline with increase in temperature were reviewed. In general, the optimum temperature for reproductive growth and development is lower than that for vegetative growth. Consequently, life cycle will progress more rapidly, especially given a shortened grain-filling duration and reduced yield as temperature rises. Furthermore, these crops are characterized by an upper failure-point temperature at which pollination and grain-set processes fail. Considering these aspects, the optimum mean temperature for grain yield is fairly low for the major agronomic crops: 18-22°C for maize, 22-24°C for soybean, 15°C for wheat, 23-26°C for rice, 25°C for sorghum, 25-26°C for cotton, 20-26°C for peanut, 23-24°C for dry bean, and 22-25°C for tomato.

Without the benefit of CO_2, the anticipated 1.2°C rise in temperature over the next 30 years is projected to decrease maize, wheat, sorghum, and dry bean yields by 4.0, 6.7, 9.4, and 8.6 percent, respectively, in their major production regions. For soybean, the 1.2°C temperature rise will increase yield 2.5 percent in the Midwest where temperatures during July, August, September average 22.5°C, but will decrease yield 3.5 percent in the South, where mean temperature during July, August, and September averages 26.7°C. Likewise, in the South, that same mean temperature will result in reduced rice, cotton, and peanut yields, which will decrease 12.0, 5.7, and 5.4 percent, respectively. An anticipated CO_2 increase from 380 to 440 ppm will increase maize and sorghum yield by only 1 percent, whereas the listed C_3 crops will increase yield by 6.1 to 7.4 percent, except for cotton, which shows a 9.2 percent increase. The response to CO_2 was developed from interpolation of extensive literature summarization of

Diversifying crops also reduces incidence of pests, diseases, and weeds, imparting resilience to the agro-ecosystem. This resilience will become increasingly important as a component of farm adaptation to climate change.

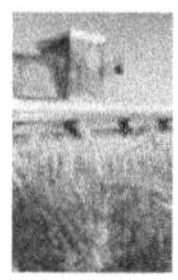

response to ambient versus doubled CO_2. The net effect of rising temperature and CO_2 on yield will be maize (-3.0 percent), soybean (Midwest, +9.9 percent; South, +3.9 percent), wheat (+0.1 percent), rice (-5.6 percent), sorghum (-8.4 percent), cotton (+3.5 percent), peanut (+1.3 percent), and dry bean (-2.5 percent). The CO_2-induced decrease in measured ET summarized from chamber and FACE studies, from 380 to 440 ppm, gives a fairly repeatable reduction in ET of 1.4 to 2.1 percent, although the 1.2°C rise in temperature would increase ET by 1.8 percent, giving an unimportant net -0.4 to +0.3 percent reduction in ET. This effect could lead to a further small -0.4 to +0.3 percent change in yield under rainfed production. A similar small change in crop water requirement will occur under irrigated production.

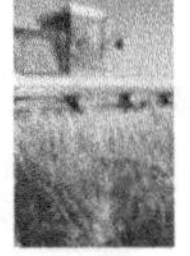

Thus, the benefits of CO_2 rise over the next 30 years mostly offset the negative effects of temperature for most C_3 crops except rice and bean, while the C_4 crop yields are reduced by rising temperature because they have little response to the CO_2 rise. The two factors also nearly balance out on crop transpiration requirements. Thus, the 30-year outlook for crop production is relatively neutral. However, the outlook for the next 100 years would not be as optimistic, if rise in temperature and CO_2 continue, because the C_3 response to rising CO_2 is reaching a saturating plateau, while the negative temperature effects will become progressively more severe. There are continual changes in the genetic resources of crop varieties and horticultural crops that will provide increases in yield due to increased resistance to water and pest stresses. These need to be considered in any future assessments of the climatic impacts; however, the genetic modifications have not altered the basic temperature response or CO_2 response of the biological system.

As temperature rises, crops will increasingly begin to experience upper failure point temperatures, especially if climate variability increases and if rainfall lessens or becomes more variable. Under this situation, yield responses to temperature and CO_2 would move more toward the negative side. Despite increased CO_2-responsiveness of photosynthesis/biomass as temperature increases, there were no published beneficial interactions of increased CO_2 upon grain yield as temperature increased because temperature effects on reproductive processes, especially pollination, are so dominant. On the other hand, there are cases of negative interactions on pollination associated with the rise in canopy temperature caused by lower stomatal conductance. For those regions and crops where climate change impairs reproductive development because of an increase in the frequency of high temperature stress events (e.g., >35°C), the potential beneficial effects of elevated CO_2 on yield may not be fully realized.

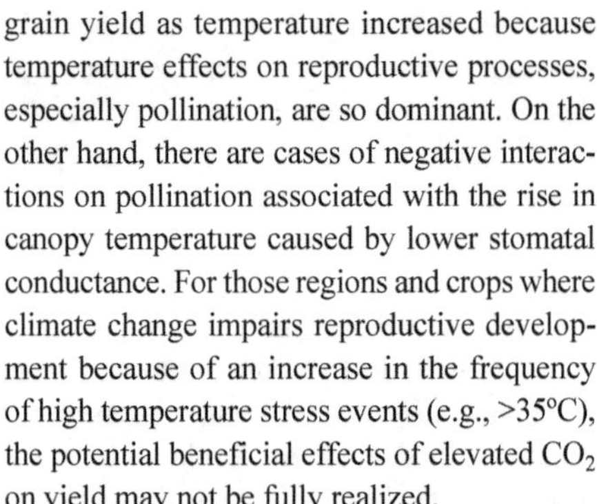

No direct conclusions were made relative to anticipated effects of rainfall change on crop production. Such assessment requires use of global climate models and the climate outputs to be directed as inputs to crop growth models to simulate production for the different crops.

2.6.1.2 Horticultural Crops

Although horticultural crops account for more than 40 percent of total crop market value in the United States (2002 Census of Agriculture), there is relatively little information on their response to CO_2, and few reliable crop simulation models for use in climate change assessments compared to that which is available for major grain and oilseed crops. The marketable yield of many horticultural crops is likely to be more sensitive to climate change than grain and oilseed crops because even short-term, minor environmental stresses can negatively affect visual and flavor quality. Perennial fruit and nut crop survival and productivity will be highly sensitive to winter, as well as summer, temperatures.

2.6.2 Weeds

The potential habitable zone of many weed species is largely determined by temperature. For example, kudzu (*Pueraria lobata, var. montana*) is an aggressive species that has a northern range currently constrained by the -20°C minimum winter temperature isocline. While other factors such as moisture and seed dispersal will affect the spread of invasive weeds such as kudzu, climate change is likely to lead to a northern migration in at least some cases.

Many weeds respond more positively to increasing CO_2 than most cash crops, particularly C_3 invasive weeds that reproduce by vegetative means (roots, stolons, etc.). Recent research also suggests that glyphosate loses its efficacy on weeds grown at elevated CO_2. While there are many weed species that have the C_4 photosynthetic pathway and therefore show a smaller response to atmospheric CO_2 relative to C_3 crops, in most agronomic situations, crops are in competition with a mix of both C_3 and C_4 weeds.

2.6.3 Insects and Disease Pests

In addition to crops and weeds, beneficial and harmful insects, invasives, microbes and other organisms present in agroecosystems will be responding to changes in CO_2 and climate. Numerous studies have already documented changes in spring arrival, over-wintering, and/or geographic range of several insect and animal species due to climate change. Disease pressure from leaf and root pathogens may increase in regions where increases in humidity and frequency of heavy rainfall events are projected, and decrease in regions projected to encounter more frequent drought.

2.6.4 Pasturelands

Today, pasturelands in the United States extend over 117 million acres; however, the area under pasturelands has experienced an 11 percent decrease over the last 25 years due mainly to expansion of urban areas. Consequently, future reductions in pastureland area will require an increase in pasture productivity in order to meet production needs.

In general, pasture species have been less studied than cropland species in terms of their response to climate change variables including atmospheric CO_2 concentration, temperature, and precipitation. Pastureland response to climate change will likely be complex because, in addition to the main climatic drivers, other plant and management factors might also influence the response (e.g., plant competition, perennial growth habits, seasonal productivity, and plant-animal interactions).

Results of studies evaluating the response of pasture species to elevated CO_2 are consistent with the general response of C_3 and C_4 type vegetation to elevated CO_2 but important exceptions exist. C_3 pasture species such as Italian ryegrass, orchardgrass, rhizoma peanut, tall fescue, and timothy have exhibited increased photosynthetic rates under elevated CO_2. Other studies suggest that Kentucky bluegrass might be at the lower end of the range in the responsiveness of C_3 grasses to elevated CO_2, especially under low nutrient conditions. Perennial ryegrass has shown a positive response in terms of photosynthetic rate, but a low or even negative response in terms of plant yield. The C_4 pasture species bahiagrass, an important pasture species in Florida, appears marginal in its response to elevated CO_2. Also, shifts in optimal temperatures for photosynthesis might be expected under elevated CO_2. Species like perennial ryegrass and tall fescue may show a downward shift in their optimal temperatures for photosynthesis.

This review has not yielded sufficient quantitative information for predicting the yield change of pastureland species under a future temperature increase of 1.2 °C. However, projected increases in temperature and the lengthening of the growing season should, in principle, extend forage production into late fall and early spring, thereby decreasing the need for accumulation of forage reserves during the winter season. In addition, water availability may play a major role in the response of pasturelands to climate change. Dallisgrass appears to better withstand conditions of moisture stress under elevated CO_2 than under ambient conditions. Simulation modeling of alfalfa yield response to climate change suggests that future alterations in precipitation will be very important in determining yields. Roughly, for every 4 mm change in annual precipitation, the models predict a 1 percent change in dryland alfalfa yields.

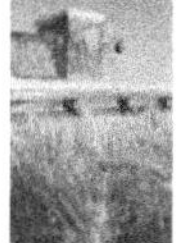

In studies using defoliation as a variable, increases in plant productivity under defoliation were only observed under ambient CO_2 while the largest response to elevated CO_2 was observed in non-defoliated plants. The effect of elevated CO_2 on pasture yield may be affected by the presence of mutualistic interactions with other organisms. Tall fescue plants infected with an endophyte fungus and exposed to elevated CO_2 showed a 15 percent higher yield response than under ambient conditions.

An improved understanding of the impacts of climate change on pastureland might be obtained through comprehensive studies that include grazing regimes, mutualistic relationships (e.g., plant roots-nematodes; N-fixing organisms), as well as the balance of carbon, nutrients and water.

2.6.5 Rangelands

The evidence from manipulative experiments, modeling exercises, and long-term observations of rangeland vegetation over the past two centuries provide indisputable evidence that warming, altered precipitation patterns, and rising atmospheric CO_2 are virtually certain to have profound impacts on the ecology and agricultural utility of rangelands.

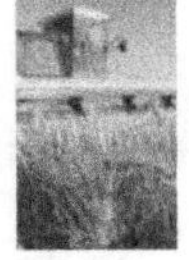

As CO_2 levels and temperatures continue to climb, and precipitation patterns change, sensitivity of different species to CO_2 will direct shifts in plant community species composition. However, lacking multiple global change experiments that incorporate CO_2, temperature, and precipitation, our knowledge about how global change factors and soil nutrient cycling will interact and affect soil N availability is limited, and reduces our ability to predict species change.

Based on current evidence, plants with the C_3 photosynthetic pathway – forbs, woody plants, and possibly legumes – seem likely to be favored by rising CO_2, although interactions of species responses with rising temperature and precipitation patterns may affect these functional group responses (Morgan 2005, 2007). (For instance, warmer temperatures and drier conditions will tend to favor C_4 species, which may cancel out the CO_2 advantage of C_3 grasses.)

There is already some evidence that climate change-induced species shifts are underway in rangelands. For instance, encroachment of woody shrubs into former grasslands is likely due to a combination of over-grazing, lack of fire, and rising levels of atmospheric CO_2. Combined effects of climate and land management change can drive species change that can have a tremendous negative impact on the range livestock industry (Bond and Midgley 2000; Morgan et al. 2007; Polley, 1997). In turn, this has altered the frequency and timing of wildfires by reducing establishment of perennial herbaceous species by pre-empting soil water early in the growing season (Young 1991). It seems likely that plant species changes will have as much or more impact on livestock operations as alterations in plant productivity.

One of our biggest concerns is in the area of how grazing animals affect ecosystem response to climate change. Despite knowledge that large grazing animals have important impacts on the productivity and nutrient cycling for rangelands (Augustine and McNaughton 2004, 2006; Semmartin et al. 2004), little global change research has addressed this particular problem. Manipulative field experiments in global change research are often conducted on plots too small to incorporate grazing animals, so these findings do not reflect the effect grazing domestic livestock can have on N cycling due to diet selectivity, species changes, and nutrient cycling, all of which can interact with CO_2 and climate (Allard et al. 2004; Semmartin et al. 2004). The paucity of data presently available on livestock-plant interactions under climate change severely compromises our ability to predict the consequences of climate change on livestock grazing.

Another important knowledge gap concerns the responses of rangelands to multiple global changes. To date, only one experiment has examined four global changes: rising CO_2, temperature, precipitation, and N deposition (Dukes et al. 2005; Zavaleta et al. 2003a). Although interactions between global change treatments on plant production were rare, strong effects on relative species abundances and functional plant group responses suggest highly complex interactions of species responses to combined global changes that may ultimately impact nutrient cycling with important implications for plant community change and C storage. Such results underscore an emerging acknowledgement that while there is certainty that rangeland ecosystems are responding to global change, our ability to understand and predict responses to future changes is limited.

Rangelands are used primarily for grazing. For most domestic herbivores, the preferred forage is grass. Other plants – including trees, shrubs, and other broadleaf species – can lessen livestock production and profitability by reducing availability of water and other resources to grasses, making desirable plants unavailable to livestock or physically complicating livestock management, or poisoning grazing animals (Dahl and Sosebee 1991).

In addition to livestock grazing, rangelands provide many other goods and services, including biodiversity, tourism, and hunting. They are also important as watershed catchments. Carbon stores are increasingly being considered as an economic product (Liebig et al. 2005; Meeting et al. 2001; Moore et al. 2001; Schuman et al. 2001). However, there is still uncertainty about the greenhouse gas sink capacity of rangelands, how it will be altered by climate change – including rising atmospheric CO_2 – and, ultimately, the economics of rangeland C sequestration (Schlesinger 2006; van Kooten 2006). While the ability to accurately predict the consequences of all aspects of climate change for rangelands is limited, a recent list of management options (Morgan 2005) suggests the types of choices ranchers and land managers will need to consider in the face of climate change (Table 2.14).

A challenge for rangeland scientists, public land managers, ranchers, and others interested in rangelands will be to understand how the dynamics of climate change and land management translate into ecological changes that impact long-term use and sustainability. Perhaps more than most occupations, ranching in the present-day United States is as much a lifestyle choice as it is an economic decision (Bartlett et al. 2002), so economics alone will not likely drive decisions that ranchers make in response to climate change. Nevertheless, ranchers are already looking to unconventional rangeland uses like tourism or C storage. In regions where vegetation changes are especially counter-productive to domestic livestock agriculture, shifts in enterprises will occur. Shifts between rangeland and more intensive agriculture may also occur, depending on the effects of climate-induced environmental changes and influence of economics that favor certain commodities. However, once a native rangeland is disturbed, whether intentionally through intensive agriculture or unintentionally through climate change, restoration can be prohibitively costly, and in some cases, impossible. Therefore, management decisions on the use of private and public rangelands will need to be made with due diligence paid toward their long-term ecological impacts.

2.6.6 Animal Production Systems

Increases in air temperature reduce livestock production during the summer season with partial offsets during the winter season. Current management systems usually do not provide as much shelter to buffer the effects of adverse weather for ruminants as for non-ruminants. From that perspective, environmental management for ruminants exposed to global warming needs to consider: 1) general increase in temperature levels, 2) increases in nighttime temperatures, and 3) increases in the occurrence of extreme events (e.g., hotter daily maximum temperature and more/longer heat waves).

In terms of environmental management needed to address global climate change, the impacts can be reduced by recognizing the adaptive ability of the animals and by proactive application of appropriate countermeasures (sunshades, evaporative cooling by direct wetting or in conjunction with mechanical ventilation, etc.). Specifically, the capabilities of livestock managers to cope with these effects are quite likely to keep up with the projected rates of change in global temperature and related climatic factors. However, coping will entail costs such as application of environmental modification techniques, use of more suitably adapted animals, or even shifting animal populations.

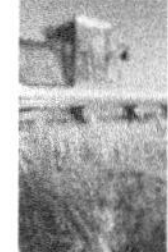

Climate changes affect certain parasites and pathogens, which could result in adverse effects on host animals. Interactions exist among temperature, humidity, and other environmental factors which, in turn, influence energy exchange. Indices or measures that reflect these interactions remain ill-defined, but research to improve them is underway. Factors other than thermal (i.e., dust, pathogens, facilities, contact surfaces, technical applications) also need better definition. Duration and intensity of potential stressors are of concern with respect to the coping and/or adaptive capabilities of an animal. Further, exposure to one type of stressor may lead to altered resistance to other types. Other interactions may exist, such that animals stressed by heat or cold may be less able to cope with other stressors (restraint, social mixing, transport, etc). Improved stressor characterization is needed to provide a basis for refinement of sensors providing input to control systems. Innovations in electronic system capabilities will undoubtedly continue to be exploited for the betterment of livestock environments with improved economic utilization of environmental measures, and mitigation strategies. There is much potential for application of improved sensors, expert systems, and electronic stockmanship. Continued progress should be closely tied to animal needs based on rational criteria, and must include further recognition of health criteria for animal caretakers as well. The ability of the animal's target tissues to respond to disruptions in normal physiological circadian rhythms may be an important indicator of stress. Also, the importance of obtaining multiple measures of stress is also becoming more apparent. However, inclusion and weighting of multiple factors (e.g., endocrine function, immune function, behavior patterns, performance measures, health status, vocalizations) is not an easy task in developing integrated stress measures. Establishing threshold limits for impaired functions that may result in reduced performance or health are essential. Improved modeling of physiological systems as our knowledge base expands will help the integration process.

Table 2.14 CO_2 and climate change responses and management options for grazing land factors. Adapted from Morgan (2005).

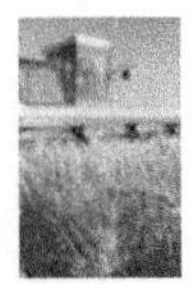

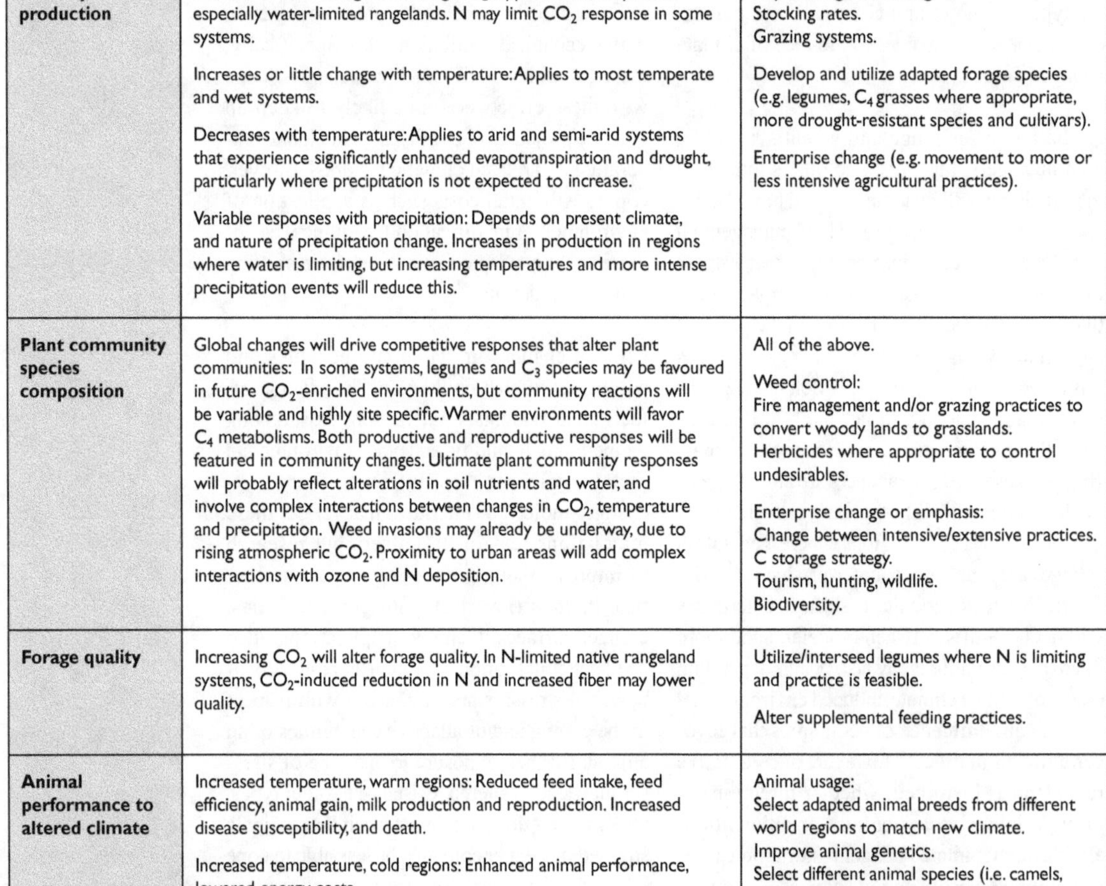

Factor	Responses to rising CO_2 and climate change	Management options
Primary production	Increase or little change with rising CO_2: Applies to most systems, especially water-limited rangelands. N may limit CO_2 response in some systems. Increases or little change with temperature: Applies to most temperate and wet systems. Decreases with temperature: Applies to arid and semi-arid systems that experience significantly enhanced evapotranspiration and drought, particularly where precipitation is not expected to increase. Variable responses with precipitation: Depends on present climate, and nature of precipitation change. Increases in production in regions where water is limiting, but increasing temperatures and more intense precipitation events will reduce this.	Adjust forage harvesting: Stocking rates. Grazing systems. Develop and utilize adapted forage species (e.g. legumes, C_4 grasses where appropriate, more drought-resistant species and cultivars). Enterprise change (e.g. movement to more or less intensive agricultural practices).
Plant community species composition	Global changes will drive competitive responses that alter plant communities: In some systems, legumes and C_3 species may be favoured in future CO_2-enriched environments, but community reactions will be variable and highly site specific. Warmer environments will favor C_4 metabolisms. Both productive and reproductive responses will be featured in community changes. Ultimate plant community responses will probably reflect alterations in soil nutrients and water, and involve complex interactions between changes in CO_2, temperature and precipitation. Weed invasions may already be underway, due to rising atmospheric CO_2. Proximity to urban areas will add complex interactions with ozone and N deposition.	All of the above. Weed control: Fire management and/or grazing practices to convert woody lands to grasslands. Herbicides where appropriate to control undesirables. Enterprise change or emphasis: Change between intensive/extensive practices. C storage strategy. Tourism, hunting, wildlife. Biodiversity.
Forage quality	Increasing CO_2 will alter forage quality. In N-limited native rangeland systems, CO_2-induced reduction in N and increased fiber may lower quality.	Utilize/interseed legumes where N is limiting and practice is feasible. Alter supplemental feeding practices.
Animal performance to altered climate	Increased temperature, warm regions: Reduced feed intake, feed efficiency, animal gain, milk production and reproduction. Increased disease susceptibility, and death. Increased temperature, cold regions: Enhanced animal performance, lowered energy costs.	Animal usage: Select adapted animal breeds from different world regions to match new climate. Improve animal genetics. Select different animal species (i.e. camels, sheep and goats for more drought-prone areas). Alter management (e.g., timing of breeding, calving, weaning) Enterprise change (above)

Land Resources: Forests and Arid Lands

Lead Authors: M.G. Ryan and S.R. Archer

Contributing Authors: R.A. Birdsey, C.N. Dahm, L.S. Heath, J.A. Hicke, D.Y. Hollinger, T.E. Huxman, G.S. Okin, R. Oren, J.T. Randerson, W.H. Schlesinger

3.1 INTRODUCTION

This synthesis and assessment report builds on an extensive scientific literature and series of recent assessments of the historical and potential impacts of climate change and climate variability on managed and unmanaged ecosystems and their constituent biota and processes. It identifies changes in resource conditions that are now being observed and examines whether these changes can be attributed in whole or part to climate change. It also highlights changes in resource conditions that recent scientific studies suggest are most likely to occur in response to climate change, and when and where to look for these changes. As outlined in the Climate Change Science Program (CCSP) Synthesis and Assessment Product 4.3 (SAP 4.3) prospectus, this chapter will specifically address climate-related issues in forests and arid lands.

In this chapter the focus is on the near-term future. In some cases, key results are reported out to 100 years to provide a larger context but the emphasis is on next 25-50 years. This nearer-term focus is chosen for two reasons. First, for many natural resources, planning and management activities already address these time scales through development of long-lived infrastructure, forest rotations, and other significant investments. Second, climate projections are relatively certain over the next few decades. Emission scenarios for the next few decades do not diverge from each other significantly because of the "inertia" of the energy system. Most projections of greenhouse gas emissions assume that it will take decades to make major changes in the energy infrastructure, and only begin to diverge rapidly after several decades have passed (30-50 years).

Forests occur in all 50 states but are most common in the humid eastern United States, the West Coast, at higher elevations in the Interior West and Southwest, and along riparian corridors in the plains states (Figure 3.1) (Zhu and Evans 1994). Forested land occupies about 740 million acres, or about one-third of the United States. Forests in the eastern United States cover 380 million acres; most of this land (83 percent) is privately owned, and 74 percent is broadleaf forest. The 360 million acres of forest land in the western United States are 78 percent conifer forests, split between public (57 percent) and private ownership (USDA Forest Service and U.S. Geological Survey 2002).

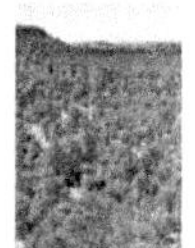

Forested land occupies about 740 million acres, or about one-third of the United States.

Forests provide many ecosystem services important to the well-being of the people of the United States: watershed protection, water quality, and flow regulation; wildlife habitat and diversity; recreational opportunities, and aesthetic and spiritual fulfillment; raw material for wood and paper products; climate regulation, carbon storage, and air quality; biodiversity conservation. While all of these services have considerable economic value, some are not easily quantified (Costanza et al. 1997; Daily et al. 2000;

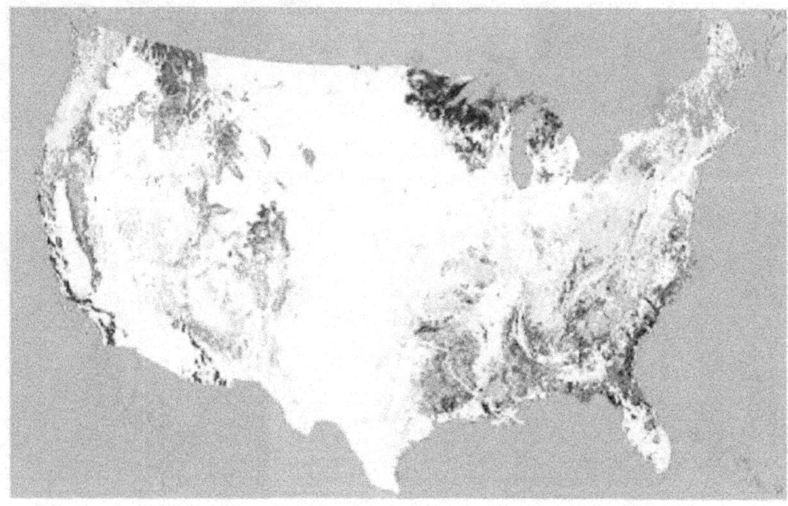

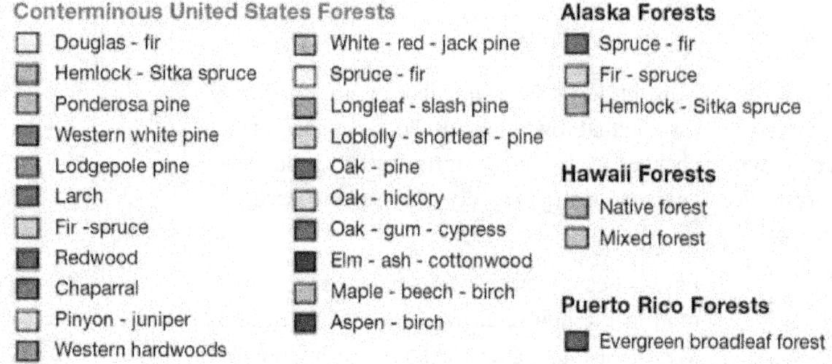

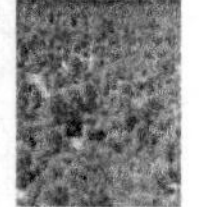

Figure 3.1 *Distribution of forest lands in the continental United States by forest type. This map was derived from Advanced Very High Resolution Radiometer (AVHRR) composite images recorded during the 1991 growing season. Each composite covered the United States at a resolution of one kilometer. Field data collected by the Forest Service were used to aid classification of AVHRR composites into forest-cover types. Details on development of the forest cover types dataset are in Zhu and Evans (1994).*

Krieger 2001; Millennium Ecosystem Assessment 2005), and many Americans are strongly attached to their forests. A changing climate will alter forests and the services they provide. Sometimes changes will be viewed as beneficial, but often they will be viewed as detrimental.

Arid lands are defined by low and highly variable precipitation, and are found in the United States in the subtropical hot deserts of the Southwest and the temperate cold deserts of the Intermountain West (Figure 3.2). Arid lands provide many of the same ecosystem services as forests (with the exception of raw materials for wood and paper products), and support a large ranching industry. These diverse environments are also valued for their wildlife habitat, plant and animal diversity, regulation of water flow and quality, opportunities for outdoor recreation, and open spaces for expanding urban environments. A changing climate will alter arid lands and their services.

Both forests and arid lands face challenges that can affect their responses to a changing climate: the legacy of historical land use, non-native invasive species, and the slow growth of many species. In forests, for instance, clearing and farming dramatically increased erosion, and the re-established forests are likely less productive as a result. In arid lands, grazing and exurban development can change plant and animal communities. Non-native invasive species are a challenge for all ecosystems, but especially so in arid lands, where non-native invasive grasses encourage fire in ecosystems where fire was historically very rare. The very slow growth of many arid land and dry forest species hinders recovery from disturbance.

Climate strongly influences both forests and arid lands. Climate shapes the broad patterns of ecological communities, the species within them, their productivity, and the ecosystem goods and services they provide. The interaction of vegetation and climate is a fundamental tenet of ecology. Many studies show how vegetation has changed with climate over the past several thousand years, so it is well understood that changes in climate will change vegetation. Given a certain climate and long enough time, resultant ecological communities can generally be predicted. However, predicting the effects of a changing climate on forests and arid lands for the next few decades is challenging, especially with regard to the rates and dynamics of change. Plants in these communities can be long lived; hence, changes in species composition may lag behind changes in climate. Furthermore, seeds and conditions for better-adapted communities are not always present.

Past studies linking climate and vegetation may also provide poor predictions for the future because the same physical climate may not occur in the future and many other factors may be changing as well. CO_2 is increasing in the atmosphere; nitrogen deposition is much greater than in the past, and appears to be increasing; ozone pollution is locally increasing; and species invasions from other ecosystems are widespread. These factors cause important changes themselves, but their interactions are difficult to predict because they represent novel combinations.

Disturbance (such as drought, storms, insect outbreaks, grazing, and fire) is part of the ecological history of most ecosystems and influences ecological communities and landscapes. Climate affects the timing, magnitude, and frequency of many of these disturbances, and a changing climate will bring changes in disturbance regimes to forests and arid lands (Dale et al. 2001). Trees and arid land vegetation can take from decades to centuries to re-establish after a disturbance. Both human-induced and natural disturbances shape ecosystems by influencing species composition, structure, and function (productivity, water yield, erosion, carbon storage, and susceptibility to future disturbance). In forests, more than 55 million acres are currently impacted by disturbance, with the largest agents being insects and pathogens (Dale et al. 2001). These disturbances cause an estimated financial loss of 3.7 billion dollars per year (Dale et al. 2001). In the past several years, scientists have learned that the magnitude and impact of these disturbances and their response to climate rivals that expected from changes in temperature and precipitation (Field et al. 2007).

Disturbance may reset and rejuvenate some ecosystems in some cases and cause enduring change in others. For example, climate may favor the spread of invasive exotic grasses into arid lands where the native vegetation is too sparse to carry a fire. When these areas burn, they typically convert to non-native monocultures and the native vegetation is lost. In another example, drought may weaken trees and make them susceptible to insect attack and death – a pattern that recently occurred in the Southwest. In these forests, drought and insects converted

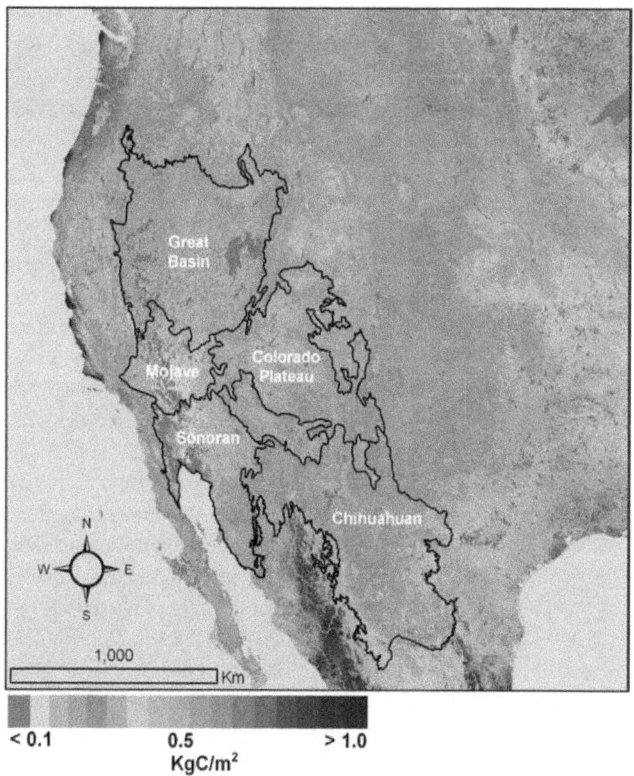

Figure 3.2 *The five major North American deserts, outlined on a 2006 map of net primary productivity (NPP). Modeled NPP was produced by the Numerical Terradynamic Simulation Group (http://www.ntsg.umt.edu/) using the fraction of absorbed photosynthetically active radiation measured by the Moderate Resolution Imaging Spectroradiometer (MODIS) satellite and land cover-based radiation use efficiency estimates Running et al. (2000). Desert boundaries based on Olson et al. (2001).*

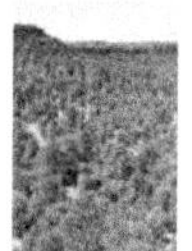

large areas of mixed pinyon-juniper forests into juniper forests. However, fire is an integral component of many forest ecosystems, and many tree species (such as the lodgepole pine forests that burned in the Yellowstone fires of 1988) depend on fire for regeneration. Climate effects on disturbance will likely shape future forests and arid lands as much as the effects of climate itself.

Arid lands are defined by low and highly variable precipitation, and are found in the United States in the subtropical hot deserts of the Southwest and the temperate cold deserts of the Intermountain West.

Disturbances and changes to the frequency or type of disturbance present challenges to resource managers. Many disturbances command quick action, public attention, and resources. Surprisingly, most resource planning in the United States does not consider disturbance, even though disturbances are common, and preliminary information exists on the frequency and areal extent of disturbances (Dale et al.

2001). Disturbances in the future may be larger and more common than those experienced historically, and planning for disturbances should be encouraged (Dale et al. 2001; Stanturf et al. 2007).

The goal of this chapter is to assess how forests and arid lands will respond to predicted anticipated changes in climate over the next few decades. It will discuss the effects of climate and its components on the structure and function of forest and arid land ecosystems. It will also highlight the effects of climate on disturbance and how these disturbances change ecosystems. Active management may increase the resiliency of forests and arid lands to respond to climate change. For example, forest thinning can reduce fire intensity, increase drought tolerance and reduce susceptibility to insect attack. Grazing management and control of invasive species can promote vegetation cover, reduce fire risk, and reduce erosion. These and other options for managing ecosystems to adapt to climate change are discussed in Synthesis and Assessment Product 4.4 (Preliminary review of adaptation options for climate-sensitive ecosystems and resources, U.S. Climate Change Science Program).

Climate strongly influences both forests and arid lands. Climate shapes the broad patterns of ecological communities, the species within them, their productivity, and the ecosystem goods and services they provide.

3.2 FORESTS

3.2.1 Brief Summary of Key Points from the Literature

Climate strongly affects forest productivity and species composition. Forest productivity in the United States has increased 2-8 percent in the past two decades, but separating the role of climate from other factors causing the increase is complicated and varies by location. Some factors that act to increase forest growth are 1) observed increases in precipitation in the Midwest and Lake States, 2) observed increases in nitrogen deposition, 3) an observed increase in temperature in the northern United States that lengthens the growing season, 4) changing age structure of forests, and 5) management practices. These factors interact, and identifying the specific cause of a productivity change is complicated by insufficient data. Even in the case of large forest mortality events, such as those associated with fire and insect outbreaks, attributing a specific event to a change in climate may be difficult because of interactions among factors. For example, in the recent widespread mortality of pinyon pine in the Southwest, intense drought weakened the trees, but generally, the Ips beetle killed them.

In addition to the direct effects of climate on tree growth, climate also affects the frequency and intensity of natural disturbances such as fire, insect outbreaks, ice storms, and windstorms. These disturbances have important consequences for timber production, water yield, carbon storage, species composition, invasive species, and public perception of forest management. Disturbances also draw management attention and resources. Because of observed warmer and drier climate in the West in the past two decades, forest fires have grown larger and more frequent during that period. Several large insect outbreaks have recently occurred or are occurring in the United States. Increased temperature and drought likely influenced these outbreaks. Fire suppression and large areas of susceptible trees (over age 50) may have also contributed.

Rising atmospheric CO_2 will increase forest productivity and carbon storage in forests if sufficient water and nutrients are available. Any increased carbon storage will be primarily in live trees. Average productivity increase for a variety of experiments was 23 percent. The response of tree growth and carbon storage to elevated CO_2 depends on site fertility, water availability, and stand age, with fertile, younger stands responding more strongly.

Forest inventories can detect long-term changes in forest growth and species composition, but they have limited ability to attribute changes to specific factors, including climate. Separating the effects of climate change from other impacts would require a broad network of indicators, coupled with a network of controlled experimental manipulations. Experiments that directly manipulate climate and observe impacts are critical components in understanding climate change impacts and in separating the effects of climate from those caused by other factors. Experiments such as free-air CO_2 enrichment, ecosystem and soil warming, and precipitation manipulation have greatly increased understanding of the direct effects of climate on ecosystems. These experiments have also attracted a large research community and fostered a thorough and integrated understanding because of their large infrastructure costs, importance and rarity. Monitoring of disturbances affecting forests is currently ineffective, fragmented, and generally unable to attribute disturbances to specific factors, including climate.

3.2.2 Observed Changes or Trends

3.2.2.1 Climate and Ecosystem Context

Anyone traveling from the lowlands to the mountains will notice that species composition changes with elevation and with it, the structure and function of these forest ecosystems. Biogeographers have mapped these different vegetation zones and linked them with their characteristic climates. The challenge facing scientists is to understand how these zones and the individual species within them will move with a changing climate, at what rate, and with what effects on ecosystem function.

Temperature, water, and radiation are the primary abiotic factors that affect forest productivity (Figure 3.3). Any response to changing climate will depend on the factors that limit production at a particular site. For example, any site where productivity is currently limited by lack of water or a short growing season will increase productivity if precipitation increases and if the growing season lengthens. Temperature controls the rate of metabolic processes for photosynthesis, respiration, and growth. Generally, plant metabolism has an optimum temperature. Small departures from this optimum usually do not change metabolism and short-term productivity, although changes in growing season length may change annual productivity. Large departures and extreme events (such as frosts in orange groves) can cause damage or tree mortality. Water controls cell division and expansion, which promote growth and stomatal opening, which regulates water loss and CO_2 uptake in photosynthesis. Productivity will generally increase with water availability in water-limited forests (Knapp et al. 2002). Radiation supplies the energy for photosynthesis, and both the amount of leaf area and incident radiation control the quantity of radiation absorbed by a forest. Nutrition and atmospheric CO_2 also strongly influence forest productivity if other factors are less limiting (Boisvenue and Running 2006), and ozone exposure can lower productivity (Hanson et al. 2005). Human activities have increased nitrogen inputs to forest ecosystems, atmospheric CO_2 concentration, and ozone levels. The effects of CO_2 are everywhere, but ozone and N deposition are common to urban areas, and forests and arid lands downwind from urban areas. The response to changes in any of these factors is likely to be complex and dependent on the other factors.

Forest trees are evolutionarily adapted to thrive in certain climates. Other factors, such as fire and competition from other plants, also regulate species presence, but if climate alone changes enough, species will adjust to suitable conditions

Temperature, water, and radiation are the primary abiotic factors that affect forest productivity.

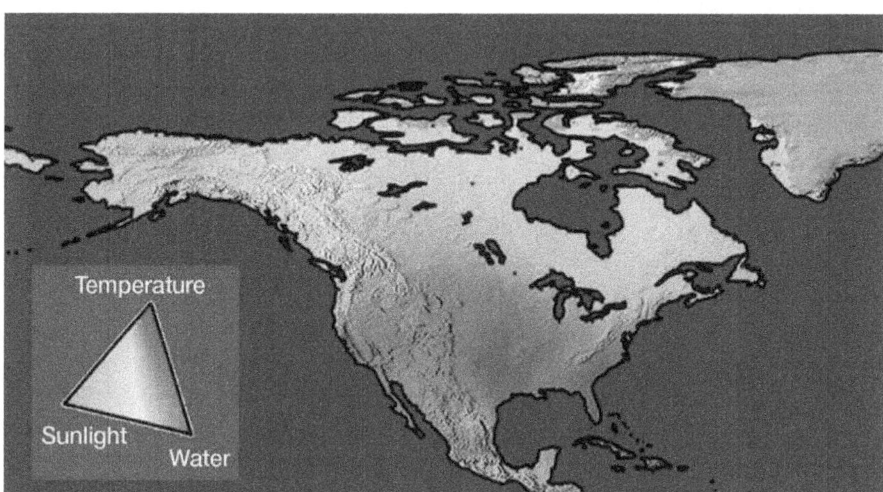

Figure 3.3 *Potential limits to vegetation net primary production based on fundamental physiological limits by sunlight, water balance, and temperature. Nutrients are also important and vary locally. From Boisvenue and Running (2006).*

or go locally extinct if suitable conditions are unavailable (Woodward 1987). One example of such a species shift is sugar maple in the northeastern United States. Suitable climate for it may move northward into Canada and the distribution will likely follow (Chuine and Beaubien 2001), assuming the species is able to disperse propagules rapidly enough to keep pace with the shifting climatic zone. Because trees live for decades and centuries, absent disturbance, it is likely that forest species composition will take time to adjust to changes in climate.

Disturbances such as forest fires, insect outbreaks, ice storms, and hurricanes also change forest productivity, carbon cycling, and species composition. Climate influences the frequency and size of disturbances. Many features of ecosystems can be predicted by forest age, and disturbance regulates forest age. After a stand-replacing disturbance, forest productivity increases until the forest fully occupies the site or develops a closed canopy, then declines to near zero in old age (Ryan et al. 1997). Carbon storage after a disturbance generally declines while the decomposition of dead wood exceeds the productivity of the new forest, then increases as the trees grow larger and the dead wood from the disturbance disappears (Kashian et al. 2006). In many forests, species composition also changes with time after disturbance. Susceptibility to fire and insect outbreaks changes with forest age, but the response of forest productivity to climate, N deposition, CO_2, and ozone differs for old and young forests is still not understood because most studies have only considered young trees or forests. Changes in disturbance prompted by climate change are likely as important as the changes in precipitation, temperature, N deposition, CO_2, and ozone for affecting productivity and species composition.

3.2.2.2 Temperature

Forest productivity in the United States has generally been increasing since the middle of the 20th century (Boisvenue and Running 2006), with an estimated increase of 2-8 percent between 1982 and 1998 (Hicke et al. 2002b), but the causes of this increase (increases in air and surface temperature, increasing CO_2, N deposition, or other factors) are difficult to isolate (Cannell et al. 1998). These effects can potentially be disentangled by experimentation, analysis of species response to environmental gradients, planting trees from seeds grown in different climates in a common garden, anomaly analysis, and other methods. Increased temperatures will affect forest growth and ecosystem processes through several mechanisms (Hughes 2000; Saxe et al. 2001) including effects on physiological processes such as photosynthesis and respiration, and responses to longer growing seasons triggered by thermal effects on plant phenology (e.g., the timing and duration of foliage growth). Across geographical or local elevational gradients, forest primary productivity has long been known to increase with mean annual temperature and rainfall (Leith 1975). This result also generally holds within a species (Fries et al. 2000) and in provenance trials where trees are found to grow faster in a slightly warmer location than that of the seed source itself (Wells and Wakeley 1966; Schmidtling 1994). In Alaska, where temperatures have warmed strongly in recent times, changes in soil processes are similar to those seen in experimental warming studies (Hinzman et al. 2005). In addition, permafrost is melting, exposing organic material to decomposition and drying soils (Hinzman et al. 2005).

Along with a general trend in warming, the length of the northern hemisphere growing season has been increasing in recent decades (Menzel and Fabian 1999; Tucker et al. 2001). Forest growth correlates with growing season length (Baldocchi et al. 2001), with longer growing seasons (earlier spring) leading to enhanced net carbon uptake and storage (Black et al. 2000; Hollinger et al. 2004). The ability to complete phenological development within the growing season is a major determinant of tree species range limits (Chuine and Beaubien 2001). However, Sakai and Weiser (1973) have also related range limits to the ability to tolerate minimum winter temperatures.

3.2.2.3 Fire and Insect Outbreaks

Westerling et al. (2006) analyzed trends in wildfire and climate in the western United States from 1974–2004. They show that both the frequency of large wildfires and fire season length increased substantially after 1985, and that these changes were closely linked with advances in the timing of spring snowmelt, and increases in spring and summer air temperatures. Much of the increase in fire activity occurred in mid-elevation forests in the northern Rocky Mountains and Sierra Nevada mountains. Earlier spring snowmelt probably contributed to greater wildfire frequency in at least two ways, by extending the period during which ignitions could potentially occur, and by reducing water availability to ecosystems in mid-summer, thus enhancing drying of vegetation and surface fuels (Westerling et al. 2006). These trends in increased fire size correspond with the increased cost of fire suppression (Calkin et al. 2005).

In boreal forests across North America, fire activity also has increased in recent decades. Kasischke and Turetsky (2006) combined fire statistics from Canada and Alaska to show that burned area more than doubled between the 1960s/70s and the 1980s/90s. The increasing trend in boreal burned-area appears to be associated with a change in both the size and number of lightning-triggered fires (>1000 km^2), which increased during this period. In parallel, the contribution of human-triggered fires to total burned area decreased from the 1960s to the 1990s (from 35.8 percent to 6.4 percent) (Kasischke and Turetsky 2006). As in the western U.S., a key predictor of burned area in boreal North America is air temperature, with warmer summer temperatures causing an increase in burned area on both interannual and decadal timescales (Gillett et al. 2004; Duffy et al. 2005; Flannigan et al. 2005). In Alaska, for example, June air temperatures alone explained approximately 38 percent of the variance of the natural log of annual burned area during 1950-2003 (Duffy et al. 2005).

Insects and pathogens are significant disturbances to forest ecosystems in the United States (Figure 3.4), costing $1.5 billion annually (Dale et al. 2001). Extensive reviews of the effects of climate change on insects and pathogens have reported many cases where climate change has affected and/or will affect forest insect species range and abundance (Ayres and Lombardero 2000; Malmström and Raffa 2000; Bale et al. 2002). This review focuses on forest insect species within the United States that are influenced by climate and attack forests that are ecologically or economically important.

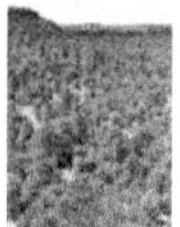

Figure 3.4 *Satellite image of the extensive attack by mountain pine beetle in lodgepole pine forests in Colorado. Pre-outbreak image taken October 2002, and post outbreak image taken August 2007. Images courtesy of DigitalGlobe, Inc. (http://digitalglobe.com/)*

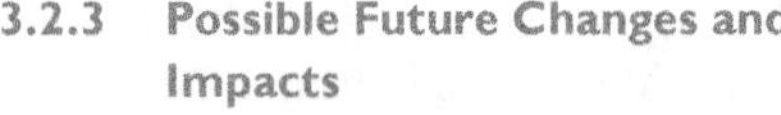

Major outbreaks in recent years include: a mountain pine beetle outbreak affected >10 million hectares (Mha) of forest in British Columbia (Taylor et al. 2006), and 267,000 ha in Colorado (Colorado State Forest Service 2007); more than 1.5 Mha was attacked by spruce beetle in southern Alaska and western Canada (Berg et al. 2006); >1.2 Mha of pinyon pine mortality occurred because of extreme drought, coupled with an Ips beetle outbreak in the Southwest (Breshears et al. 2005); and millions of hectares were affected by southern pine beetle, spruce budworm, and western spruce budworm in recent decades in southeastern, northeastern, and western forests, respectively (USDA Forest Service 2005). Ecologically important whitebark pine is being attacked by mountain pine beetle in the northern and central Rockies (Logan and Powell 2001). For example, almost 70,000 ha, or 17 percent, of whitebark pine forest in the Greater Yellowstone Ecosystem is infested by mountain pine beetle (Gibson 2006). Evident from these epidemics is the widespread nature of insect outbreaks in forests throughout the United States.

For the projected temperature increases over the next few decades, most studies support the conclusion that a modest warming of a few degrees Celsius will lead to greater tree growth in the United States.

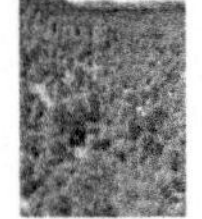

Climate plays a major role in driving, or at least influencing, infestations of these important forest insect species in the United States (e.g., Holsten et al. 1999; Logan et al. 2003a; Carroll et al. 2004; Tran et al. in press), and these recent large outbreaks are likely influenced by observed increases in temperature. Temperature controls life cycle development rates, influences synchronization of mass attacks required to overcome tree defenses, and determines winter mortality rates (Hansen et al. 2001b; Logan and Powell 2001; Hansen and Bentz 2003; Tran et al. in press). Climate also affects insect populations indirectly through effects on hosts. Drought stress, resulting from decreased precipitation and/or warming, reduces the ability of a tree to mount a defense against insect attack (Carroll et al. 2004; Breshears et al. 2005), though this stress may also cause some host species to become more palatable to some types of insects (Koricheva et al. 1998). Fire suppression and large areas of susceptible trees (a legacy from logging in the late 1800s and early 1900s (Birdsey et al. 2006)), may also play a role.

3.2.3 Possible Future Changes and Impacts

3.2.3.1 Warming

A review of recent experimental studies found that rising temperatures would generally enhance tree photosynthesis (Saxe et al. 2001), as a result of increased time operating near optimum conditions, and because rising levels of atmospheric CO_2 increase the temperature optimum for photosynthesis (Long 1991). Warming experiments, especially for trees growing near their cold range limits, generally increase growth (Bruhn et al. 2000; Wilmking et al. 2004; Danby and Hik 2007). The experimental warming of soils alone has been found to stimulate nitrogen mineralization and soil respiration (Rustad et al. 2001). An important concern for all experimental manipulations is that the treatments occur long enough to determine the full suite of effects. It appears that the large initial increases in soil respiration observed at some sites decrease with time back toward pretreatment levels (Rustad et al. 2001; Melillo et al. 2002). This result may come about from changes in C pool sizes, substrate quality (Kirschbaum 2004; Fang et al. 2005), or other factors (Davidson and Janssens 2006).

A general response of leaves, roots, or whole trees to short-term increases in plant temperature is an approximate doubling of respiration with a 10°C temperature increase (Ryan et al. 1994; Amthor 2000). Over the longer term, however, there is strong evidence for temperature acclimation (Atkin and Tjoelker 2003; Wythers et al. 2005), which is probably a consequence of the linkage of respiration to the production of photosynthate (Amthor 2000). One negative consequence of warming for trees is that it can increase the production of isoprene and other hydrocarbons in many tree species (Sharkey and Yeh 2001) – compounds that may lead to higher levels of surface ozone and increased plant damage. Physiologically, the overall result of the few degrees of warming expected over the next few decades is likely a modest increase in photosynthesis and tree growth (Hyvonen et al. 2007). However, where increased temperature coincides with decreased precipitation (western Alaska, Interior West, Southwest), forest growth is expected to be lower (Hicke et al. 2002b).

For the projected temperature increases over the next few decades, most studies support the conclusion that a modest warming of a few degrees Celsius will lead to greater tree growth in the United States. There are many causes for this enhancement including direct physiological effects, a longer growing season, and potentially greater mineralization of soil nutrients. Because different species may respond somewhat differently to warming, the competitive balance of species in forests may change. Trees will probably become established in formerly colder habitats (more northerly, higher altitude) than at present.

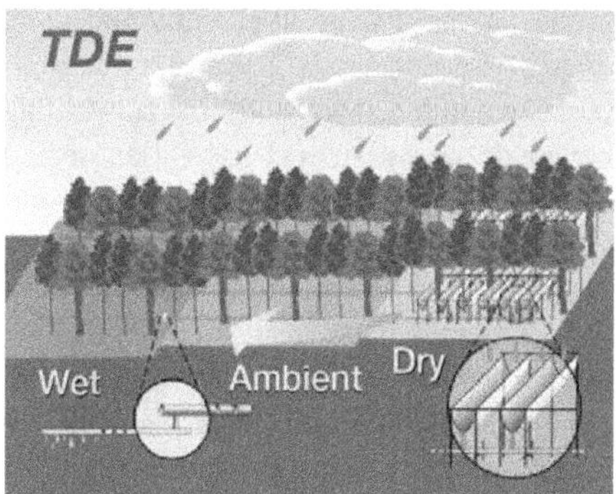

Figure 3.5 *Direct manipulation of precipitation in the Throughfall Displacement experiment (TDE) at Walker Branch (Paul Hanson, Oak Ridge National Laboratory).*

3.2.3.2 Changes in Precipitation

Relationships between forest productivity and precipitation have been assessed using continental gradients in precipitation (Webb et al. 1983; Knapp and Smith 2001), interannual variability within a site (Hanson et al. 2001), and by manipulating water availability (Hanson et al. 2001). Forest productivity varies with annual precipitation across broad gradients (Webb et al. 1983; Knapp and Smith 2001), and with interannual variability within sites (Hanson et al. 2001). Some of these approaches are more informative than others for discerning climate change effects.

Gradient studies likely poorly predict the response to changes in precipitation, because site-specific factors such as site fertility control the response to precipitation (Gower et al. 1992; Maier et al. 2004). The response of forest productivity to interannual variability also likely poorly predicts response to precipitation changes, because forests have the carbohydrate storage and deep roots to offset drought effects over that time, masking any effects that might be apparent over a longer-term trend.

The effects of precipitation on productivity will vary with air temperature and humidity. Warmer, drier air will evaporate more water and reduce water availability faster than cooler, humid air. Low humidity also promotes the closure of stomata on leaves, which reduces photosynthesis and lowers productivity even where soil water availability is abundant.

Manipulation of water availability in forests allows an assessment of the direct effects of precipitation (Figure 3.5). Two experiments where water availability was increased through irrigation showed only modest increases in forest production (Gower et al. 1992; Maier et al. 2004), but large increases with a combination of irrigation and nutrients. In contrast, forest productivity did not change when precipitation was increased or reduced 33 percent, but with the same timing as natural precipitation (Hanson et al. 2005). Tree growth in this precipitation manipulation experiment also showed strong interannual variability with differences in annual precipitation. Hanson et al. (2005) conclude that "differences in seasonal patterns of rainfall within and between years have greater impacts on growth than percentage changes in rainfall applied to all rainfall events."

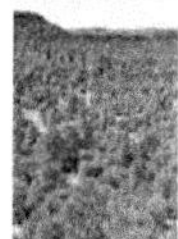

No experiments have assessed the effect of changes in precipitation on forest tree species composition. Hanson et al. (2005) showed that growth and mortality changed in response to precipitation manipulation for some smaller individuals, but we do not know if these changes will lead to composition changes. However, one of the best understood patterns in ecology is the variation of species with climate and site water balance. So, if precipitation changes substantially, it is highly likely that species composition will change (Breshears et al. 2005). However, limited studies exist with which to predict the rate of change and the relationship with precipitation amount.

Drought is a common feature of all terrestrial ecosystems (Hanson and Weltzin 2000), and generally lowers productivity in trees. Drought events can have substantial and long-lasting effects on ecosystem structure, species composition, and function by differentially killing certain species or sizes of trees (Hanson and Weltzin 2000; Breshears et al. 2005), weakening trees to make them more susceptible to insect attacks (Waring 1987), or by increasing the incidence and intensity of forest fires (Westerling et al. 2006). Forest management by thinning trees can improve water available to the residual trees. (Donner and Running 1986; Sala et al. 2005).

If existing trends in precipitation continue, forest productivity will likely decrease in the Interior West, the Southwest, eastern portions of the Southeast, and Alaska. Forest productivity will likely increase in the northeastern United States, the Lake States, and in western portions of the Southeast. An increase in drought events will very likely reduce forest productivity wherever these events occur.

3.2.3.3 Elevated Atmospheric CO_2 and Carbon Sequestration

The effects of increasing atmospheric CO_2 on carbon cycling in forests are most realistically observed in Free Air CO_2 Enrichment (FACE) experiments (Figure 3.6). These experiments have recently begun to provide time-series sufficiently long for assessing the effect of CO_2 projected for the mid-21st century on some components of the carbon cycle. The general findings from a number of recent syntheses using data from the three American and one European FACE sites (King et al. 2004; Norby et al. 2005; McCarthy et al. 2006a; Palmroth et al. 2006) show that North American forests will absorb more CO_2 and might retain more carbon as atmospheric CO_2 increases. The increase in the rate of carbon sequestration will be highest (mostly in wood) on nutrient-rich soils with no water limitation and will decrease with decreasing fertility and water supply. Several yet unresolved questions prevent a definitive assessment of the effect of elevated CO_2 on other components of the carbon cycle in forest ecosystems:

- Although total carbon allocation to belowground increases with CO_2 (King et al. 2004; Palmroth et al. 2006), there is only equivocal evidence of CO_2-induced increase in soil carbon (Jastrow et al. 2005; Lichter et al. 2005).

Figure 3.6 *FACE ring at the Duke Forest FACE, Durham, North Carolina. (Photo courtesy Duke University.)*

- Older forests can be strong carbon sinks (Stoy et al. 2006), and older trees absorb more CO_2 in an elevated CO_2 atmosphere, but wood production of these trees show limited or only transient response to CO_2 (Körner et al. 2005).

- When responding to CO_2, trees require and obtain more nitrogen (and other nutrients) from the soil. Yet, despite appreciable effort, the soil processes supporting such increased uptake have not been identified, leading to the expectation that nitrogen availability may increasingly limit the response to elevated CO_2 (Finzi et al. 2002; Luo et al. 2004; de Graaff et al. 2006; Finzi et al. 2006; Luo et al. 2006).

To understand the complex processes controlling ecosystem carbon cycling under elevated CO_2 and solve these puzzles, longer time series are needed (Walther 2007), yet the three FACE studies in the U.S. forest ecosystems are slated for closure in 2007-2009.

Major findings on specific processes leading to these generalities

Net primary production (NPP) is defined as the balance between canopy photosynthesis and plant respiration. Canopy photosynthesis increases with atmospheric CO_2, but less than expected based on physiological studies because of negative feedbacks in leaves (biochemical down-regulation) and canopies (reduced light, and conductance with increasing leaf area index (LAI); (Saxe et al. 2001; Schäfer et al. 2003; Wittig et al. 2005). On the other hand, plant respiration increases only in proportion to tree growth and amount of living biomass – that is, tissue-specific respiration does not change under elevated CO_2 (Gonzalez-Meler et al. 2004). The balance between these processes, NPP, increases in stands on moderately fertile and fertile soils. The short-term (<10 years), median response among the four "forest" FACE experiments was an increase of 23±2 percent (Norby et al. 2005). Although the average response was similar among these sites that differed in productivity (Norby et al. 2005), the within-site variability in the response to elevated CO_2 can be large (<10 percent to >100 percent). At the Duke FACE site, this within-site variability was related to nitrogen availability (Oren et al. 2001; Finzi et al. 2002; Norby et al. 2005). The absolute magnitude of the additional carbon sink varies greatly among years. At the Duke FACE, much of this variability is caused by droughts and disturbance events (McCarthy et al. 2006a).

The enhancement of NPP at low LAI is largely driven by an enhancement in LAI, whereas at high LAI, the enhancement reflects increased light-use efficiency (Norby et al. 2005; McCarthy et al. 2006a). The sustainability of the NPP response and the partitioning of carbon among plant components may depend on soil fertility (Curtis and Wang 1998; Oren et al. 2001; Finzi et al. 2002). NPP in intermediate fertility sites may undergo several phases of transient response, with CO_2-induced enhancement of stemwood production dominating initially, followed by fine-root production after several years (Oren et al. 2001; Norby et al. 2004). In high fertility plots, the initial response so far appears sustainable (Körner 2006).

Carbon partitioning to pools with different turnover times is highly sensitive to soil nutrient availability. Where nutrient availability is low, increasing soil nutrient supply promotes higher LAI. Under elevated CO_2 and increased nutrient supply, LAI becomes increasingly greater than that of stands under ambient CO_2. This response affects carbon allocation to other pools. Aboveground NPP increases with LAI (McCarthy et al. 2006a) with no additional effects of elevated CO_2. The fraction of Aboveground NPP allocated to wood, a moderately slow turnover pool, increases with LAI in broadleaf FACE experiments (from ~50 percent at low LAI, to a maximum of 70 percent at mid-range LAI), with the effect of elevated CO_2 on allocation entirely accounted for by changes in LAI. In pines, allocation to wood decreased with increasing LAI (from ~65 percent to 55 percent), but was higher (averaging ~68 percent versus 58 percent) under elevated CO_2 (McCarthy et al. 2006a). Despite the increased canopy photosynthesis, there is no evidence of increased wood production in pines growing on very poor, sandy soils (Oren et al. 2001).

Total carbon allocation belowground and CO_2 efflux from the forest floor decrease with increasing LAI, but the enhancement under elevated CO_2 is approximately constant (~22 percent) over the entire range of LAI (King et al. 2004; Palmroth et al. 2006). About a third of the extra carbon allocated belowground under elevated CO_2 is retained in litter and soil storage at the U.S. FACE sites (Palmroth et al. 2006). At Duke FACE, a third of the incremental carbon sequestration is found in the forest floor. The CO_2-induced enhancement in fine root and mycorrhizal fungi turnover has not translated to a significant net incremental storage of carbon in the mineral soil (Schlesinger and Lichter 2001; Jastrow et al. 2005; Lichter et al. 2005). A recent meta-analysis (Jastrow et al. 2005), incorporating data from a variety of studies in different settings, estimated a median CO_2-induced increase in the rate of soil C sequestration of 5.6 percent (+19 g C m^{-2} y^{-1}). Because soil C is highly variable and a large fraction of ecosystem carbon, a long time-series is necessary to statistically detect changes at any one site (McMurtrie et al. 2001).

3.2.3.4 Forests and Carbon Sequestration

Forest growth and long-lived wood products currently offset about 20 percent of annual U.S. fossil fuel carbon emissions (U.S. Climate Change Science Program Synthesis and Assessment Product 2.2 2007). Because a large forested landscape should be carbon neutral over long periods of time (Kashian et al. 2006), the presence of this large forest carbon sink is either a legacy of past land use (regrowth after harvest or reforestation of land cleared for pasture or crops) or a response to increased CO_2 and nitrogen deposition, or both (Canadell et al. 2007). This carbon sink is an enormous ecosystem service by forests, and its persistence will be important to any future mitigation strategy. If the sink primarily results from past land use, it will diminish through time. If not, it may continue until the effects of CO_2 and N diminish (Canadell et al. 2007).

To understand whether forest growth and long-lived forest products will continue their important role in offsetting a fraction of U.S. carbon emissions, significant unknowns and uncertainties would have to be addressed. The scale of the problem is large: Jackson and Schlesinger (2004) estimate that for afforestation to offset an additional 10 percent of U.S. emissions, immediate conversion of one-third of current croplands to forests would be required. Some of the unknowns and uncertainties are: 1) the economics of sequestration (Richards and Stokes 2004); 2) the timeline for valuing carbon stored in forests – should carbon stored today be worth more than carbon stored later (Fearnside 2002); 3) the permanence of stored carbon and its value if not permanent (Kirschbaum 2006); 4) the ability to permanently increase forest carbon stores in the face of changes in climate that may change species (Bachelet et al. 2001) and increase disturbance (Westerling et al. 2006), and change the process of carbon storage itself (Boisvenue and Running 2006); 5) how much carbon can be counted as "additional" given the self-replacing nature of forests; 6) identification of methods for increasing carbon sequestration in a variety of ecosystems and management goals; 7) how to account for carbon storage "gained" from management or avoided losses in fire; 8) identification of uniform methods and policies for validating carbon storage; 9) vulnerability of sequestered carbon to fire, windthrow or other disturbance; 10) "leakage" or displacement of carbon storage on one component of the landscape to carbon release on another (Murray et al. 2004); 11) will saturation of the carbon sink in North America work against forest C sequestration (Canadell et al. 2007)? 12) the impacts of carbon sequestration on the health of forest ecosystems and the climate system itself; and 13) the impacts of increasing carbon storage on other forest values such as biodiversity and water yield.

3.2.3.5 Interactive Effects Including Ozone, Nitrogren Deposition, and Forest Age

Ozone is produced from photochemical reactions of nitrogen oxides and volatile organic compounds. Ozone can damage plants (Ashmore 2002) and lower productivity, and these responses have been documented for U.S. forests (Matyssek and Sandermann 2003; Karlsson et al. 2004). In the United States, controls on emissions of nitrogen oxides and volatile organic compounds are expected to reduce the peak ozone concentrations that currently cause the most plant damage (Ashmore 2005). However, background tropospheric concentrations may be increasing as a result of increased global emissions of nitrogen oxides (Ashmore 2005).

These predicted increases in background ozone concentrations may reduce or negate the effects of policies to reduce ozone concentrations (Ashmore 2005). Ozone pollution will modify the effects of elevated CO_2 and any changes in temperature and precipitation (Hanson et al. 2005), but these interactions are difficult to predict because they have been poorly studied.

Nitrogen deposition in the eastern United States and California can exceed 10 kg N ha^{-1} yr^{-1} and likely has increased 10-20 times above pre-industrial levels (Galloway et al. 2004). Forests are generally limited by nitrogen availability, and fertilization studies show that this increased deposition will enhance forest growth and carbon storage in wood (Gower et al. 1992; Albaugh et al. 1998; Adams et al. 2005). There is evidence that chronic nitrogen deposition also increases carbon storage in surface soil over large areas (Pregitzer et al. 2008). Chronic nitrogen inputs over many years could lead to "nitrogen saturation" (a point where the system can no longer use or store nitrogen), a reduction in forest growth, and increased levels of nitrate in streams (Aber et al. 1998; Magill et al. 2004), but observations of forest ecosystems under natural conditions have not detected this effect (Magnani et al. 2007). Experiments and field studies have shown that the positive effect of elevated CO_2 on productivity and carbon storage can be constrained by low nitrogen availability, but in many cases elevated CO_2 causes an increase in nitrogen uptake (Finzi et al. 2006; Johnson 2006; Luo et al. 2006; Reich et al. 2006). For nitrogen-limited ecosystems, increased nitrogen availability from nitrogen deposition enhances the productivity increase from elevated CO_2 (Oren et al. 2001) and the positive effects of changes in temperature and precipitation. Overall, there is strong evidence that the effects of nitrogen deposition on forest growth and carbon storage are positive and might exceed those of elevated CO_2 (Körner 2000; Magnani et al. 2007).

Forest growth changes with forest age (Ryan et al. 1997), likely because of reductions in photosynthesis (Ryan et al. 2004). Because of the link of forest growth with photosynthesis, the response to drought, precipitation, nitrogen availability, ozone, and elevated CO_2 may also change with forest age. Studies of elevated CO_2 on trees have been done with young trees (which show a positive growth response), but the one study on mature trees showed no growth response (Körner et al. 2005). This is consistent with model results found in an independent study (Kirschbaum 2005). Tree size or age may also affect ozone response and response to drought, with older trees possibly more resistant to both (Grulke and Miller 1994; Irvine et al. 2004).

3.2.3.6 Fire Frequency and Severity

Several lines of evidence suggest that large, stand-replacing wildfires will likely increase in frequency over the next several decades because of climate warming (Figure 3.7). Chronologies derived from fire debris in alluvial fans (Pierce et al. 2004) and fire scars in tree rings (Kitzberger et al. 2007) provide a broader temporal context for interpreting contemporary changes in the fire regime. These longer-term records unequivocally show that warmer and drier periods during the last millennium are associated with more frequent and severe wildfires in western forests. GCM projections of future climate during 2010-2029 suggest that the number of low humidity days (and high fire danger days) will increase across much of the western U.S., allowing for more wildfire activity with the assumption that fuel densities and land management strategies remain the same (Flannigan et al. 2000; Brown et al. 2004). Flannigan et al. (2000) used two GCM simulations of future climate to calculate a seasonal severity rating related to fire intensity and difficulty of fire control. Depending on the GCM used, forest fire seasonal severity rating in the Southeast is projected to increase from 10 to 30 percent and 10 to 20 percent in the Northeast by 2060. Other biome models used with a variety of GCM climate projections simulate a larger increase in fire activity and biomass loss in the Southeast, sufficient to convert the southernmost closed-canopy Southeast forests to savannas (Bachelet et al. 2001). Forest management options to reduce fire size and intensity are discussed in Synthesis and Assessment Product 4.4 (Preliminary review of adaptation options for climate-sensitive ecosystems and resources, U.S. Climate Change Science Program).

Future increases in fire emissions across North America will have important consequences for climate forcing agents, air quality, and ecosystem services.

By combining climate-fire relationships derived from contemporary records with GCM simula-

Figure 3.7 *Ponderosa pine after the Hayman fire in Colorado, June 2002. While no one fire can be related to climate or changes in climate, research shows that the size and number of Western forest fires has increased substantially since 1985, and that these increases were linked with earlier spring snowmelt and higher spring and summer air temperature. Photo courtesy USDA Forest Service.*

tions of future climate, Flannigan et al. (2005) estimated that future fire activity in Canadian boreal forests will approximately double by the end of this century for model simulations in which fossil fuel emissions were allowed to increase linearly at a rate of 1 percent per year. Both Hadley Center and Canadian GCM simulations projected that fuel moisture levels will decrease and air temperatures will increase within the continental interior of North America because of forcing from greenhouse gases and aerosols.

Santa Ana winds and human-triggered ignitions play an important role in shaping the fire regime of Southern California shrublands and forests (Keeley and Fotheringham 2001; Westerling et al. 2004). Santa Ana winds occur primarily during fall and winter and are driven by large-scale patterns of atmospheric circulation (Raphael 2003; Conil and Hall 2006). Using future predictions from GCMs, Miller and Schlegel (2006) assessed that the total number of annual Santa Ana events would not change over the next 30 years. One of the GCM simulations showed a shift in the seasonal cycle, with fewer Santa Ana events occurring in September and more occurring in December. The implication of this change for the fire regime was unknown.

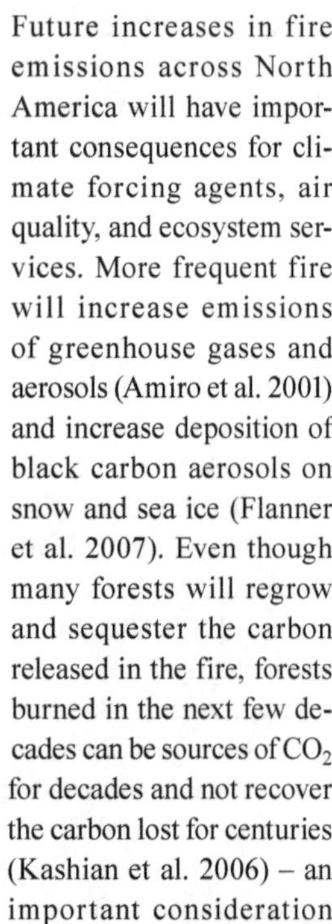

Future increases in fire emissions across North America will have important consequences for climate forcing agents, air quality, and ecosystem services. More frequent fire will increase emissions of greenhouse gases and aerosols (Amiro et al. 2001) and increase deposition of black carbon aerosols on snow and sea ice (Flanner et al. 2007). Even though many forests will regrow and sequester the carbon released in the fire, forests burned in the next few decades can be sources of CO_2 for decades and not recover the carbon lost for centuries (Kashian et al. 2006) – an important consideration for slowing the increase in atmospheric CO_2. In boreal forests, the warming effects from fire-emitted greenhouse gases may be offset at regional scales by increases in surface albedo caused by a shift in the stand age distribution (Randerson et al. 2006). Any climate driven changes in boreal forest fires in Alaska and Canada will have consequences for air quality in the central and eastern United States because winds often transport carbon monoxide, ozone, and aerosols from boreal fires to the south (McKeen et al. 2002; Morris et al. 2006; Pfister et al. 2006). Increased burning in boreal forests and peatlands also has the potential to release large stocks of mercury currently stored in cold and wet soils (Turetsky et al. 2006). These emissions may exacerbate mercury toxicities in northern hemisphere food chains caused by coal burning.

3.2.3.7 Insect Outbreaks

Rising temperature is the aspect of climate change most influential on forest insect species through changes in insect survival rates, increases in life cycle development rates, facilitation of range expansion, and effects on host plant capacity to resist attack (Ayres and Lombardero 2000; Malmström and Raffa 2000; Bale et al. 2002). Future northward range expansion attributed to warming temperatures has been predicted for mountain pine beetle (Logan and Powell 2001) and southern pine beetle (Ungerer et al. 1999). Future range expansion of mountain pine beetle has the potential of invading jack pine, a suitable host that extends across the boreal forest of North America (Logan and Powell 2001). Increased probability of spruce beetle outbreak (Logan et al. 2003a) as well as increase in climate suitability for mountain pine beetle attack in high-elevation ecosystems (Hicke et al. 2006) has been projected in response to future warming. The combination of higher temperatures with reduced precipitation in the Southwest has led to enhanced tree stress, and also affected Ips beetle development rates; continued warming, as predicted by climate models, will likely maintain these factors (Breshears et al. 2005).

Indirect effects of future climate change may also influence outbreaks. Increasing atmospheric CO_2 concentrations may lead to increased ability of trees to recover from attack (Kruger et al. 1998). Enhanced tree productivity in response to favorable climate change, including rises in atmospheric CO_2, may lead to faster recovery of forests following outbreaks, and thus a reduction in time to susceptibility to subsequent attack (Fleming 2000). Although eastern spruce budworm life cycles are tightly coupled to host tree phenology even in the presence of climate change, enemy populations that are significant in governing epidemic dynamics are not expected to respond to climate change in a synchronized way (Fleming 2000). Changing fire regimes in response to climate change (Flannigan et al. 2005) will affect landscape-scale forest structure, which influences susceptibility to attack (Shore et al. 2006).

Nonnative invasive species are also significant disturbances to forests in the United States. Although little has been reported on climate influences on these insects, a few studies have illustrated climate control. The hemlock woolly adelgid is rapidly expanding its range in the eastern United States, feeding on several species of hemlock (Box 1). The northern range limit of the insect in the United States is currently limited by low temperatures (Parker et al. 1999), suggesting range expansion in the event of future warming. In addition, the hemlock woolly adelgid has evolved greater resistance to cold conditions as it has expanded north (Butin et al. 2005). The introduced gypsy moth has defoliated millions of hectares of forest across the eastern United States, with great efforts expended to limit its introduction to other areas (USDA Forest Service 2005). Projections of future climate and gypsy moth simulation modeling reveal substantial increases in probability of establishment in the coming decades (Logan et al. 2003a).

As important disturbances, insect outbreaks affect many forest ecosystem processes. Outbreaks alter tree species composition within stands, and may result in conversion from forest to herbaceous vegetation through lack of regeneration (Holsten et al. 1995). Carbon stocks and fluxes are modified through a large decrease in living biomass and a corresponding large increase in dead biomass, reducing carbon uptake by forests as well as enhancing decomposition fluxes. In addition to effects at smaller scales, widespread outbreaks have significant effects on regional carbon cycling (Kurz and Apps 1999; Hicke et al. 2002a). Other biogeochemical cycles, such as nitrogen, are affected by beetle-caused mortality (Throop et al. 2004). Defoliation, for example as related to gypsy moth outbreaks, facilitates nitrogen movement from forest to aquatic ecosystems, elevating stream nitrogen levels (Townsend et al. 2004).

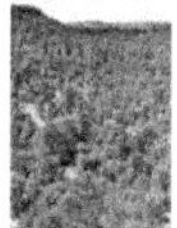

Significant changes to the hydrologic cycle occur after a widespread insect epidemic, including increases in annual water yield, advances in the annual hydrograph, and increases in low flows (Bethlahmy 1974; Potts 1984). Water quantity is enhanced through reductions in transpiration, in addition to reductions in snow interception, and subsequent redistribution and sublimation. These effects can last for many years following mortality (Bethlahmy 1974).

Interactions of outbreaks and fire likely vary with time, although observational evidence is limited to a few studies (Romme et al. 2006). In central Colorado, number of fires, fire extent, and fire severity were not enhanced following outbreaks of spruce beetle (Bebi et al. 2003; Bigler et al. 2005; Kulakowski and Veblen in press). Other studies of the 1988 Yellowstone

BOX 1: The Eastern Hemlock and its Woolly Adelgid

Outbreaks of insects and diseases affect forest structure and composition, leading to changes in carbon, nutrients, biodiversity, and ecosystem services. The hemlock woolly adelgid (HWA), native to Asia, was first recorded in 1951 in Virginia, and has since spread, causing a severe decline in vitality and survival of eastern hemlock in North American forests (Maps 3.1 & 3.2, Stadler et al. 2006). Roads, major trails, and riparian corridors provide for long-distance dispersal of this aphid-like insect, probably by humans or birds (Koch et al. 2006). Although HWA is consumed by some insect predators (Flowers et al. 2006), once it arrives at a site, complete hemlock mortality is inevitable (Orwig et al. 2002; Stadler et al. 2005).

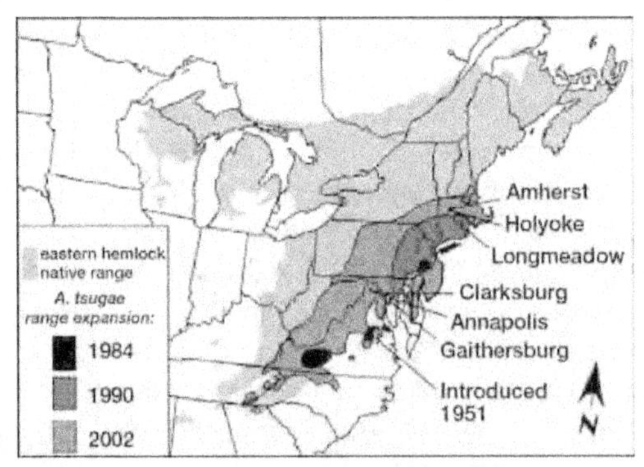

Map 3.1 *Sample sites and range expansion of* Adelges tsugae *relative to the native range of eastern hemlock in North America. Map from Butin et al. 2005 (redrawn from USDA Forest Service and Little, 1971).*

HWA will change biodiversity and species composition. Hemlock seedlings are readily attacked and killed by the HWA, so damaged hemlock stands are replaced by stands of black birch, black oaks, and other hardwoods (Brooks 2004; Small et al. 2005; Sullivan and Ellison 2006). After HWA attack, plant biodiversity increases in the canopy and in the understory; invasive shrubs and woody vines of several species also expand in response to the improved light conditions (Goslee et al. 2005; Small et al. 2005; Eschtruth et al. 2006). Four insectivorous bird species have high affinity for hemlock forest type, and two of these, the blue-headed vireo and Blackburnian warbler, are specialists in the hemlock habitat. Expansion of HWA could negatively impact several million pairs of these birds by eliminating their habitat (Tingley et al. 2002; Ross et al. 2004).

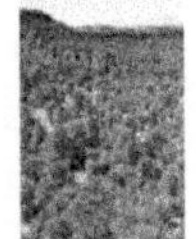

Changes in canopy attributes upon replacement of hemlock with deciduous broadleaf species alter the radiation regime, hydrology, and nutrient cycling (Cobb et al. 2006; Stadler et al. 2006), and result in greater temperature fluctuations and longer periods of times in which streams are dry (Snyder et al. 2002). These conditions reduce habitat quality for certain species of fish. Brook trout and brown trout were two to three times as prevalent in hemlock than hardwood streams (Ross et al. 2003). Low winter temperature is the main factor checking the spread of HWA (Skinner et al. 2003). However, the combination of increasing temperature and the capacity of HWA to evolve greater resistance to cold shock as it has expanded its range northward (Butin et al. 2005) means that stands that have been relatively protected by cold temperatures (Orwig et al. 2002) may fall prey to the insect in the not-so-distant future (Map 3.3).

Map 3.2 *Counties in the range of eastern hemlock that are uninfected, newly infected, and infected. From Onken and Reardon.(2005).*

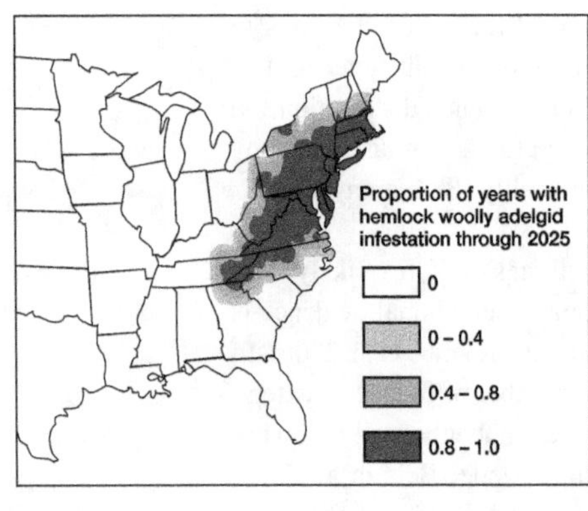

Map 3.3 *Hemlock woolly adelgid spread map prepared by Randall Marin, Northeastern Research Station, U.S. Forest Service (Souto et al. 1996).*

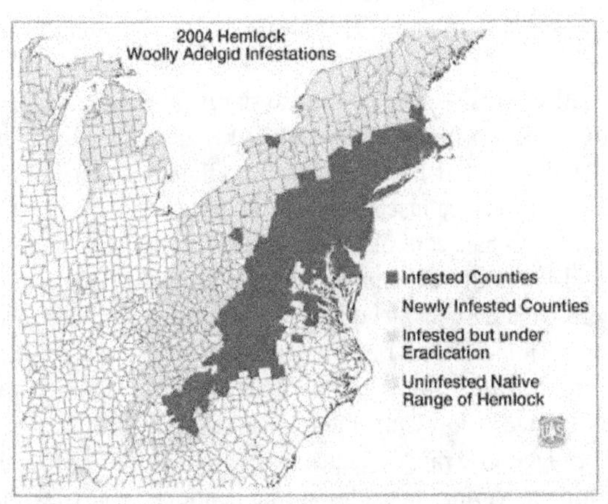

fire that followed two mountain pine beetle epidemics found mixed fire effects, depending on outbreak severity and time since outbreak (Turner et al. 1999; Lynch et al. 2006). A quantitative modeling study of fire behavior found mixed fire effects following an outbreak of western spruce budworm (Hummel and Agee 2003); more modeling studies that incorporate complete effects are needed to explore other situations.

Multiple socioeconomic impacts follow severe insect outbreaks. Timber production and manufacturing and markets may not be able to take advantage of vast numbers of killed trees (Ferguson 2004), and beetle-killed timber has several disadvantages from a manufacturing perspective (Byrne et al. 2006). Perceived enhanced fire risk and views about montane aesthetics following beetle-cause mortality influence public views of insect outbreaks, which could drive future public policy. Threats to ecologically important tree species may have ramifications for charismatic animal species (e.g., influences of whitebark pine mortality on the grizzly bear) (Logan and Powell 2001). Impacts are enhanced as human population, recreation, and tourism increase in forested regions across the nation.

3.2.3.8 Storms (Hurricanes, Ice Storms, Windstorms)

Predictions of forest carbon (C) sequestration account for the effect of fires (e.g., Harden et al. 2000), yet other wide-ranging and frequent disturbances, such as hurricanes (Figure 3.8) and ice storms, are seldom explicitly considered. Both storm types are common in the southeastern United States, with an average return time of six years for ice storms (Bennett 1959) and two years for hurricanes (Smith 1999). These, therefore, have the potential for significant impact on C sequestration in this region, which accounts for ~20 percent of annual C sequestration in conterminous U.S. forests (Birdsey and Lewis 2002; Bragg et al. 2003). Recent analysis demonstrated that a single Category 3 hurricane or severe ice storm could each transfer to the decomposable pool the equivalent of 10 percent of the annual U.S. C sequestration, with subsequent reductions in sequestration caused by direct stand damage (McNulty 2002; McCarthy et al. 2006b). For example, Hurricanes

Figure 3.8 *Forest damage from Hurricane Katrina. Dr. Jeffrey Q. Chambers, Tulane University.*

Rita and Katrina together damaged a total of 2,200 ha and 63 million m^3 of timber volume (Stanturf et al. 2007) which, when decomposed over the next several years, will release a total of 105 teragrams (Tg) of C into the atmosphere, roughly equal to the annual net sink for U.S. forests (Chambers et al. 2007).

Common forest management practices such as fertilization and thinning, forest type, and increasing atmospheric CO_2 levels can change wood and stand properties, and thus may change vulnerability to ice storm damage. A pine plantation experienced a ~250 g C m^{-2} reduction in living biomass during a single ice storm, equivalent to ~30 percent of the annual net ecosystem carbon exchange of this ecosystem. In this storm at the Duke FACE, nitrogen fertilization had no effect on storm damage; conifer trees were more than twice as likely to be killed by ice storm damages as leafless deciduous-broadleaf trees; and thinning increased broken limbs or trees threefold. Damage in the elevated CO_2 stand was one third as much as in the ambient CO_2 stand. (McCarthy et al. 2006b). Although this result suggests that forests may suffer less damage in a future ice storm when atmospheric CO_2 is higher, future climate may create conditions leading to greater ice storm frequency, extent and severity (da Silva et al. 2006), which may balance the decreased sensitivity to ice damage under elevated CO_2. All of these predictions are very uncertain (Cohen et al. 2001).

3.2.3.9 Changes in Overstory Species Composition

Several approaches can predict changes in biomes (major vegetation assemblages such as conifer forests and savanna/woodland) and changes in species composition or overstory species communities (Hansen et al. 2001a). These approaches use either rules that define the water balance, temperature, seasonality, etc. required for a particular biome, or statistically link species distributions or community composition with climate envelopes. These efforts have mostly focused on equilibrium responses to climate changes over the next century (Hansen et al. 2001a), so predictions for the next several decades are unavailable.

Bachelet et al. (2001) used the Mapped Atmosphere-Plant-Soil System (MAPPS) model with the climate predictions generated by seven different global circulation models to predict how biome distributions would change with a doubling of CO_2 by 2100. Mean annual temperature of the United States increased from 3.3 to 5.8 °C for the climate predictions. Predicted forest cover in 2100 declined by an average of 11 percent (range for all climate models was +23 percent to -45 percent). The MAPPS model coupled to the projected future climates predicts that biomes will migrate northward in the East and to higher elevations in the West. For example, mixed conifer and mixed hardwood forests in the Northeast move into Canada, and decline in area by 72 percent (range: -14 to -97 percent), but are replaced by eastern hardwoods. In the Southeast, grasslands or savannas displace forests and move their southern boundaries northward, particularly for the warmer climate scenarios. In the West, forests displace alpine ecosystems, and the wet conifer forests of the Northwest decline in area 9 percent (range: 54 to + 21 percent), while the area of interior western pines changes little. Species predictions for the eastern United States using a statistical approach showed that most species moved north 60-300 miles (Hansen et al. 2001a).

Authors of these studies cautioned that these equilibrium approaches do not reflect the transient and species-specific nature of the community shifts that are projected to occur. Success in moving requires suitable climate, but also dispersal that may lag behind changes in climate (Hansen et al. 2001a). Some species will be able to move quicker than others, and some biomes and communities may persist until a disturbance allows changes to occur (Hansen et al. 2001a). Because trees are long-lived and may tolerate growing conditions outside of their current climate envelopes, they may be slower to change than modeled (Loehle and LeBlanc 1996). The authors of these studies agreed that while climate is changing, novel ecosystems will arise – novel because some species will persist in place, some species will depart, and new species will arrive.

3.2.4 Indicators and Observing Systems

3.2.4.1 Characteristics of Observing Systems

Many Earth observing systems (Bechtold and Patterson 2005; Denning 2005) are designed to allow for integration of multiple kinds of observations using a hierarchical approach that takes advantage of the characteristics of each. A typical system uses remote sensing to obtain a continuous measurement over a large area, coupled with statistically-designed field surveys to obtain more detailed data at a finer resolution. Statistically, this approach (known as "multi-phase" sampling) is more efficient than sampling with just a single kind of observation or conducting a complete census (Gregoire and Valentine, in press). Combining observed data with models is also common because often the variable of interest cannot be directly observed, but observation of a closely-related variable may be linked to the variable of interest with a model. Model-data synthesis is often an essential component of Earth observing systems (Raupach et al. 2005).

To be useful, the system must observe a number of indicators more than once over a period, and also cover a large enough spatial scale to detect a change. The length of time required to detect a change with a specified level of precision depends on the variability of the population being sampled, the precision of measurement, and the number of samples (Smith 2004). Non-climatic local factors, such as land use change, tend to dominate vegetation responses at small scales, masking the relationship with climate (Parmesan and Yohe 2003). A climate signal is therefore more likely to be revealed by analyses that can identify trends across large geographic regions (Walther et al. 2002).

The relationship between biological observations and climate is correlational; thus, it is difficult to separate the effects of climate change from other possible causes of observed effects (Walther et al. 2002). Inference of causation can be determined with carefully controlled experiments that complement the observations. Yet, observation systems can identify some causal relationships and therefore have value in developing climate impact hypotheses. Schreuder and Thomas (1991) determined that if both the potential cause and effect variables were measured at inventory sample plots, a relationship could be established if the variables are measured accurately, estimated properly, and based on a large enough sample. But, in practice, additional information is often needed to strengthen a case – for example, a complementary controlled experiment to verify the relationship.

3.2.4.2 Indicators of Climate Change Effects

The species that comprise communities respond both physiologically and competitively to climate change. One scheme for assessing the impacts of climate change on species and communities is to assess the effects on: (1) the physiology of photosynthesis, respiration, and growth; (2) species distributions; and (3) phenology, particularly life cycle events such as timing of leaf opening. There may also be effects on functions of ecosystems such as hydrologic processes.

Effects on physiology

Net primary productivity is closely related to indices of "greenness" and can be detected by satellite over large regions (Hicke et al. 2002b). Net ecosystem productivity (NEP) can be measured on the ground as changes in carbon stocks in vegetation and soil (Boisvenue and Running 2006). Root respiration and turnover are sensitive to climate variability and may be good indicators of climate change if measured over long enough time periods (Atkin et al. 2000; Gill and Jackson 2000). Gradient studies show variable responses of growth to precipitation changes along elevational gradients (Fagre et al. 2003). Climate effects on growth patterns of individual trees is confounded by other factors such as increasing CO_2 and N deposition, so response of tree growth is difficult to interpret without good knowledge of the exposure to many possible causal variables. For example, interannual variability in NPP, which can mask long-term trends, can be summarized from long-term ecosystem studies and seems to be related to interactions between precipitation gradients and growth potential of vegetation (Knapp and Smith 2001).

Effects on species distributions

Climate change affects forest composition and geographical distribution, and these changes are observable over time by field inventories, remote sensing, and gradient studies. Both range expansions and retractions are important to monitor (Thomas et al. 2006), and population extinctions or extirpations are also possible. Changes in the range and cover of shrubs and trees have been observed in Alaska by field studies and remote sensing (Hinzman et al. 2005). Detecting range and abundance shifts in wildlife populations may be complicated by changes in habitat from other factors (Warren et al. 2001).

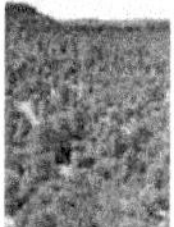

Effects on phenology

Satellite and ground systems can document onset and loss of foliage, with the key being availability of long-term data sets (Penuelas and Filella 2001). Growing season was found significantly longer in Alaska based on satellite normalized difference vegetation index (NDVI) records (Hinzman et al. 2005). Schwartz et al. (2006) integrated weather station observations of climate variables with remote sensing and field observations of phenological changes using Spring Index phenology models. However, Fisher et al. (2007) concluded that species or community compositions must be known to use satellite observations for predicting the phenological response to climate change.

Effects on natural disturbances and mortality

Climate change can affect forests by altering the frequency, intensity, duration, and timing of natural disturbances (Dale et al. 2001). The correlation of observations of changes in fire frequency and severity with changes in climate are well documented (e.g., Flannigan et al.

2000; Westerling et al. 2006), and the inference of causation in these studies is established by in situ studies of fire and vegetation response, and fire behavior models. Similar relationships hold for forest disturbance from herbivores and pathogens (Ayres and Lombardero 2000; Logan et al. 2003b). Tree mortality may be directly caused by climate variability, such as in drought (Gitlin et al. 2006).

Effects on hydrology

Climate change will affect forest water budgets. These changes have been observed over time by long-term stream gauge networks and research sites. Changes in permafrost and streamflow in the Alaskan Arctic region are already apparent (Hinzman et al. 2005). There is some evidence of a global pattern (including in the United States) in response of streamflow to climate from stream-gauge observations (Milly et al. 2005). Inter-annual variation in transpiration of a forest can be observed by sap flow measurements (Phillips and Oren 2001; Wullschleger et al. 2001).

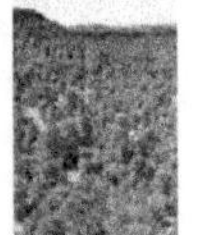

Causal variables

It is important to have high-quality, spatially-referenced observations of climate, air pollution, deposition, and disturbance variables. These are often derived from observation networks using spatial statistical methods (e.g., Thornton et al. 2000).

3.2.4.3 Current Capabilities and Needs

There are strengths and limitations to each kind of observation system: intensive monitoring sites such as Long Term Ecological Research (LTER) sites and protected areas; extensive observation systems such as Forest Inventory and Analysis (FIA) or the U.S. Geological Survey (USGS) stream gauge network; and remote sensing. A national climate observation system may be improved by integration under an umbrella program such as the National Ecological Observatory Network (NEON), or Global Earth System of Systems (GEOSS) (see Table 3.1). Also, increased focus on "sentinel" sites could help identify early indicators of climate effects on ecosystem processes, and provide observations of structural and species changes (NEON 2006).

Intensive monitoring sites measure many of the indicators that are likely to be affected by climate change, but have mostly been located independently and so do not optimally represent either (1) the full range of forest condition variability, or (2) forest landscapes that are most likely to be affected by climate change (Hargrove et al. 2003). Forest inventories are able to detect long-term changes in composition and growth, but they are limited in ability to attribute observed changes to climate, because they were not designed to do so. Additions to the list of measured variables and compiling potential causal variables would improve the inventory approach (The Heinz Center 2002; USDA 2003). Remote sensing, when coupled with models, can detect changes in vegetation-response to climate variability (Running et al. 2004; Turner et al. 2004). Interpretation of remote sensing observations is greatly improved by associating results with ground data (Pan et al. 2006).

Maintaining continuity of remote sensing observations at appropriate temporal and spatial scales must be a high priority. NASA's Earth Science division cannot support continued operations of all satellites indefinitely, so it becomes a challenge for the community using the measurements to identify long-term requirements for satellites, and provide a long-term framework for critical Earth science measurements and products (NASA Office of Earth Science 2004).

Another high-priority need is to improve ability to monitor the effects of disturbance on forest composition and structure, and to attribute changes in disturbance regimes to changes in climate. This will involve a more coordinated effort to compile maps of disturbance events from satellite or other observation systems, to follow disturbances with in situ observations of impacts, and to keep track of vegetation changes in disturbed areas over time. There are several existing programs that could be augmented to achieve this result, such as intensifying the permanent sample plot network of the FIA program for specific disturbance events, or working with forest regeneration and restoration programs to install long-term monitoring plots.

3.2.5 How Changes in One Resource can Affect Other Resources

Disturbances in forests such as fire, insect outbreaks, and hurricanes usually kill some or all of the trees and lower leaf area. These reductions in forest cover and leaf area will likely change the hydrology of the disturbed areas. Studies of forest harvesting show that removal of the tree canopy or transpiring surface will increase water yield, with the increase proportional to the amount of tree cover removed (Stednick 1996). The response will vary with climate and species, with wetter climates showing a greater response of water yield to tree removal. For all studies, average water yield increased 2.5 mm for each 1 percent of the tree basal-area removed (Stednick 1996). High-severity forest fires can increase sediment production and water yield as much as 10 to 1000 times, with the largest effects occurring during high-intensity summer storms (see review in Benavides-Solorio and MacDonald 2001). An insect epidemic can increase annual water yield, advance the timing of peak runnoff, and increase base flows (Bethlahmy 1974; Potts 1984). Presumably, the same effects would occur for trees killed in windstorms and hurricanes.

Disturbances can also affect native plant species diversity, by allowing opportunities for establishment of non-native invasives, particularly if the disturbance is outside of the range of variability for the ecosystem (Hobbs and Huenneke 1992). Areas most vulnerable to invasion by non-natives are those areas that support the highest plant diversity and growth (Stohlgren et al. 1999). In the western United States, these are generally the riparian areas (Stohlgren et al. 1998). We expect that disturbances that remove forest litter or expose soil (fire, trees tipped over by wind) will have the highest risk for admitting invasive non-native plants.

Table 3.1 Current and Planned Observation Systems for Climate Effects on Forests

Observation System	Characteristics	Reference
Forest Inventory and Analysis (U.S. Forest Service)	Annual and periodic measurements of forest attributes at a large number (more than 150,000) of sampling locations. Historical data back to 1930s in some areas.	Bechtold and Patterson 2005
AmeriFlux (Department of Energy and other Agencies)	High temporal resolution (minutes) measurements of carbon, water, and energy exchange between land and atmosphere at about 50 forest sites. A decade or more of data available at some of the sites.	http://public.ornl.gov/ameriflux/
Long Term Ecological Research network (National Science Foundation)	The LTER network is a collaborative effort involving more than 1,800 scientists and students investigating ecological processes over long temporal and broad spatial scales. The 26 LTER Sites represent diverse ecosystems and research emphases	http://www.lternet.edu/
Experimental Forest Network (U.S. Forest Service)	A network of 77 protected forest areas where long-term monitoring and experiments have been conducted.	Lugo 2006
National Ecological Observation Network	The NEON observatory is specifically designed to address central scientific questions about the interactions of ecosystems, climate, and land use.	http://www.neoninc.org/
Global Terrestrial Observing System (GTOS)	GTOS is a program for observations, modeling, and analysis of terrestrial ecosystems to support sustainable development.	http://www.fao.org/gtos/

3.3 ARID LANDS

3.3.1 Brief Summary of Key Points from the Literature

Plants and animals in arid lands live near their physiological limits, so slight changes in temperature and precipitation will substantially alter the composition, distribution, and abundance of species, and the products and services that arid lands provide. Observed and projected decreases in the frequency of freezing temperatures, lengthening of the frost-free season, and increased minimum temperatures will alter plant species ranges and shift the geographic and elevational boundaries of the Great Basin, Mojave, Sonoran, and Chihuahuan deserts. The extent of these changes will also depend on changes in precipitation and fire. Increased drought frequency will likely cause major changes in vegetation cover. Losses of vegetative cover coupled with increases in precipitation intensity and climate-induced reductions in soil aggregate stability will dramatically increase potential erosion rates. Transport of eroded sediment to streams coupled with changes in the timing and magnitude of minimum and maximum flows will affect water quality, riparian vegetation, and aquatic fauna. Wind erosion will have continental-scale impacts on downwind ecosystems, air quality, and human populations.

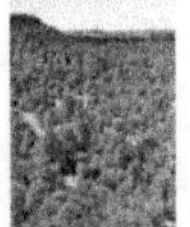

Plants and animals in arid lands live near their physiological limits, so slight changes in temperature and precipitation will substantially alter the composition, distribution, and abundance of species, and the products and services that arid lands provide.

The response of arid lands to climate change will be strongly influenced by interactions with non-climatic factors at local scales. Climate effects should be viewed in the context of these other factors, and simple generalizations should be viewed with caution. Climate will strongly influence the impact of land use on ecosystems and how ecosystems respond. Grazing has traditionally been the most extensive land use in arid regions. However, land use has significantly shifted to exurban development and recreation in recent decades. Arid land response to climate will thus be influenced by environmental pressures related to air pollution and N-deposition, energy development, motorized off-road vehicles, feral pets, and horticultural invasives, in addition to grazing.

Non-native plant invasions will likely have a major impact on how arid land ecosystems respond to climate and climate change. Exotic grasses generate large fuel loads that predispose arid lands to more frequent and intense fire than historically occurred with sparser native fuels. Such fires can radically transform diverse desert scrub, shrub-steppe, and desert grassland/savanna ecosystems into monocultures of non-native grasses. This process is well underway in the cold desert region, and is in its early stages in hot deserts. Because of their profound impact on the fire regime and hydrology, invasive plants in arid lands may trump direct climate impacts on native vegetation.

Given the concomitant changes in climate, atmospheric CO_2, nitrogen deposition, and species invasions, novel wildland and managed ecosystems will likely develop. In novel ecosystems, species occur in combinations, and relative abundances that have not occurred previously in a given biome. In turn, novel ecosystems present novel challenges for conservation and management.

3.3.2 Observed and Predicted Changes or Trends

3.3.2.1 INTRODUCTION

Arid lands occur in tropical, subtropical, temperate, and polar regions and are defined based on physiographic, climatic, and floristic features. Arid lands are characterized by low (typically <400 mm), highly variable annual precipitation, along with temperature regimes where potential evaporation far exceeds precipitation inputs. In addition, growing season rainfall is often delivered via intense convective storms, such that significant quantities of water run off before infiltrating into soil; precipitation falling as snow in winter may sublimate or run off during snowmelt in spring while soils are frozen. As a result of these combined factors, production per unit of precipitation can be low. Given that many organisms in arid lands are near their physiological limits for temperature and water stress tolerance, slight changes in temperature and precipitation that affect water availability and water requirements could have substantial ramifications. Thus, predicted transitions toward more arid conditions (e.g., higher temperatures that elevate potential evapotranspiration and more intense thunderstorms that generate more run off; Seager et al. 2007) have the real potential to alter species composition and abundance, and the ecosystem goods and services that arid lands can provide for humans (Field et al. 2007).

The response of arid lands to climate and climate change is contingent upon the net effect of non-climatic factors interacting with climate at local scales (Figure 3.9). Some of these factors may reinforce and accentuate climate effects (e.g., livestock grazing); others may constrain, offset or override climate effects (e.g., soils, atmospheric CO_2 enrichment, fire, non-native species). Climate effects should thus be viewed in the context of other factors, and simple generalizations regarding climate effects should be viewed with caution. A literature review of the relationship between climate change and land use indicate land use change has had a much greater effect on ecosystems than has climate change; and that the vast majority of land use changes have little to do with climate or climate change (Dale 1997). Today's arid lands reflect a legacy of historic land uses, and future land use practices will arguably have the greatest impact on arid land ecosystems in the next two to five decades. In the near-term, climate fluctuation and change will be important primarily as it influences the impact of land use on ecosystems, and how ecosystems respond to land use.

3.3.2.2 Bio-Climatic Setting

Arid lands of the continental United States are represented primarily by the subtropical hot deserts of the Southwest, and the temperate cold deserts of the Intermountain West (Figure 3.2). The hot deserts differ primarily with respect to precipitation seasonality (Figure 3.10). The Mojave desert is dominated by winter precipitation (thus biological activity in the cool season), whereas the Chihuahuan desert is dominated by summer precipitation (thus biological activity during hotter conditions). The hottest of the three deserts, the Sonoran, is the intermediate, receiving both winter and summer precipitation. The cold deserts are also dominated by winter precipitation, much of which falls as snow, owing to the more temperate latitudes and higher elevations (West 1983). These arid land formations are characterized by unique plants and animals, and if precipitation seasonality changes, marked changes in species and functional group composition and abundance would be expected.

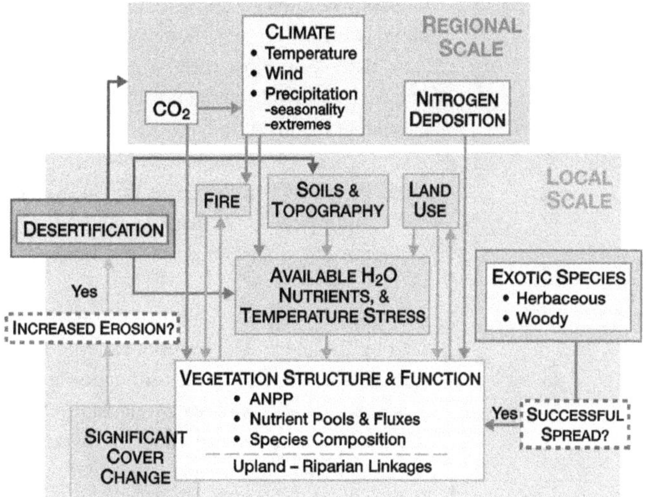

Figure 3.9 *Organizational framework for interpreting climate and climate change effects on arid land ecosystems.*

There is broad consensus among climate models that the arid regions of the southwestern United States will dry in the 21st century and that a transition to a more arid climate is already underway. In multimodel ensemble means reported by Seager et al. (2007), there is a transition to a sustained drier climate that begins in the late 20th and early 21st centuries. Both precipitation and evaporation are expected to decrease, but precipitation is expected to decrease more than evaporation leading to an overall drier climate. The increasing aridity is primarily reduced in winter, when precipitation decreases and evaporation remains unchanged or slightly reduced. The projected ensemble

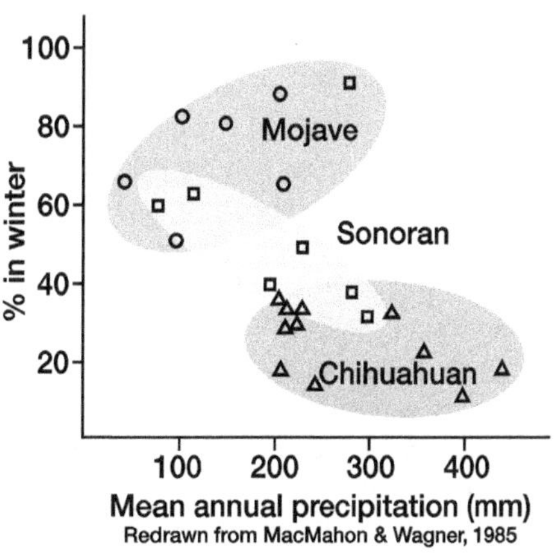

Figure 3.10 *Mean annual precipitation and its seasonality in three hot deserts (from MacMahon and Wagner 1985).*

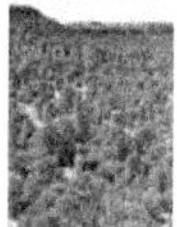

median reduction in precipitation reaches 0.1 mm/day in mid-century, though several models show that the decrease could occur in the early 21st century. A substantial portion of the mean circulation contribution, especially in winter, is explained by the change in zonal mean flow alone, indicating that changes in the Hadley Cell and extratropical mean meridional circulation are important to the climate of this region.

The Great Basin is a cold desert characterized by limited water resources and periodic droughts in which a high proportion of the year's precipitation falls as winter snow (Wagner 2003). Snow-derived runoff provides the important water resources to maintain stream and river channels that support riparian areas and human utilization of this region. In the last century, the Great Basin warmed by 0.3° to 0.6°C and is projected to warm by an additional 5° to 10°C in the coming century (Wagner 2003). In the last half-century, total precipitation has increased 6-16 percent and this increase is projected to continue in the future (Baldwin et al. 2003). The increase in total precipitation is offset partially by the decrease in snowpack, which in the Great Basin is among the largest in the nation (Mote et al. 2005). The onset of snow runoff is currently 10–15 days earlier than 50 years ago, with significant impacts on the downstream utilization of this water (Cayan et al. 2001; Baldwin et al. 2003; Stewart et al. 2004). Increased warming is likely to continue to accelerate spring snowmelt. Warmer temperatures are also likely to lead to more precipitation falling as rain which would further reduce overall snowpack and spring peak flow.

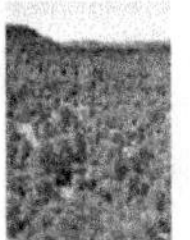

The vegetation growing season, as defined by continuous frost-free air temperatures, has increased by on average about two days/decade since 1948 in the conterminous United States, with the largest changes occurring in the West.

Throughout the dry western United States, extreme temperature and precipitation events are expected to change in the next century. Warm extremes will generally follow increases in the mean summertime extremes, while cold extremes will warm faster than warm extremes (Kharin et al. 2007). As a result, what is currently considered an unusually high temperature (e.g., 20-year return interval) will become very frequent in the desert Southwest, occurring every couple of years. On the other hand, unusually low temperatures will become increasingly uncommon. As a result winters will be warmer, leading to higher evapotranspiration and lower snowfall. Changes in precipitation are also expected. Precipitation events that are currently considered extreme (20-year return interval) are also expected to occur roughly twice as often as they currently do, consistent with general increases in rainstorm intensity (Kharin et al. 2007).

Changes in species and functional group composition might first occur in the geographic regions where biogeographic formations and their major subdivisions interface. Extreme climatic events are major determinants of arid ecosystem structure and function (Holmgren et al. 2006). Thus, while changes in mean temperature will affect levels of physiological stress and water requirements during the growing season, minimum temperatures during winter may be a primary determinant of species composition and distribution. In the Sonoran Desert, warm season rainfall and freezing temperatures strongly influence distributions of many plant species (Turner et al. 1995). The vegetation growing season, as defined by continuous frost-free air temperatures, has increased by on average about two days/decade since 1948 in the conterminous United States, with the largest changes occurring in the West (Easterling 2002; Feng and Hu 2004). A recent analysis of climate trends in the Sonoran Desert (1960-2000) also shows a decrease in the frequency of freezing temperatures, lengthening of the frost-free season, and increased minimum temperatures (Weiss and Overpeck 2005). With warming expected to continue throughout the 21st century, potential ecological responses may include contraction of the overall boundary of the Sonoran Desert in the southeast and expansion northward, eastward, and upward in elevation, and changes to plant species ranges. Realization of these changes will be co-dependent on what happens with precipitation and disturbance regimes (e.g., fire).

The biotic communities that characterize many U.S. arid lands are influenced by basin and range topography. Thus, within a given bioclimatic zone, communities transition from desert scrub and grassland to savanna, woodland, and forest along strong elevation gradients (Figure 3.11). Changes in climate will affect the nature of this zonation, with arid land communities potentially moving up in elevation in response to warmer and drier conditions. Experimental data suggest shrub recruitment at woodland-grassland ecotones will be favored by increases in summer precipitation, but unaffected by increases in winter precipitation (Weltzin and McPherson 2000). This suggests that increases in summer precipitation would favor the downslope shifts in this ecotone. In the Great Basin, favorable climatic conditions at the turn of the last century enabled expansion of woodlands into sagebrush steppe (Miller and Rose 1999; Miller et al. 2005) and ongoing expansion is significantly increasing fuel loads and creating conditions for catastrophic fire. Plant composition and ecosystem processes (e.g., plant growth, water and nutrient use, herbivory) change along these elevation gradients in a manner that broadly mimics changes in ecosystem structure and function along continental-scale latitudinal gradients (Whittaker 1975). Changes along local elevation gradients may therefore be early indicators of regional responses to climate change (Peters 1992).

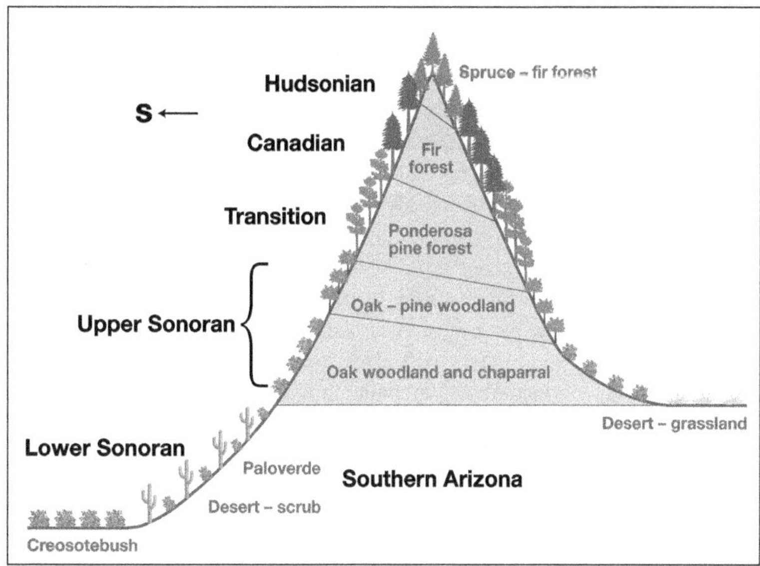

Figure 3.11 *Elevation life zones along an arid land elevation gradient (from Brown 1994).*

3.3.2.3 Climate Influences at Local Scales

Climate and atmospheric CO_2 influence communities at broad spatial scales, but topography, soils, and landform control local variation in ecosystem structure and function within a given elevation zone, making local vegetation very complex. Topography influences water balance (south-facing slopes are drier), air drainage and night temperatures, and routing of precipitation. Soil texture and depth affect water capture, water storage, and fertility (especially nitrogen). These factors strongly interact with precipitation to limit plant production and control species composition. Plants that can access water in deep soil or in groundwater depend less on precipitation for growth and survival, but such plants may be sensitive to precipitation changes that affect the recharge of deep water stores. If the water table increases with increases in rainfall or decreased plant cover, soil salinity may increase and adversely affect vegetation in some bottomland locations (McAuliffe 2003). To predict vegetation response to climate change, it is necessary to understand these complex relationships among topography, soil, soil hydrology, and plant response.

3.3.2.4 Climate and Disturbance

Disturbances such as fire and grazing are superimposed against the backdrop of climate variability, climate change, and spatial variation in soils and topography. The frequency and intensity of a given type of disturbance will determine the relative abundance of annual, perennial, herbaceous, and woody plants on a site. Extreme climate events such as drought may act as triggers to push arid ecosystems experiencing chronic disturbances, such as grazing, past desertification 'tipping points' (CCSP 4.2 2008; Gillson and Hofffman 2007). An increase in the frequency of climate trigger events would make arid systems increasingly susceptible to major changes in vegetation cover. Climate is also a key factor dictating the effectiveness of resource management plans and restoration efforts (Holmgren and Scheffer 2001). Precipitation (and its interaction with temperature) plays a major role in determining how plant communities are impacted by, and how they respond to, a given type and intensity of disturbance. It is generally accepted that effects of grazing in arid lands may be somewhat mitigated in years of good rainfall and accentuated in drought years. However, this generalization is context

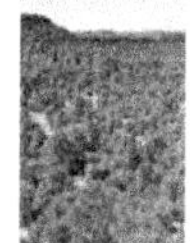

dependent. Landscape-scale factors such as rainfall and stocking rate affect grass cover in pre- and post-drought periods, but grass dynamics before, during, and after drought varies with species-specific responses to local patch-scale factors (e.g., soil texture, micro-topographic redistribution of water) (Yao et al. 2006). As a result, a given species may persist in the face of grazing and drought in some locales and be lost from others. Spatial context should thus be factored in to assessments of how changes in climate will affect ecosystem stability: their ability to maintain function in the face of disturbance (resistance), and the rate and extent to which they recover from disturbance (resilience). Advances in computing power, geographic information systems, and remote sensing now make this feasible.

Chronic disturbance will also affect rates of ecosystem change in response to climate change because it reduces vegetation resistance to slow, long-term changes in climate (Cole 1985; Overpeck et al. 1990). Plant communities dominated by long-lived perennials may exhibit considerable biological inertia, and changes in community composition may lag behind significant changes in climate. Conditions required for seed germination are largely independent of conditions required for subsequent plant survival (Miriti 2007). Species established under previous climate regimes may thus persist in novel climates for long periods of time. Indeed, it has been suggested that the desert grasslands of the Southwest were established during the cooler, moister Little Ice Age but have persisted in the warmer, drier climates of the 19th and 20th centuries (Neilson 1986). Disturbances can create opportunities for species better adapted to the current conditions to establish. In the case of desert grasslands, livestock grazing subsequent to Anglo-European settlement may have been a disturbance that created opportunities for desert shrubs such as mesquite and creosote bush to increase in abundance. Rates of ecosystem compositional change in response to climate change may therefore depend on the type and intensity of disturbance, and the extent to which fundamental soil properties (especially depth and fertility) are altered by disturbance.

3.3.2.5 Desertification

Precipitation and wind are agents of erosion. Wind and water erosion are primarily controlled by plant cover. Long-term reductions in plant cover by grazing and short-term reductions caused by fire create opportunities for accelerated rates of erosion; loss of soils feeds back to affect species composition in ways that can further reduce plant production and cover. Disturbances in arid lands can thus destabilize sites and quickly reduce their ability to capture and retain precipitation inputs. This is the fundamental basis for desertification, a long-standing concern (Van de Koppel et al. 2002). Desertification involves the expansion of deserts into semi-arid and subhumid regions, and the loss of productivity in arid zones. It typically involves loss of ground cover and soils, replacement of palatable, mesophytic grasses by unpalatable xerophytic shrubs, or both (Figure 3.12). There has been long-standing controversy in determining the relative contribution of climatic and anthropogenic factors as drivers of desertification. Local fence line contrasts argue for the importance of land use (e.g., changes in grazing, fire regimes); vegetation change in areas with no known change in land use argue for climatic drivers.

Grazing has traditionally been the most pervasive and extensive climate-influenced land use in arid lands (with the exception of areas where access to ground or surface water allows agriculture; see Chapter 2). Large-scale, unregulated livestock grazing in the 1800s and early 1900s is widely regarded as contributing to widespread desertification (Fredrickson et al. 1998). Grazing peaked around 1920 on public lands in the West, and by the 1970s had been reduced by approximately 70 percent (Holechek et al. 2003). These declines reflect a combination of losses in carrying capacity (ostensibly associated with soil erosion and reductions in the abundance of palatable species); the creation of federally funded experimental ranges in the early 1900s (e.g., the Santa Rita Experimental Range in Arizona, and the Jornada Experimental Range in New Mexico, which are charged with developing stocking rate guidelines); the advent of the science of range management; federal legislation intended to regulate grazing (Taylor Grazing Act of 1934) and combat soil

There has been long-standing controversy in determining the relative contribution of climatic and anthropogenic factors as drivers of desertification.

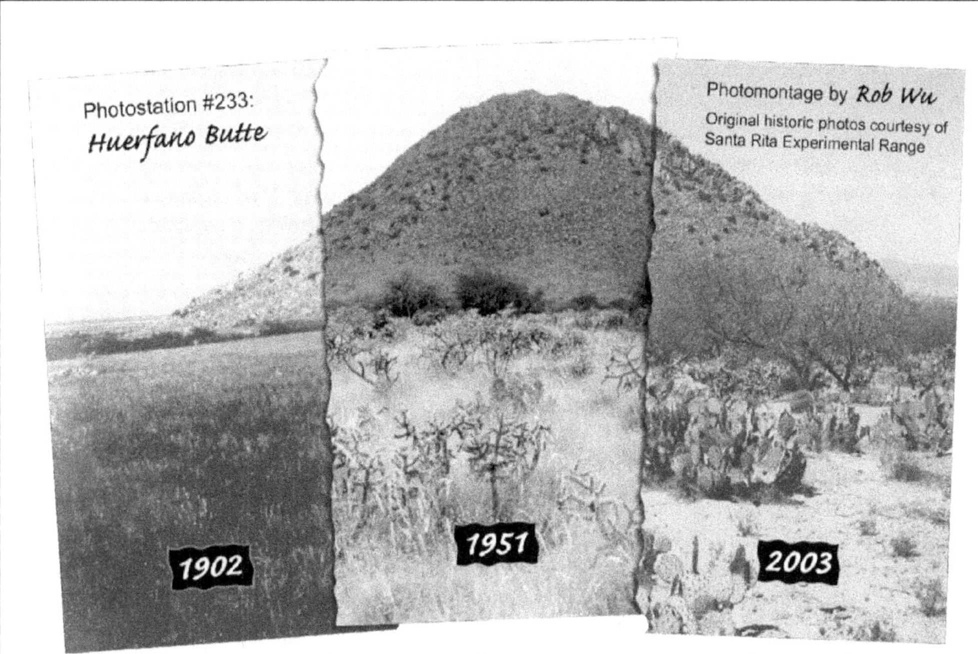

Figure 3.12 *Desertification of desert grassland (Santa Rita Experimental Range [SRER] near Tucson, AZ). Collage developed by Rob Wu from photos in the SRER photo archives (http://ag.arizona.edu/SRER/).*

erosion (Soil Erosion Act 1935); and the shift of livestock production operations to higher rainfall regions.

Arid lands can be slow to recover from livestock grazing impacts. Anderson and Inouye (2001) found that at least 45 years of protection was required for detectable recovery of herbaceous perennial understory cover in cold desert sagebrush steppe. Development of warmer, drier climatic conditions would be expected to further slow rates of recovery. On sites where extensive soil erosion or encroachment of long-lived shrubs occurs, recovery from grazing may not occur over time frames relevant to ecosystem management. While livestock grazing remains an important land use in arid lands, there has been a significant shift to exurban development and recreation, reflecting dramatic increases in human population density since 1950 (Hansen and Brown 2005). Arid land response to future climate will thus be mediated by new suites of environmental pressures such as air pollution and N-deposition, energy development, motorized off-road vehicles, feral pets, and invasion of non-native horticultural plants and grazing.

3.3.2.6 Biotic Invasions

Arid lands of North America were historically characterized by mixtures of shrublands, grasslands, shrub-steppe, shrub-savanna, and woodlands. Since Anglo-European settlement, shrubs and trees have increased at the expense of grasses (Archer 1994). Causes for this shift in plant-life-form abundance are the topic of active debate, but center around climate change, atmospheric CO_2 enrichment, nitrogen deposition, and changes in grazing and fire regimes (Archer et al. 1995; Van Auken 2000). In many cases, increases in woody plant cover reflect the proliferation of native shrubs or trees (mesquite, creosote bush, pinyon, juniper); in other cases, non-native shrubs have increased in abundance (tamarix). Historically, the displacement of grasses by woody plants in arid lands was of concern due to its perceived adverse impacts on stream flow and ground water recharge (Wilcox 2002; Owens and Moore 2007) and livestock production. More recently, the effects of this change in land cover has been shown to have implications for carbon sequestration, and land surface-atmosphere interactions (Schlesinger et al. 1990; Archer et al. 2001; Wessman et

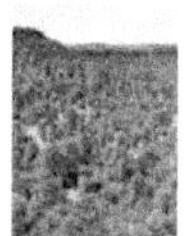

al. 2004). Warmer, drier climates with more frequent and intense droughts are likely to favor xerophytic shrubs over mesophytic native grasses, especially when native grasses are preferentially grazed by livestock. However, invasions by non-native grasses are markedly changing the fire regime in arid lands and impacting shrub cover.

In arid lands of the United States, non-native grasses often act as "transformer species" (Richardson et al. 2000; Grice 2006) in that they change the character, condition, form or nature of a natural ecosystem over substantial areas. Land use and climate markedly influence the probability, rate, and pattern of alien species invasion, and future change for each of these drivers will interact to strongly impact scenarios of plant invasion and ecosystem transformation (Sala et al. 2000; Walther et al. 2002; Hastings et al. 2005). Plant invasions are strongly influenced by seed dispersal and resource availability, but disturbance and abrupt climatic changes also play key roles (Clarke et al. 2005). Changes in ecosystem susceptibility to invasion by non-native plants may be expected with changes in climate (Ibarra et al. 1995; Mau-Crimmins et al. 2006), CO_2 (Smith et al 2000; Nagel et al. 2004) and nitrogen deposition (Fenn et al. 2003). Invasibility varies across elevation gradients. For cheatgrass, a common exotic annual in the Great Basin, invasibility is related to temperature at higher elevations and soil water availability at lower elevations. Increased variability in soil moisture and reductions in perennial herbaceous cover also increased susceptibility of low elevation sites to cheatgrass invasion (Chambers et al 2007). In a 45-year study of cold desert sagebrush steppe that included the major drought of the 1950s, abundance of native species was found to be an important factor influencing community resistance to invasion (Anderson and Inouye 2001). Thus, maintenance of richness and cover of native species should be a high management priority in the face of climate change (see also Chapter 5, this report).

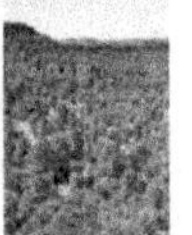

Non-native plant invasions, promoted by enhanced nitrogen deposition (Fenn et al. 2003) and increased anthropogenic disturbance (Wisdom et al. 2005), will have a major impact on how arid land ecosystems respond to climate and climate change. Once established, non-native annual and perennial grasses can generate massive, high-continuity fine-fuel loads that predispose arid lands to fires more frequent and intense than those with which they evolved (Figure 3.13). The result is the potential for desert scrub, shrub-steppe, and desert grassland/savanna biotic communities to be quickly and radically transformed into monocultures of invasive grasses over large areas. This is already well underway in the cold desert region (Knapp 1998) and is in its early stages in hot deserts (Williams and Baruch 2000; Kupfer and Miller 2005; Salo 2005; Mau-Crimmins 2006). By virtue of their profound impact on the fire regime and hydrology, invasive plants in arid lands will very likely trump direct climate impacts on native vegetation where they gain dominance (Clarke et al. 2005). There is a strong climate-wildfire synchrony in forested ecosystems of western North America (Kitzberger et al. 2007); longer fire seasons and more frequent episodes of extreme fire weather are predicted (Westerling et al. 2006). As the areal extent of fire-prone exotic grass communities increases, low elevation arid ecosystems will likely experience similar climate-fire synchronization where none previously existed, and spread of low elevation fires upslope may constitute a new source of ignition for forest fires. Exurban development (Nelson 1992; Daniels 1999) has been and will continue to be a major source for both ignitions (Keeley et al. 1999) and exotic species introductions by escape from horticulture.

3.3.2.7 A Systems Perspective

As reviewed in the preceding sections, the response of arid lands to climate and climate change is contingent upon the net outcome of several interacting factors (Fig 3.9). Some of these factors may reinforce and accentuate climate effects (e.g., soils, grazing); others may constrain, offset or override climate effects (soils, atmospheric CO_2 enrichment, fire, exotic species). Furthermore, extreme climatic events can themselves constitute disturbance (e.g., soil erosion and inundation associated with high intensity rainfall events and flooding; burial abrasion and erosion associated with high winds, mortality caused by drought and extreme temperature stress). Climate effects should thus be viewed in the context of other factors, and simple generalizations regarding climate effects should be viewed with caution. This is not to say, however, that there is insufficient data and theory to guide prediction of future outcomes. Today's arid lands reflect a legacy of historic land uses, and future land use practices will arguably have the greatest impact on arid land ecosystems in the next two to five decades. In the near-term, climate fluctuation and change will be important primarily as it influences the impact of land use on ecosystems and how ecosystems respond to land use. Given the concomitant changes in climate, disturbance frequency/intensity, atmospheric CO_2, nitrogen deposition, and species invasions, it also seems likely that novel wildland and managed ecosystems will develop (Hobbs et al. 2006). Communities that are compositionally unlike any found today have occurred in the late-glacial past (Williams and Jackson 2007). In climate simulations for the IPCC emission scenarios, novel climates arise by 2100 AD. These future novel climates (which are warmer than any present climates, with spatially variable shifts in precipitation) increase the risk of species reshuffling into future novel communities and other ecological surprises (Williams and Jackson 2007). These novel ecosystems will present novel challenges and opportunities for conservation and management.

The following sections will address specific climate/land use/land cover issues in more detail. Section 3.10 will discuss climate and climate change effects on species distributions and community dynamics. Section 3.11 will review the consequences for ecosystem processes. Section 3.12 will focus on climate change implications for structure and function of riparian and aquatic ecosystems in arid lands. Implications for wind and water erosion will be reviewed in Section 3.13.

3.3.3 Species Distributions and Community Dynamics

3.3.3.1 Climate-Fire Regimes

The climate-driven dynamic of the fire cycle is likely to become the single most important feature controlling future plant distributions in U.S. arid lands. Rising temperatures, decreases in precipitation and a shift in seasonality and variability, and increases in atmospheric CO_2

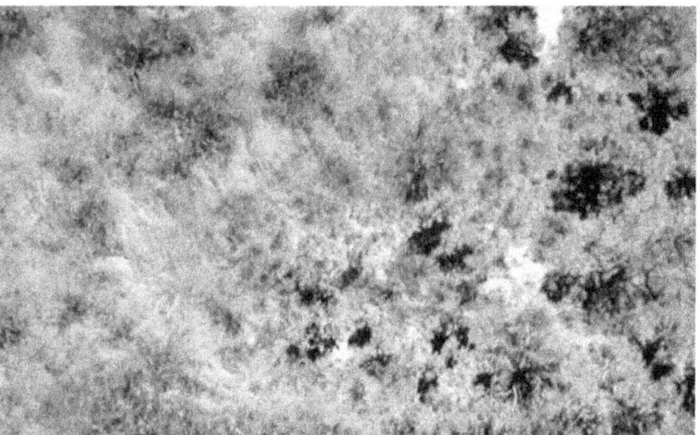

Figure 3.13 *Top-down view of native big sagebrush steppe (right) invaded by cheatgrass, an exotic annual grass (left). Photo: Steve Whisenant.*

and nitrogen deposition (Sage 1996), coupled with invasions of exotic grasses (Brooks et al. 2004; Brooks and Berry 2006) will accelerate the grass-fire cycle in arid lands and promote development of near monoculture stands of invasive plants (D'Antonio and Vitousek 1992). The frequency of fire in the Mojave Desert has dramatically increased over the past 20 years and effected a dramatic conversion of desert shrubland to degraded annual-plant landscapes (Bradley et al. 2006; Brooks and Berry 2006). Given the episodic nature of desert plant establishment and the high susceptibility of the new community structure to additional fire, it will be exceedingly difficult to recover native plant dominance. A similar conversion has occurred in many Great Basin landscapes (Knapp 1995, 1996), and given the longer period of non-native annual plant presence (Novak and Mack 2001), the pattern is much more advanced and has lowered ecosystem carbon storage (Bradley et al. 2006). Contemporary patterns in natural settings (Wood et al. 2006) and field experiments (Smith et al. 2000) suggest non-native response to climate change will be extremely important in the dynamics of arid land fire cycle, and changes in climate that promote fires will exacerbate land cover change in arid and semi-arid ecosystems.

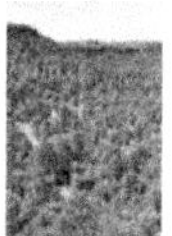

There is some debate as to how climate contributed to a non-native component of this vegetation-disturbance cycle over the first half of the 20th century. For the upper elevations in the Sonoran Desert, Lehmann lovegrass, a perennial C_4 African grass introduced for cattle forage and erosion control, has spread aggressively and independently of livestock grazing (McClaran 2003). Its success relative to native grasses appears related to its greater seedling drought tolerance and its ability to more effectively utilize winter moisture. Relatively wet periods associated with the Pacific Decadal Oscillation appear to have been more important than increases in N-deposition or CO_2 concentrations in the spread of these species (Salo 2005).

More recently, warm, summer-wet areas in northern Mexico (Sonora) and the southwestern United States have become incubators for perennial C_4 African grasses such as buffelgrass, purposely introduced to improve cattle forage in the 1940s. These grasses escape plantings on working ranches and, like exotic annual grasses, initiate a grass-fire cycle (Williams and Baruch 2000). In the urbanized, tourism-driven Sonoran Desert of southern Arizona, buffelgrass invasion is converting fireproof and picturesque desert scrub communities into monospecific, flammable grassland. Buffelgrass, like other neotropical exotics, is sensitive to low winter temperatures. The main invasion of buffelgrass in southern Arizona happened with warmer winters beginning in the 1980s, and its range will extend farther north and upslope as minimum temperatures continue to increase (Arriaga et al. 2004). This is complicated further by ongoing germplasm research seeking to breed more drought- and cold-resistant varieties. For example, a cold-resistant "Frio" variety of buffelgrass recently released by USDA-Agricultural Research Service has been planted 40 km south of the Arizona border near Cananea, Mexico. Escape of "Frio" north of the United States-Mexico border may extend the potential niche of buffelgrass a few hundred meters in elevation and a few hundred kilometers to the north.

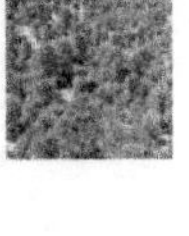

3.3.3.2 Drought and Vegetation Structure

Over the past 75 years, the drought of the 1950s and the drought of the early 2000s represent two natural experiments for understanding plant community response to future environmental conditions. While both had similar reductions in precipitation, the 1950s drought was typical of many Holocene period droughts throughout the Southwest, whereas the drought that spanned the beginning of the 21st century was relatively hot (with both greater annual temperatures and greater summer maximum temperatures) (Mueller et al. 2005; Breshears et al. 2005). The 1950s drought caused modest declines in the major shrubs in the Sonoran Desert, whereas the 2000s drought caused much higher mortality rates in numerous species, including the long-lived C_3 creosote bush, which had shown essentially no response to the 1950s drought (Bowers 2005). A similar pattern was seen in comparing the two time periods for perennial species in the Mojave Desert, where dry periods close to the end of the 20th century were associated with reductions in C_3 shrubs and both C_3 and C_4 perennial grass species (Hereford et

Figure 3.14 *Buffelgrass invasion of saguaro stand in the Tucson Mountains, Arizona (left); fire-damaged saguaro (right). Photos courtesy Ben Wilder.*

al. 2006). Thus, the greater temperatures and higher rates of evapotranspiration predicted to co-occur with drought portend increased mortality for the dominant woody vegetation typical of North American deserts, and open the door for establishment of non-native annual grasses. These patterns are mostly driven by changes in winter precipitation, but in systems where summer rainfall is abundant, woody plant-grass interactions may also be important. Given the projected increases in the frequency of these "global warming type" droughts (e.g., Breshears et al. 2005), increases in summer active, non-native C_4 grasses (such as buffelgrass in the Sonoran Desert (Franklin et al. 2006)), and the increased probability of fire (Westerling et al. 2006), a similar pattern of a wide-spread woody vegetation conversion to degraded non-native grasslands can be anticipated for the hot deserts of North America – a pattern similar to that already seen in the Great Basin (Bradley et al. 2006).

3.3.3.3 Plant Functional Group Responses

Annual plants are a major source of plant diversity in the North American deserts (Beatley 1967), but exotic annuals are rapidly displacing native annuals. The density of both native and non-native desert annuals in the Sonoran Desert, at Tumamoc Hill in Tucson, AZ, has been reduced by an order of magnitude since 1982 (from ~2,000 plants/m^{-2} to ~150/plants m^{-2}) (Venable and Pake 1999). Similar reductions have been recorded for the Nevada Test Site (Rundel and Gibson 1996). At the same time, there has been an increase in the number of non-native annual species (Hunter 1991; Salo et al. 2005; Schutzenhofer and Valone 2006). High CO_2 concentrations appear to benefit non-native grasses and "weeds" more so than native species (Huxman and Smith 2001; Ziska 2003; Nagel et al. 2004). Thus, when rainfall is relatively high in the Mojave Desert, non-natives comprise about 6 percent of the flora and ~66 percent of the community biomass, but when rainfall is restricted, they comprise ~27 percent of the flora and >90 percent of the biomass (Brooks and Berry 2006). Competition between annuals and perennials for soil nitrogen during relatively wet periods can be intense (Holzapfel and Mahall 1999). At the western fringe of the Mojave and Sonoran Deserts, nitrogen deposition is tipping the balance toward the annual plant community (typically non-native) with the resulting loss of woody native species (Wood et al. 2006).

Based on theory and early experiments, rising atmospheric CO_2 and increasing temperature are predicted to increase the competitive ability of C_3 versus C_4 plants in water-limited systems, potentially reducing the current pattern of C_4 dominance in many warm season semi-arid ecosystems (Long 1991; Ehleringer et al. 1997; Poorter and Navas 2003). Photosynthesis and stomatal conductance of leaves of plants in mixed C_3/C_4 communities often show a greater proportional response in C_3 as compared to C_4 species at elevated CO_2 (Polley et al. 2002). However, community composition and productivity do not always reflect leaf level patterns and more sophisticated experiments show complex results. It is likely that whole-system water budgets are significantly altered and more effectively influence competitive interactions between C_3 and C_4 species as compared to any direct CO_2 effects on leaf function (Owensby et al. 1993; Polley et al. 2002). In the Great Basin, which is dominated by C_3 plants, CO_2 enrichment favors non-native annual cheatgrass over native C_3 plants (Smith et al. 2000; Ziska et al. 2005).

Based on theory and early experiments, rising atmospheric CO_2 and increasing temperature are predicted to increase the competitive ability of C_3 versus C_4 plants in water-limited systems, potentially reducing the current pattern of C_4 dominance in many warm season semi-arid ecosystems.

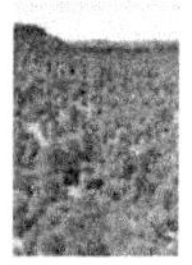

Where C_3 species have increased in abundance in elevated CO_2 experiments (Morgan et al. 2007), the photosynthetic pathway variation also reflected differences in herbaceous (C_4) and woody (C_3) life forms. CO_2 enhancement of C_3 woody plant seedling growth, as compared to growth of C_4 grasses, may facilitate woody plant establishment (Polley et al. 2003). Reduced transpiration rates from grasses under higher CO_2 may also allow greater soil water recharge to depth, and favor shrub seedling establishment (Polley et al. 1997). Changes in both plant growth and the ability to escape the seedling-fire-mortality constraint are critical for successful shrub establishment in water-limited grasslands (Bond and Midgley 2000). However, interactions with other facets of global change may constrain growth form and photosynthetic pathway responses to CO_2 fertilization. Increased winter temperatures would lengthen the C_4 growing season. Greater primary production at elevated CO_2 combined with increased abundance of non-native grass species may alter fire frequencies (see sections 3.2.2.6 and 3.3.3.1 and 3.3.4.1). Nitrogen deposition may homogenize landscapes, favoring grassland physiognomies over shrublands (Reynolds et al. 1993). Changes in the occurrence of episodic drought may alter the relative performance of these growth forms in unexpected ways (Ward et al. 1999). Predicting changes in C_3 versus C_4 dominance, or changes in grass versus shrub abundance in water-limited ecosystems, will require understanding of multifactor interactions of global change the scientific community does not yet possess.

3.3.3.4 Charismatic Mega Flora

Saguaro density is positively associated with high cover of perennial vegetation and mean summer precipitation, but total annual precipitation and total perennial cover are the best predictors of reproductive stem density (Drezner 2006). Because of how these drivers co-vary in the southwestern United States, the drier western regions have lower saguaro densities than the southeastern region where summer rainfall is greater. Additionally, the Northeast and Southeast both have very high reproductive stem densities relative to the West. These patters reflect the interaction between summer rainfall and the frequency of episodic freezing events that constrain the species' northern range boundary. Despite predicted reductions in the number of freezing events (Weiss and Overpeck 2005), predicted increases in annual temperature, loss of woody plant cover from a greater frequency of "global warming-type" droughts, and increasing fire resulting from non-native grass invasions (Figure 3.14) suggest a restriction of the Saguaro's geographic range and reductions in abundance within its historic range.

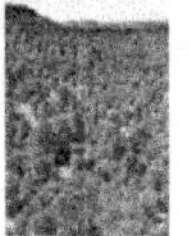

The direct effects of rising CO_2 on climatic tolerance and growth of Joshua trees also suggest important shifts in this Mojave Desert species' range (Dole et al. 2003). Growth at elevated CO_2 improves the ability of seedlings to tolerate periods of cold temperature stress (Loik et al. 2000). When applied to downscale climate outputs and included in the rules that define species distribution, this direct CO_2 effect suggests the potential for a slight increase in geographic range. However, like all long-lived, large-statured species in the North American deserts, the frequency of fire will be a primary determinant of whether this potential will be realized.

3.3.4 Ecosystem Processes

3.3.4.1 Net Primary Production and Biomass

Semi-arid and arid ecosystems of the western United States are characterized by low plant growth (NPP), ranging from 20 to 60 g/m^2/yr in the Mojave Desert of Nevada (Rundel and Gibson 1996) to 100 to 200 g/m^2/yr (aboveground) in the Chihuahuan Desert of New Mexico (Huenneke et al. 2002). In most studies, the belowground component of plant growth is poorly characterized, but observations of roots greater than 9 meters deep suggest that root production could be very large and perhaps underestimated in many studies (Canadell et al. 1996).

With water as the primary factor limiting plant growth, it is not surprising that the variation in plant growth among desert ecosystems, or year-to-year variation within arid ecosystems, is related to rainfall. High spatial and interannual variation make it difficult to identify trends in aboveground net primary production (ANPP) over time, especially when disturbances such as livestock gazing co-occur as an additional confounding factor. In their comparison of cold desert sagebrush steppe vegetation structure and production during two 10-year studies from the late 1950s to the late 1960s and three years in the 1990s, West and Yorks (2006) noted high coefficients of variation in aboveground plant production associated with five-fold differences in precipitation at a given locale, sometimes in consecutive years. In the Chihuahuan Desert, shrub encroachment into desert grassland has increased the spatial heterogeneity of ANPP and soil nutrients (Schlesinger and Pilmanis 1998; Huenneke et al. 2002). Although grasslands tended to support higher ANPP than did shrub-dominated systems, grasslands demonstrated higher interannual variation. Projected increases in precipitation variability coupled with changes in species composition would be expected to further increase the already substantial variation in arid land plant production. Other factors, such as soil texture and landscape position, also affect soil moisture availability and determine plant growth in local conditions

(Schlesinger and Jones 1984; Wainwright et al. 2002). Increases in temperature and changes in the amount and seasonal distribution of precipitation in cold deserts (Wagner 2004) and hot deserts (Seager et al. 2007) can be expected to have a dramatic impact on the dominant vegetation, NPP, and carbon storage in arid lands.

Jackson et al. (2002) found that plant biomass and soil organic matter varied systematically in mesquite-dominated ecosystems across west Texas and eastern New Mexico, demonstrating some of the changes that can be expected with future changes in rainfall regimes. The total content of organic matter (plant + soil) in the ecosystem was greatest at the highest rainfall, but losses of soil carbon in the driest sites were compensated by increases in plant biomass, largely mesquite. Despite consistent increases in aboveground carbon storage with woody vegetation encroachment, a survey of published literature revealed no correlation between mean annual rainfall and changes in soil organic carbon pools subsequent to woody plant encroachment (Asner and Archer in press). Differences in soil texture, topography, and historical land use across sites likely confound assessments of precipitation influences on soil organic carbon pool responses to vegetation change.

3.3.4.2 Soil Respiration

Soil respiration includes the flux of CO_2 from the soil to the atmosphere from the combined activities of plant roots and their associated mycorrhizal fungi and heterotrophic bacteria and fungi in the soil. It is typically measured by placing small chambers over replicated plots of soil or estimated using eddy-covariance measurements of changes in atmospheric properties, particularly at night. Soil respiration is the dominant mechanism that returns plant carbon dioxide to Earth's atmosphere, and it is normally seen to increase with increasing temperature. Mean soil respiration in arid and semi-arid ecosystems is 224 g C/m^2/yr (Raich and Schlesinger 1992; Conant et al. 1998), though in individual sites, it can be expected to vary with soil moisture content during and between years.

Intensification of the hydrologic cycle due to atmospheric warming is expected reduce rainfall frequency, but increase the intensity and/or size of individual precipitation events. A change in the size-class distribution of precipitation has important implications for instantaneous fluxes of carbon dioxide from soils and the potential for ecosystems to sequester carbon (Austin et al. 2004; Huxman et al. 2004a; Jarvis et al. 2007). This is due to differences in the way soil microbial populations and plants respond to moisture entering the soil following rainfall events of different sizes. Larger rainfall events that increase the wetting depth in the soil profile should increase the number of periods within a year where substantial plant activity and carbon storage can occur (Huxman et al. 2004b; Pereira et al. 2007; Kurc and Small 2007; Patrick et al. 2007). However, reducing the frequency of wet-dry cycles in soils will retard microbial activity and nutrient cycling, likely introducing a long-term nitrogen limitation to plant growth (Huxman et al. 2004a). For winter rainfall ecosystems, these shifts in wet-dry cycles can cause reductions in productivity and soil carbon sequestration (Jarvis et al. 2007; Pereira et al. 2007).

Given the low NPP of arid lands, they are likely to result in only small amounts of carbon sequestration.

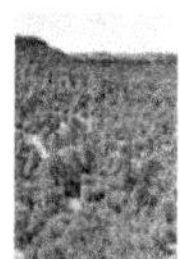

3.3.4.3 Net Carbon Balance

The net storage or loss of carbon in any ecosystem is the balance between carbon uptake by plants (autotrophic) and the carbon released by plant respiration and heterotrophic processes. Although elegant experiments have attempted to measure these components independently, the difference between input and output is always small and thus measurement errors can be proportionately large. It is usually easier to estimate the accumulation of carbon in vegetation and soils on landscapes of known age. This value, NEP, typically averages about 10 percent of NPP in forested ecosystems. Arid soils contain relatively little soil organic matter, and collectively make only a small contribution to the global pool of carbon in soils (Schlesinger 1977; Jobbagy and Jackson 2002). Given the low NPP of arid lands, they are likely to result in only small amounts of carbon sequestration. Since soil organic matter is inversely related to mean annual temperature in many arid regions (Schlesinger 1982; Nettleton and Mays 2007), anticipated increases in regional temperature will lead to a loss of soil carbon to the atmosphere, exacerbating increases in atmospheric carbon dioxide. Recent measurements of NEP by micrometeorological techniques, such as eddy covariance, across relatively large spatial scales confirm this relatively low carbon uptake for arid lands (Grunzweig et al. 2003), but point to the role of life-form (Unland et al. 1996), seasonal rainfall characteristics (Hastings et al. 2005; Ivans et al. 2006), and potential access to groundwater as important modulators of the process (Scott et al. 2006).

Several scientists have suggested that arid lands might be managed to sequester carbon in soils and mitigate future climate change (Lal 2001). The prospects for such mitigation are limited by the low sequestration rates of organic and inorganic carbon that are seen in arid lands under natural conditions (Schlesinger 1985, 1990), the tendency for warmer soils to store lesser amounts of soil organic matter, and the small increases in net productivity that might be expected in these lands in a warmer, drier future climate. Moreover, when desert lands are irrigated, there can be substantial releases of carbon dioxide from the fossil fuels used to pump irrigation water (Schlesinger 2001). Globally, the greatest potential for soil carbon sequestration is found in soils that are cold and/or wet, not in soils that are hot and dry.

In many areas of desert, the amount of carbon stored in inorganic soil carbonates greatly exceeds the amount of carbon in vegetation and soil organic matter, but the formation of such carbonates is slow and not a significant sink for carbon in its global cycle (Schlesinger 1982; Monger and Martinez-Rios 2000). Some groundwater contains high (supersaturated) concentrations of carbon dioxide, which is released to the atmosphere when this water is brought to the Earth's surface for irrigation, especially when carbonates and other salts precipitate (Schlesinger 2000). Thus, soil carbonates are unlikely to offer significant potential to sequester atmospheric carbon dioxide in future warmer climates.

3.3.4.4 Biogeochemistry

Arid-land soils often have limited supplies of nitrogen, such that nitrogen and water can "co-limit" the growth of vegetation (Hooper and Johnson 1999). These nitrogen limitations normally appear immediately after the receipt of seasonal rainfall. The nitrogen limitations of arid lands stem from small amounts of N received by atmospheric deposition and nitrogen fixation and rather large losses of N to wind erosion and during microbial transformations of soil N that result in the losses of ammonia (NH_3), nitric oxide (NO), nitrous oxide (N_2O), and nitrogen gas (N_2) to the atmosphere (Schlesinger et al. 2006). These microbial processes are all stimulated by seasonal rainfall, suggesting that changes in the rainfall regime as a result of climate change will alter N availability and plant growth. N deposition is spatially variable, being greater in areas downwind from major urban centers such as Los Angeles, increasing the abundance of herbaceous vegetation and potentially increasing the natural fire regime in the Mojave Desert (Brooks 2003).

In arid lands dominated by shrub vegetation, the plant cycling of N and other nutrients is often heterogeneous, with most of the activity focused in the soils beneath shrubs (Schlesinger et al. 1996). It remains to be seen how these local nutrient hot spots will influence vegetation composition and ecosystem function in future environments. In cold desert shrub steppe, non-native cheatgrass is often most abundant under shrubs, resulting in rapid consumption of the shrub during fire and mortality of native plants and seed banks; the higher available resources on the fertile island enables greater biomass and seed production of cheatgrass in the post-fire period (Chambers et al. 2007). Thus, the rate and extent of invasion of cold desert sagebrush-steppe by cheatgrass may initially be a function of the cover and density of sagebrush plants and the fertile islands they create.

3.3.4.5 Trace Gases

In addition to significant losses of N trace gases, some of which confer radiative forcing on the atmosphere (e.g., N_2O), deserts are also a minor source of methane, largely resulting from activities of some species of termites, and volatile organic compounds (VOC) and non-methane hydrocarbon gaseous emissions from vegetation and soils (Geron et al. 2006). Isoprene, produced by many woody species and a few herbaceous species, is the dominant VOC released by vegetation; the ability to produce significant amounts of isoprene may or may not be shared by members of the same plant family or genus (Harley et al. 1999). No phylogenetic pattern is obvious among the angiosperms, with the trait widely scattered and present (and absent) in both primitive and derived taxa, although confined largely to woody species. VOCs can serve as precursors to the formation of tropospheric ozone and organic aerosols, thus influencing air pollution. Emissions of such gases have increased as a result of the invasion of grasslands by desert shrubs during the past 100 years (Guenther et al. 1999), and emissions of isoprene are well known to increase with temperature (Harley et al. 1999). The flux of these gases from arid lands is not well studied, but is known to be sensitive to temperature, precipitation, and drought stress. For example, total annual VOC emissions in deserts may vary threefold between dry and wet years, and slight increases in daily leaf temperatures can increase annual desert isoprene and monoterpene fluxes by 18 percent and 7 percent, respectively (Geron et al. 2006). Thus, changes in VOC emissions from arid lands can be expected to accompany changes in regional and global climate.

3.3.5 Arid Land Rivers and Riparian Zones

Springs, rivers and floodplain (riparian) ecosystems commonly make up less than 1 percent of the landscape in arid regions of the world. Their importance, however, belies their small areal extent (Fleischner 1994; Sada et al. 2001; Sada and Vinyard 2002). These highly productive ecosystems embedded within much lower productivity upland ecosystems attract human settlement and are subjected to a variety of land uses. They provide essential wildlife habitat for migration and breeding, and these environments are critical for breeding birds, threatened and endangered species, and arid-land vertebrate species. Riparian vegetation in arid lands can occur at scales from isolated springs to ephemeral and intermittent watercourses, to perennial rivers (Webb and Leake 2006). The rivers and riparian zones of arid lands are dynamic ecosystems that react quickly to changing hydrology, geomorphology, human utilization, and climate change. Certain types of springs can also be highly responsive to these changes. As such, spring, river and riparian ecosystems will likely prove to be responsive components of arid landscapes to future climate change.

Effects of climate change on aquatic organisms in arid lands are not well known. Introductions of non-native fish and habitat modification have caused the extinction of numerous endemic species, subspecies and populations of fishes, mollusks and insects since the late 1800s. Declines in abundance or distribution have been attributed to (in order of decreasing importance) water flow diversions, competitive or predatory interactions with non-native species, livestock grazing, introductions for sport fisheries management, groundwater pumping, species hybridization, timber harvest, pollution, recreation and habitat urbanization (reviewed by Sada and Vinyard 2002). Most taxa were influenced by multiple factors. It is likely that projected climate changes will exacerbate these existing threats via effects on water temperature, sedimentation, and flows.

Global climate change can potentially impact river and riparian ecosystems in arid regions through a wide variety of mechanisms and pathways (Regab and Prudhomme 2002). Three pathways in which riverine corridors in arid lands are highly likely to be affected are particularly important. The first is the impact of climate change on water budgets. Both sources of water and major depletions will be considered. The second is competition between native and non-native species in a changing climate. The potential importance of thresholds in these interactions will be explicitly considered. The third mechanism pertains to the role of extreme climate events (e.g., flood and droughts) in a changing climate. Extreme events have always shaped ecosystems, but the interactions of a warmer climate with a strengthened and more variable hydrologic cycle are likely to be significant structuring agents for riverine corridors in arid lands.

Global climate change can potentially impact river and riparian ecosystems in arid regions through a wide variety of mechanisms and pathways.

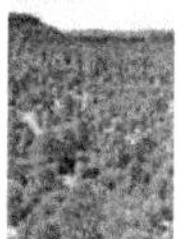

3.3.5.1 Water Budgets

Analysis of water budgets under a changing climate is one tool for assessing the impact of climate change on arid-land rivers and riparian zones. Christensen et al. (2004) have produced a detailed assessment of the effects of climate change on the hydrology and water resources of the Colorado River basin. Hydrologic and water resources scenarios were evaluated through coupling of climate models, hydrologic models, and projected greenhouse gas scenarios for time periods 2010-2039, 2040-2069, and 2070-2099. Average annual temperature changes for the three periods were 1.0°C, 1.7°C, and 2.4°C, respectively, and basin-average annual precipitation was projected to decrease by 3, 6, and 3 percent for the three periods, respectively. These scenarios produced annual runoff decreases of 14, 18, and 17 percent from historical conditions for the three designated time periods. Such decreases in runoff will have substantial effects on human populations and river and riparian ecosystems, particularly in the lower elevation arid land compartments of this heavily appropriated catchment (e.g., Las Vegas and southern California).

Changing climate also can have a significant effect on major depletions of surface waters in arid regions. Dahm et al. (2002) examined major depletions along a 320-km reach of the Rio Grande in central New Mexico. Major depletions were reservoir evaporation, riparian zone evapotranspiration, agriculture, groundwater recharge, and urban/suburban use. All of these depletions are sensitive to climate warming. Reservoir evaporation is a function of temperature, wind speed, and atmospheric humidity. Riparian zone evapotranspiration is sensitive to the length of the growing season, and climate warming will lengthen the period of time that riparian plants will be actively respiring (Goodrich et al. 2000; Cleverly et al. 2006), and also increase the growing season for agricultural crops dependent on riparian water. Temperature increases positively affect groundwater recharge rates from surface waters through changes in viscosity (Constantz and Thomas 1997; Constantz et al. 2002). The net result of climate warming is greater depletion of water along the riverine corridor (Figure 3.15). Global warming will place additional pressure on the major depletions of surface water in arid regions, in addition to likely effects on the supply side of the equation.

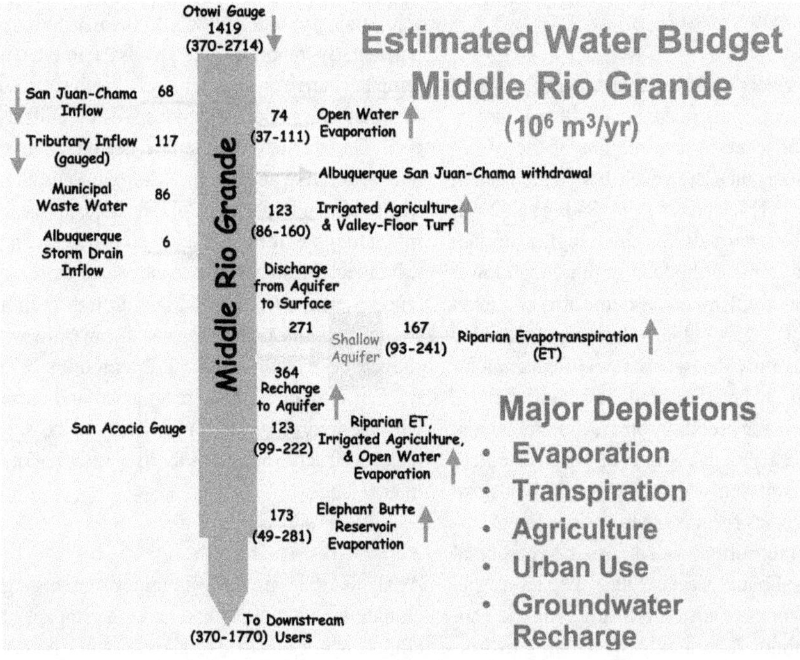

Figure 3.15 *A water budget for a 320-km segment of the Middle Rio Grande of New Mexico, USA, with water sources on the left and top, depletions on the right, and downstream output on the bottom (Dahm et al. 2002). The red arrows indicate the direction of change for various water sources and depletions predicted with a warmer climate. Otowi Guage values are a 26 year mean with range; releases from Elephant Butte dam are ranges only, because releases vary depending on delivery requirements and because releases sometimes include storage water (dam volume is being drawn down) or is much less than inflow (water going into storage). Ranges reflect both interannual variability and measurement uncertainty. The budget balances, but only coarsely, because of the large ranges.*

3.3.5.2 Native and Non-Native Plant Interactions

Competition between native and non-native species in a changing climate is a second area where climate change is predicted to have a substantial effect on riparian zones of arid lands. Riparian zones of arid lands worldwide are heavily invaded by non-native species of plants and animals (Prieur-Richard and Lavorel 2000; Tickner et al. 2001). Salt cedar and Russian olive are particularly effective invaders of the arid land riparian zones of the western United States (Figure 3.16) (Brock 1994; Katz and Shafroth 2003). Shallow ground water plays an important role in structuring riparian plant communities (Stromberg et al. 1996) and groundwater level decline, both by human depletions and intensified drought in a changing climate, will alter riparian flora. Stromberg et al. (1996) describe riparian zone "desertification" from a lowered water table whereby herbaceous species and native willows and cottonwoods are negatively impacted. Horton et al. (2001a, b) describe a threshold effect where native canopy dieback occurs when depth to ground water exceeds 2.5-3.0 meters. Non-native salt cedar (Tamarix chinensis), however, are more drought tolerant when water tables drop, and readily return to high rates of growth when water availability again increases. Plant responses like these are predicted to shift the competitive balance in favor of non-native plants and promote displacement of native plants in riparian zones under a warmer climate.

Another example of a threshold effect on river and riparian ecosystems in arid lands is the persistence of aquatic refugia in a variable or changing climate. Hamilton et al. (2005) and Bunn et al. (2006) have shown the critical importance of waterhole refugia in the maintenance of biological diversity and ecosystem productivity in arid-land rivers. Arid regions worldwide, including this example from inland Australia, are dependent on the persistence of these waterholes during drought. Human appropriation of these waters or an increase in the duration and intensity of drought due to climate change would dramatically affect aquatic biodiversity and the ability of these ecosystems to respond to periods of enhanced water availability. For example, most waterhole refugia throughout the entire basin would be lost if drought persisted for more than two years in the Cooper Creek basin of Australia, or if surface diversions of flood waters reduced the available water within refugia in the basin (Hamilton et al. 2005; Bunn et al. 2006). Desiccation of waterholes could become more common if climate change increases annual evapotranspiration rates or if future water withdrawals reduce the frequency and intensity of river flows to waterholes. Roshier et al. (2001) pointed out that temporary wetland habitats throughout arid lands in Australia are dependent upon infrequent, heavy rainfalls and are extremely vulnerable to any change in frequency or magnitude. Climate change that induces drying or reduced frequency of large floods would

Figure 3.16 *Non-native salt cedar (right) has invaded and displaced native cottonwood and poplar forests (left) in many southwestern riparian corridors. Photo credits: Jim Thibault and James Cleverly.*

deleteriously impact biota, particularly water birds that use these temporary arid land habitats at broad spatial scales.

3.3.5.3 Extreme Events

The role of extreme events (e.g., flood and droughts) in a changing climate is predicted to increase with a warmer climate (IPCC 2007). Extreme climatic events are thought to strongly shape arid and semi-arid ecosystems worldwide (Holmgren et al. 2006). Climate variability, such as associated with the El Niño Southern Oscillation (ENSO) phenomenon, strongly reverberates through food webs in many arid lands worldwide. Fluvial systems and riparian vegetation are especially sensitive to the timing and magnitude of extreme events, particularly the timing and magnitude of minimum and maximum flows (Auble et al. 1994). GCMs do not yet resolve likely future regional precipitation regimes or future temperature regimes. A stronger overall global hydrologic cycle, however, argues for more extreme events in the future (IPCC 2007). The ecohydrology of arid-land rivers and riparian zones will certainly respond to altered precipitation patterns (Newman et al. 2006), and the highly variable climate that characterizes arid lands is likely to become increasingly variable in the future.

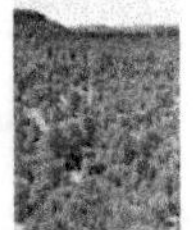

3.3.6 Water and Wind Erosion

Due to low and discontinuous cover, there is a strong coupling between vegetation in arid lands and geomorphic processes such as wind and water erosion (Wondzell et al. 1996). Erosion by wind and water has a strong impact on ecosystem processes in arid regions (Valentin et al. 2005; Okin et al. 2006). Erosion impacts the ability of soils to support plants and can deplete nutrient-rich surface soils, thus reducing the probability of plant establishment and recruitment. Although erosion by water has received by far the most attention in the scientific literature, the few studies that have investigated both wind and water erosion have shown that they can be of similar magnitude under some conditions (Breshears et al. 2003).

3.3.6.1 Water Erosion

Water erosion primarily depends on the erosivity of precipitation events (rainfall rate, storm duration, and drop size) and the erodibility of the surface (infiltration rate, slope, soil, and vegetation cover). Climate change may impact all of these except slope. For instance, it is well established that the amount of soil that is detached (and hence eroded) by a particular depth of rain is related to the intensity at which this rain falls. Early studies suggest soil splash rate is related to rainfall intensity and raindrop fall velocity (Ellison 1944; Bisal 1960). It is also well established that the rate of runoff depends on soil infiltration rate and rainfall intensity. When rainfall intensity exceeds rates of infiltration, water can run off as inter-rill flow, or be channeled into rills, gullies, arroyos, and streams. The intensity of rainfall is a function of climate, and therefore may be strongly impacted by climate change. The frequency of heavy precipitation events has increased over most land areas, including the United States, which is consistent with warming and observed increases in atmospheric water vapor (IPCC 2007). Climate models predict additional increases in the frequency of heavy precipitation, and thus highly erosive events. Warming climates may also be responsible for changes in surface soils themselves, with important implications for the erodibility of soils by water. In particular, higher temperatures and decreases in soil moisture, such as those predicted in many climate change scenarios, have been shown to decrease the size and stability of soil aggregates, thus increasing their susceptibility to erosion (Lavee et al. 1998).

By far the most significant impact of climate change on water erosion is via its effects on vegetation cover. Vegetation conversion to annual grasses or weedy forbs can result in loss of soil nutrients, siltation of streams and rivers, and increased susceptibility to flooding (Knapp 1996). Although some fireproof shrublands in the Southwest have been invaded by non-native grasses, thus changing the fire ecology and endangering those ecosystems (Knapp 1996; Bradley et al. 2006), many other areas have experienced the loss of native perennial grasses, which have been replaced by shrubs (van Auken 2000; sections 3.9.4 and 3.9.5). This widespread conversion of grasslands to shrublands throughout the desert Southwest has resulted in significantly greater erosion, though research on natural rainfall events to quantify the total amounts of erosion is ongoing. Flow and erosion plots in the Walnut Gulch Experimental

Watershed in Arizona and the Jornada LTER site in New Mexico have demonstrated significant differences in water erosion between grasslands and shrublands (Wainwright et al. 2000). For instance, greater splash detachment rates (Parsons et al. 1991, 1994, 1996), and interrill erosion rates (Abrahams et al. 1988) are observed in shrublands compared to grasslands; shrubland areas are more prone to develop rills, which are responsible for significant increases in overall erosion rates (Luk et al. 1993). Episodes of water erosion are often associated with decadal drought-interdrought cycles because depressed vegetation cover at the end of the drought makes the ecosystem vulnerable to increased erosion when rains return (McAuliffe et al. 2006). No study to date has used climate models to estimate how the periodicity of these cycles might change in the future.

U.S. arid regions have already experienced dramatic increases in erosion rates due to widespread losses of vegetation cover. These changes have created conditions where anticipated increases in precipitation intensity, coupled with reductions in soil aggregate stability due to net warming and drying, will likely increase potential erosion rates dramatically in coming decades.

3.3.6.2 Wind Erosion

As with water erosion, the magnitude of wind erosion is related to both the erosivity of the wind and the erodibility of the surface. However, the impact of increased wind erosion in deserts can have continental-scale impacts because the resulting dust can travel long distances with significant impacts to downwind ecosystems, air quality, and populations. Both hemispheres have experienced strengthening of mid-latitude westerly winds since the 1960s (IPCC 2007). This trend is likely to continue into the near future. Thus, desert regions of the United States are likely to experience more erosive conditions in the near future.

The susceptibility of soil to erosion by wind is determined by both the erodibility of the surface soil and the amount of vegetation present to disrupt wind flows and shelter the surface from erosion. Anticipated net aridification in the desert Southwest (Seager et al. 2007) is likely to lead to a decrease in soil aggregate size and stability. Increased temperatures and drought occurrence will result in lower relative humidity in arid lands. Because the top few millimeters of soil are in equilibrium with soil moisture in the overlying air, the decrease in relative humidity may result in soils that require less wind power to initiate erosion (Ravi et al. 2006). Increased drought occurrence throughout the western United States can further lead to lower soil moisture content, which can also increase the erodibility of the soil (Bisal 1960; Cornelis et al. 2004).

Short-term changes in vegetation cause significant changes in the wind erodibility of the surface. For instance, Okin and Reheis (2002) and Reheis (2006) have shown that annual variation in wind erosion on a regional scale is related to variation in precipitation. There appears to be a one-year lag in this effect, with low precipitation one year resulting in significant wind erosion and dust emission the following year. This lag is hypothesized to be due to the fact that the effect of low precipitation must propagate through the system by first affecting vegetation cover. This one-year lag effect has been observed in other arid systems (Zender and Kwon 2005). In addition, dust emission from dry lakes or playas in the desert Southwest also appears to occur after years of particularly intense rainfall. This phenomenon seems to result from the increased delivery of fine-grained sediment to these playas during especially wet years or years with intense rainfall events. Anticipated climatic changes in the coming decades include both increased drought frequency and also increased precipitation intensity during rain events (IPCC 2007). Both of these effects are likely to increase wind erosion and dust emission in arid regions due to, in the first case, suppression of vegetation and, in the second case, greater water erosion resulting in increased delivery of sediment to dry lakes.

Long-term and ongoing vegetation changes in arid regions, namely the conversion of grasslands to shrublands, have dramatically increased wind erosion and dust production due to increased bare areas in shrublands compared to the grasslands they replaced. Measurements of aeolian sediment flux in the Chihuahuan Desert have shown nearly ten-fold greater rates of wind erosion and dust emission in mesquite-

Long-term and ongoing vegetation changes in arid regions, namely the conversion of grasslands to shrublands, have dramatically increased wind erosion and dust production due to increased bare areas in shrublands compared to the grasslands they replaced.

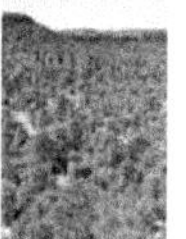

dominated shrublands compared to grasslands on similar soils (Gillette and Pitchford 2004). Large-scale conversion of grasslands to shrublands, coupled with anticipated changes in climate in the coming decades, and increases in wind speed, temperature, drought frequency, and precipitation intensity, contribute to greater wind erosion and dust emission from arid lands.

3.3.6.3 Impacts of Water and Wind Erosion

Dust can potentially influence global and regional climate by scattering and absorbing sunlight (Sokolik and Toon 1996) and affecting cloud properties (Wurzler et al. 2000), but the overall effect of mineral dusts in the atmosphere is likely to be small compared to other human impacts on the Earth's climate system (IPCC 2007). Desert dust is thought to play a major role in ocean fertilization and CO_2 uptake (Duce and Tindale 1991; Piketh et al. 2000; Jickells et al. 2005), terrestrial soil formation, and nutrient cycling (Swap et al. 1992; Wells et al. 1995; Chadwick et al. 1999), and public health (Leathers 1981; Griffin et al. 2001). In addition, desert dust deposited on downwind mountain snowpack has been shown to decrease the albedo of the snowpack, thus accelerating melt by as much as 20 days (Painter et al. 2007).

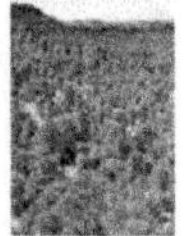

In arid regions, erosion has been shown to increase sediment delivery to large rivers (e.g., the Rio Grande), and can change the flow conditions of those rivers (Jepsen et al. 2003). Transport of eroded sediment to streams can change conditions in waterways, impacting water quality, riparian vegetation, and water fauna (Cowley 2006).

3.3.7 Indicators and Observing Systems

3.3.7.1 Existing Observing Systems

A summary of arid land sites with inventory and monitoring programs is given in Table 3.2. Data from such sites will be important for helping track the consequences of climate change, but unfortunately, most sites do not have this as an explicit part of their mission. Furthermore, there is virtually no coordination among these sites with respect to the variables being monitored, the processes being studied, the methodologies being used or the spatial and temporal scales over which change is occurring. Lack of coordination and standardization across these existing sites, programs and networks constitutes a missed opportunity.

Repeat photography is a valuable tool for documenting changes in vegetation and erosion. Hart and Laycock (1996) present a bibliography listing 175 publications using repeat photography and information on the ecosystems photographed, where they are located, number of photographs, and dates when the photographs were taken. More recent publications have added to this list (e.g., Webb 1996; McClaran 2003; Webb et al. 2007), and Hall (2002) has published a handbook of procedures. Time-series aerial photographs dating back to the 1930s and 1940s are also a useful source for quantifying landscape-scale changes in land cover (e.g., Archer 1996; Asner et al. 2003b; Bestelmeyer et al. 2006; Browning et al. 2008).

3.3.7.2 Observing Systems Required for Detecting Climate Change Impacts

While the deserts of North America have been the site of many important ecological studies, there have been relatively few long-term monitoring sites at an appropriate spatial representation that allow us the means to access changes in ecosystem structure and function in response to global change. Coordinated measurements of plant community composition in plots across the North American deserts would enhance our ability to detect change and relate that to aspects of climate. Several important data sets stand as benchmarks – the long-term photographic record at the Santa Rita Experimental Range, the long-term vegetation maps and livestock management records at the Jornada Experimental Range, the long-term perennial plant and winter annual plant studies at Tumamoc Hill, the long-term data collected from large-scale ecosystem manipulations at Portal Arizona, and the new Mojave Desert Climate Change Program. Greater spatial representation of such efforts is important in future assessment of change in these biomes.

Table 3.2 Arid land sites with research and monitoring systems.

Monitoring system	Site	Location	Source
Free-Air CO_2 Enrichment (FACE) Site	Nevada Desert FACE	Nevada Test Site, NV	http://www.unlv.edu/Climate_Change_Research
International Biome Project (IBP) Sites	Rock Valley	Nevada Test Site, NV	archived at University of California, Los Angeles, CA
	Silverbell	Arva Valley, AZ	archived at University of Arizona, Tucson, AZ
Land-Surface Flux Assessment Sites	Audubon Ranch	Ameriflux Sites	http:/public.ornl.gov/ameriflux/
	USDA Agricultural Research Service Flux Tower Network		http://edcintl.cr.usgs.gov/carbon_cycle/FluxesReseachActivities.html
	Semi-arid Ecohydrology Array (SECA)		http://eebweb.arizona.edu/faculty/huxman/seca/
Research Sites (some with long-term data sets)	Desert Experimental Range	Pine Valley, UT	http://www.fs.fed.us/rmrs/experimental-forests/desert-experimental-range/
	V Bar V Ranch	Rimrock, AZ	http://ag.arizona.edu/aes/vbarv/
	Valles Caldera National Preserve	Jemez Springs, NM	http://www.vallescaldera.gov/
	Sweeny Granite Mountains Desert Research Center	Kelso, CA	http://nrs.ucop.edu/Sweeney-Granite.htm
	Boyd Deep Canyon Desert Research Center	Palm Desert, CA	http://deepcanyon.ucnrs.org/
	Great Basin Experimental Range	Manti-LaSal National Forest, UT	http://www.fs.fed.us/rm/provo/great_basin/great_basin.shtml
	Indio Mountains Research Station	Van Horn, TX	http://www.utep.edu/indio/
	Jornada Experimental Range	Jornada Basin, NM	http://usda-ars.nmsu.edu/
	USA National Phenology Network		http://www.usanpn.org
	The Portal Project	Portal, AZ	http://biology.unm.edu/jhbrown/Portal-LTREB/PortalFront.htm
	Reynolds Creek Experimental Watershed	Boise, ID	http://ars.usda.gov/main/site_main.htm?modecode=53-42-45-00
	San Juaquin Experimental Range	O'Neals, CA	http://www.fs.fed.us/psw/ef/san_joaquin
	Santa Rita Experimental Range	Green Valley, AZ	http://ag.arizona.edu/SRER/
	UA Desert Laboratory at Tumamoc Hill Permanent Plots	Tucson, AZ	http://www.paztcn.wr.usgs.gov/home.html
	U.S. Sheep Experiment Station	Snake River Plain, ID	http://ars.usda.gov/main/site_main.htm?modecode=53-64-00-00
	Walnut Gulch Experimental Watershed	Tombstone, AZ	http://ars.usda.gov/PandP/docs.htm?docid=10978&page=2
Long-Term Ecological Research (LTER) Sites	Jornada Basin	Las Cruces, NM	http://jornada-www.nmsu.edu/
	Sevilleta	Albuquerque, NM	http://sev.lternet.edu/
	Central Arizona-Phoenix	Phoenix, AZ	http://caplter.asu.edu/
National Ecological Observatory Network (NEON)	Santa Rita Experimental Range	Tucson, AZ	http://www.sahra.arizona.edu/santarita/
	Onaqui-Benmore	Salt Lake City, UT	http://www.neoninc.org
National Park Service Inventory & Monitoring Program	The NPS has recently initiated Inventory & Monitoring programs at many of its Parks and Monuments in arid lands		http://science.nature.nps.gov
TRENDS Project	Synthesis of long-term data from 44 research sites		http://fire.lternet.edu/Trends/
Rainfall Manipulations	USDA Agricultural Research Service Rainout Shelter	Burns, OR	Svejcar et al. 2003
	Nevada Global Change Experiment	Nevada Test Site, NV	http://www.unlv.edu/Climate_Change_Research/

BOX 2: Ecosystem "tipping points."

There is widespread recognition that ecosystems may exist in "alternate states" (e.g., perennial grassland state vs. annual grassland state; grassland state vs. shrubland state). Within a given state, ecosystems may tolerate a range of climate variability, stress and disturbance and exhibit fluctuation in structure (e.g., species composition) and function (e.g., rates of primary production and erosion). However, there may be "tipping points" that occur where certain levels of stress, resource availability, or disturbance are exceeded, causing the system transition to an alternate state (Archer and Stokes 2000). Thus, change in ecosystem structure and function in response to changes in stress levels or disturbance regimes may be gradual and linear up to a certain point(s), and then change dramatically and profoundly. Once in an alternate state, plant cover, composition and seedbanks, and soil physical properties, nutrient status and water holding capacity, etc. may have been altered to the point that it is difficult for the system to revert to its previous state even if the stresses or disturbance causing the change are relaxed.

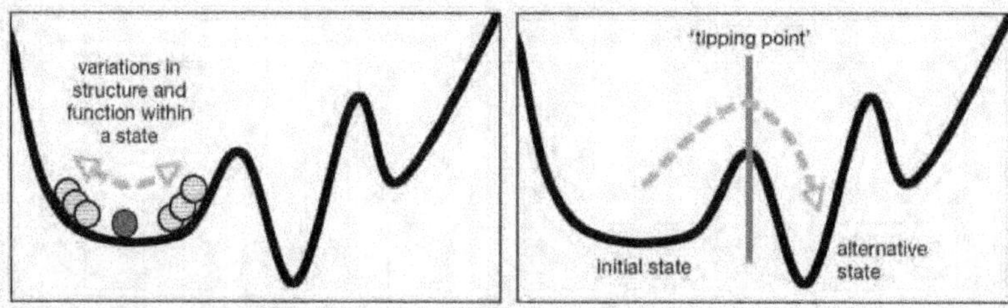

In arid lands, threshold examples include shifts from grassland states to shrubland states (Archer 1989) and desertification (Schlesinger et al. 1990). It appears that these state-transitions occur as result of various combinations of vegetation-fire, soil, hydrology, animal, and climate feedbacks (e.g., Thurow 1991; Wainwright et al. 2002; Okin et al. 2006, D'Odorico et al. 2006).

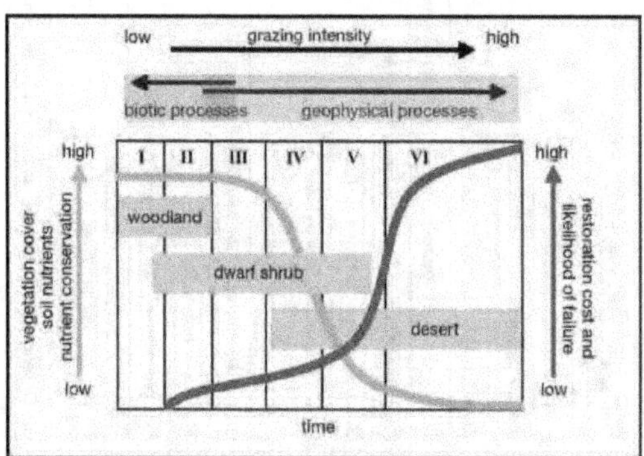

While there is substantial observational evidence for these threshold phenomena, our quantitative understanding is limited and many questions remain:

- How far, and under what circumstances, can an ecosystem be pushed before entering into an alternate state?
- What changes in ecosystem properties and feedbacks are involved in these state-transitions?
- What variables could be monitored to predict when a system is nearing a 'tipping point'?
- How do climate factors influence the risk of exceeding state-transition thresholds?

Soil moisture is a key indicator and integrator of ecological and hydrological processes. However, as noted in the Water Resources chapter (Chapter 4), there is a dearth of information on the long-term patterns and trends in this important variable. Even on well-instrumented watersheds in arid lands (e.g., Lane and Kidwell 2003; NWRC 2007; SWRC 2007) soil moisture records are only erratically collected over time and are limited in their spatial coverage and depth. Thus, there is a pressing need for a distributed network of soil moisture sensors in arid lands that would be a component of a network of monitoring precipitation, evapotranspiration, and temperature. Ideally, such a network would also include collection of plant, soil and precipitation samples for determination of the stable isotope composition of C, O, and H. Such isotope data would provide important clues regarding when and where plants were obtaining soil moisture and how primary production and water use efficiency are being affected by environmental conditions (e.g., Boutton et al. 1999; Roden et al. 2000; Williams and Ehleringer 2000).

Effects of climate change will be most easily observed in relatively few arid land springs. Springs that dry periodically are relatively poor candidates, as long periods of record will be required to determine "baseline" conditions. Similarly, springs supported by large, regional aquifers are also poor candidates, as transmissivity is low and surface discharge is primarily ancient water (Mifflin 1968; Hershey and Mizell 1995; Thomas et al. 2001; Knochenmus et al. 2007). The USGS maintains quantitative historic Web-based records of surface water discharge from springs. These records could provide a "baseline" discharge, but the effect of climate change on such springs will not be evident for decades or much, much longer. Persistent springs fed by aquifers with moderate transmissivity are good candidates to assess effects of climate change. In the arid western United States, these geologically persistent springs are characterized by crenobiontic macroinvertebrates, including aquatic insects and springsnails. They occur on bajadas, at the base of mountains, and sometimes on valley floors (Taylor 1985; Hershler and Sada 2001; Polhemus and Polhemus 2001). While discharge from these springs fluctuates, they have not dried. Transmissivity through aquifers supporting these springs is relatively high, hence their response to changes in precipitation will be relatively rapid and measurable (Plume and Carlton 1988; Thomas et al. 1996). An existing database, consisting of surveys of >2000 springs (mostly Great Basin and in the northwestern United States) over the past 15 years, includes hundreds of springs that would qualify as potential climate change monitoring sites (Sada and Hershler 2007).

Most land-surface exchange research has focused on forested systems. There is, however, a need for understanding the seasonal carbon dynamics, biomass, annual productivity, canopy structure, and water use in deserts (Asner et al. 2003a, b; Farid et al. 2006; Sims et al. 2006). Studies to date do not yet yield clear generalizations. For example, shifts from grass to shrub domination may show no net effects on evapotranspiration due to offsetting changes in radiant energy absorption and the evaporative fraction in the contrasting cover types (Kurc and Small 2004). However, this may depend upon the type of shrubs (Dugas et al. 1996). Although net changes in evapotranspiration may not occur with this land cover change, ecosystem water use efficiency may be significantly reduced (Emmerich 2007). Part of the challenge in predicting functional ecosystem dynamics in arid lands derives from our relatively poor understanding of non-equilibrium processes driven by highly episodic inputs of precipitation (Huxman et al. 2004). Part derives from the importance of the strong, two-way coupling between vegetation phenology and the water cycle, which is critical for predicting how climate variability influences surface hydrology, water resources, and ecological processes in water-limited landscapes (e.g., Scanlon et al. 2005). Shifts in phenology represent an integrated vegetation response to multiple environmental factors, and understanding of vegetation phenology is prerequisite to interannual studies and predictive modeling of land surface responses to climate change (White et al. 2005). Along these lines, the ability to detect ecosystem stress and impacts on vegetation structure will be requisite to understanding regional aspects of drought (Breshears et al. 2005) that result in substantial land use and land cover changes.

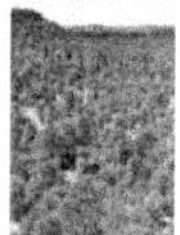

In regions where the eroded surfaces are connected to the regional hydraulic systems (i.e., not in closed basins), sediment delivery to streams and streambeds is an excellent indicator of integrated erosion in the catchment when coupled with stream gauging and precipitation data. USGS gauges are few and far between in arid lands and many have been or are being decommissioned due to lack of funds (as is also the case for watersheds on U.S. Forest Service lands). There is currently no integrated monitoring system in place for the measurement of bedload, but the USGS National Water Information System does collect water quality data that could inform sediment loads. Unfortunately, there are very few sites in the arid United States that are monitored continuously. Additional arid region rivers could be instrumented and sampled to provide further monitoring of stream flow as well as water erosion. In closed basins, or the upland portion of open basins, the development and expansion of rills and gullies is the clearest indicator of water erosion. There is no system in place for the monitoring of these features (Ries and Marzolff 2003), but high-resolution remote sensing (~1-meter resolution) might be used to monitor the largest of these features.

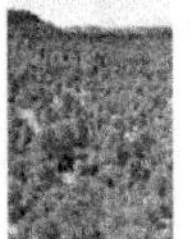

The most important indicator of wind erosion is the dust that it produces. Because dust is transported long distances, even a sparse network of monitoring sites can identify dust outbreaks. For instance, Okin and Reheis (2002) have used meteorological data collected as part of the National Climatic Data Center's network of cooperative meteorological stations (the COOP network) to identify dust events and to correlate them to other meteorological variables. The expansion of this network to include observations in more locations, and especially at locations downwind of areas of concern, would be a significant improvement to monitoring wind in the arid portions of the United States. This existing observation network might also be integrated with data from NASA's Aerosol Robotic Network (AERONET) on aerosol optical depth and radar or lidar systems deployed throughout the region, but particularly near urban centers and airports. In addition, there are several remote sensing techniques that can be used to identify the spatial extent and timing of dust outbreaks (Chomette et al. 1999; Chavez et al. 2002; Miller 2003), though there is no system in place to integrate or track the evolution of dust sources through time.

Novel communities (with a composition unlike any found today) have occurred in the late-glacial past and will develop in the greenhouse world of the future (Williams and Jackson 2007). Most ecological models are at least partially parameterized from modern observations and so may fail to accurately predict ecological responses to novel climates occurring in conjunction with direct plant responses to elevated atmospheric CO_2 and nitrogen deposition. There is a need to test the robustness of ecological models to conditions outside modern experience.

3.4 FINDINGS AND CONCLUSIONS

3.4.1 Forests

Climate strongly influences forest productivity, species composition, and the frequency and magnitude of disturbances that impact forests. The effect of climate change on disturbances such as forest fire, insect outbreaks, storms, and severe drought will command public attention and place increasing demands on management resources. Other effects, such as increases in temperature, the length of the growth season, CO_2, and nitrogen deposition may be more incremental and subtle, but may have equally dramatic long-term effects.

Climate change has very likely increased the size and number of forest fires, insect outbreaks, and tree mortality in the interior west, the Southwest, and Alaska, and will continue to do so. An increased frequency of disturbance is at least as important to ecosystem function as incremental changes in temperature, precipitation, atmospheric CO_2, nitrogen deposition, and ozone pollution. Disturbances partially or completely change forest ecosystem structure and species composition, cause short-term productivity and carbon storage loss, allow better opportunities for invasive alien species to become established, and command more public and management attention and resources.

Rising CO_2 will very likely increase photosynthesis for forests, but the increased photosynthesis will likely only increase wood production in young forests on fertile soils. Where nutrients are not limiting, rising CO_2 increases photosynthesis and wood production. But on infertile soils the extra carbon from increased photosynthesis will be quickly respired. The response of older forests to CO_2 is uncertain, but possibly will be lower than the average of the studied younger forests.

Nitrogen deposition and warmer temperatures have very likely increased forest growth where water is not limiting and will continue to do so in the near future. Nitrogen deposition has likely increased forest growth rates over large areas, and interacts positively to enhance the forest growth response to increasing CO_2. These effects are expected to continue in the future as N deposition and rising CO_2 continue.

The combined effects of expected increased temperature, CO_2, nitrogen deposition, ozone, and forest disturbance on soil processes and soil carbon storage remain unclear. Soils hold an important, long-term store of carbon and nutrients, but change slowly. Long-term experiments are needed to identify the controlling processes to inform ecosystem models.

3.4.2 Arid Lands

Disturbance and land use on arid lands will control their response to climate change. Many plants and animals in arid ecosystems are near their physiological limits for tolerating temperature and water stress. Thus, even slight changes in stress will have significant consequences. Projected climate changes will increase the sensitivity of arid lands to disturbances such as grazing and fire. Invasion of non-native grasses will increase fire frequency. In the near-term, fire effects will trump climate effects on ecosystem structure and function. These factors cause important changes themselves, but the outcome of their interactions are difficult to predict in the context of increased concentrations of atmospheric CO_2 and nitrogen deposition. This is particularly so because these interactions represent novel combinations.

Higher temperatures, increased drought and more intense thunderstorms will very likely increase erosion and promote invasion of exotic grass species. Climate change will create physical conditions conducive to wildfire, and the proliferation of exotic grasses will provide fuel, thus causing fire frequencies to increase in a self-reinforcing fashion (Figure 3.17). In arid regions where ecosystems have not coevolved with a fire cycle, the probability of loss of iconic, charismatic mega flora such as saguaro cacti and Joshua trees will be greatly increased.

Arid lands very likely do not offer a large capacity to serve as a "sink" for atmospheric CO_2 and will likely lose carbon as climate-induced disturbance increases. Climate-induced changes in vegetation cover and erosion will reduce the availability of nitrogen in dryland soils, which (after water) is an important control of primary productivity and carbon cycling.

Arid land river and riparian ecosystems will very likely be negatively impacted by decreased streamflow, increased water removal, and greater competition from non-native species. Dust deposition on alpine snow pack will accelerate the spring delivery of montane water sources and potentially contribute

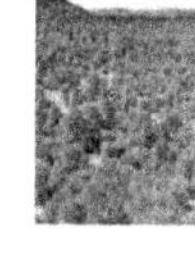

Figure 3.17 *Mojave Desert scrub near Las Vegas, NV (foreground); and area invaded by the exotic annual grass red brome (background) following a fire that carried from desert floor upslope into pinyon-juniper woodlands. Photo: T.E. Huxman.*

to earlier seasonal drought conditions in lower stream reaches. Riparian ecosystems will likely contract, and in the remainder, aquatic ecosystems will be less tolerant of stress. The combination of increased droughts and floods, land use and land cover change, and human water demand will amplify these impacts and promote sedimentation.

Changes in temperature and precipitation will very likely decrease the cover of vegetation that protects the ground surface from wind and water erosion. More intense droughts and floods will accelerate fluvial erosion and higher frequencies of dust storms. Higher intensity rainfall will result in greater sheet erosion. All of these factors will periodically increase the sediment load in water and the atmosphere and decrease air and water quality.

3.4.3 Observing Systems for Forests and Arid Lands

Current observing systems are very likely inadequate to separate the effects of changes in climate from other effects. The major findings in the Land Resources Chapter relied on publications that used data assembled from diverse sources, generally for that specific study. In many cases, finding, standardizing, and assembling the data was the primary task in these studies. This was particularly the case for studies relating climate and disturbance. Findings on the effects of CO_2 and nitrogen deposition were largely based on short-term, small-scale experimental manipulations. Those for the interaction of climate and invasive species with vegetation, riparian ecosystems, and erosion generally came from long-term monitoring and survey studies. The NOAA weather network was invaluable for climate information, but most studies needed to extrapolate weather data to create uniform coverage across the United States. This was and remains a considerable task, and is particularly problematic in arid lands where precipitation is highly localized and varies significantly across short distances. Separating the effects of climate change from other impacts would require a broad network of indicators, coupled with a network of controlled experimental manipulations.

There is no coordinated national network for monitoring changes associated with disturbance and land cover and land use change. Because of the spatial heterogeneity of insect outbreaks and other disturbances, new sampling and monitoring approaches are needed to provide a comprehensive assessment of how climate is affecting the disturbance regime of forest ecosystems and changes in forest soils.

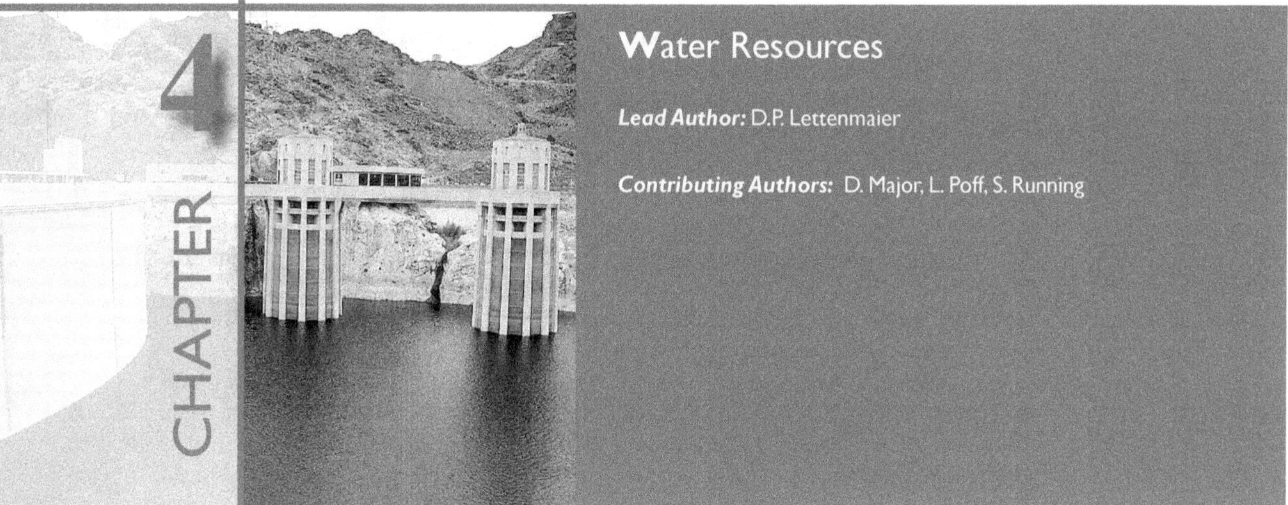

Water Resources

Lead Author: D.P. Lettenmaier

Contributing Authors: D. Major, L. Poff, S. Running

4.1 INTRODUCTION

This synthesis and assessment report builds on an extensive scientific literature and series of recent assessments of the historical and potential impacts of climate change and climate variability on managed and unmanaged ecosystems and their constituent biota and processes. It identifies changes in resource conditions that are now being observed, and examines whether these changes can be attributed in whole or part to climate change. It also highlights changes in resource conditions that recent scientific studies suggest are most likely to occur in response to climate change, and when and where to look for these changes. As outlined in the Climate Change Science Program (CCSP) Synthesis and Assessment Product 4.3 (SAP 4.3) prospectus, this chapter will specifically address climate-related issues in freshwater supply and quality. In this chapter the focus is on the near-term future. In some cases, key results are reported out to 100 years to provide a larger context, but the emphasis is on the next 25-50 years. This nearer-term focus is chosen for two reasons. First, for many natural resources, planning and management activities already address these time scales through development of long-lived infrastructure, forest or crop rotations, and other significant investments. Second, climate projections are relatively certain over the next few decades. Emission scenarios for the next few decades do not diverge from each other significantly because of the "inertia" of the energy system. Most projections of greenhouse gas emissions assume that it will take decades to make major changes in the energy infrastructure, and only begin to diverge rapidly after several decades have passed (30-50 years).

Water is essential to life and is central to society's welfare and to sustainable economic growth. Plants, animals, natural and managed ecosystems, and human settlements are sensitive to variations in the storage, fluxes, and quality of water at the land surface – notably storage in soil moisture and groundwater, snow, and surface water in lakes, wetlands, and reservoirs, and precipitation, runoff, and evaporative fluxes to and from the land surface, respectively. These, in turn, are sensitive to climate change.

Water is essential to life and is central to society's welfare and to sustainable economic growth.

Water managers have long understood the implications of variability in surface water supplies at time scales ranging from days to months and years on the reliability of water resource systems, and many sophisticated methods (e.g. Jain and Singh, 2003) have been developed to simulate and respond to such variability in water resource system design and operation. The distinguishing feature of all such methods, however, is that they assume that an observed record of streamflow, on which planning is based, is statistically stationary – that is, the probability distribution(s) from which the observations are drawn does not change with time. As noted by

Arnell (2002), Lettenmaier (2003), and NRC (1998), in the era of climate change this assumption is no longer tenable. In this vein, Milly et al. (2008) argue that "stationarity is dead," and advocate the urgent need for a major new initiative at the level of the Harvard Water Program of the 1960s (Maass et al.1962) to develop more applicable methods for water planning as climate changes. These new paradigms would provide the basis for assessing plausible ranges of future conditions for purposes of hydrologic design and operation. Such assessments are also needed to understand how changes in the availability and quality of water will affect animals, plants, and ecosystems.

This chapter briefly reviews the current status of U.S. water resources, both in terms of characteristics of the physical system(s), trends in water use, and observed space-time variability in the recent past. It then examines changes to the natural hydrologic systems (primarily stream flow, but also evapotranspiration and snow water storage) over recent decades for six regions of the United States (the West, Central, Northeast, and South and Southeast, as well as Alaska and Hawaii, which are defined as aggregates of U.S. Geological Survey (USGS) hydrologic regions). Finally, recent studies based on climate model projections archived for the 2007 IPCC report, which project the implications of climate change for these six major U.S. regions, are reviewed.

4.1.1 Hydroclimatic Variability in the United States

The primary driver of the land surface hydrologic system is precipitation. Figure 4.1 shows variations in mean annual precipitation and its variability (expressed as the coefficient of variation, defined as the standard deviation divided by the mean) across the continental United States. The semi-humid conditions of the eastern United States yield to drier conditions to the west, with the increasing dryness eventually interrupted by the Rocky Mountains. The driest climates, however, exist in the Intermountain West and the Southwest, which give way as one proceeds west and north to more humid conditions on

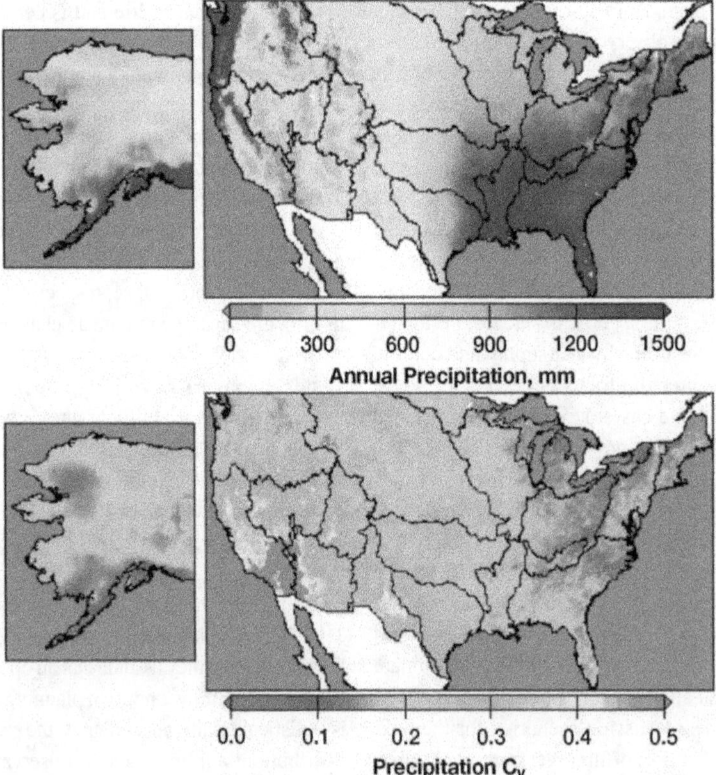

Figure 4.1 *Mean and coefficient of variation of annual precipitation in the continental U.S. and Alaska. Data replotted from Maurer et al. (2002).*

the upslope areas of the Cascade and coastal mountain ranges, especially in the Pacific Northwest. The bottom panel of Figure 4.1, which shows the coefficient of variation of precipitation, indicates that precipitation variability generally is lowest in the humid areas, and highest in the arid and semi-arid West, with a tendency toward lower variability in the Pacific Northwest, which is more similar to that of the East than the rest of the West.

Figure 4.2 (upper panel) shows that runoff patterns, for the most part, follow those of precipitation. The runoff ratio (annual runoff divided by annual precipitation; second panel in Figure 4.2) generally decreases from east to west, but the decline in runoff from east to west is sharper than it is for precipitation. The runoff ratio increases in headwaters regions of the mountainous source areas of the West, and more generally in the Pacific Northwest. This increase in runoff ratio with elevation is critical to the hydrology of the West, where a large fraction of runoff originates in a relatively small fraction of the area – much more so than in the semi-humid East and Southeast, where runoff generation is relatively uniform spatially. The bottom panel in Figure 4.2 shows the ratio of maximum annual snow accumulation to annual runoff, and can be considered an index to the relative fraction of runoff that is derived from snowmelt. This panel emphasizes the critical role of snow processes to the hydrology of the western United States, and to a more limited extent, in the northern tier of states.

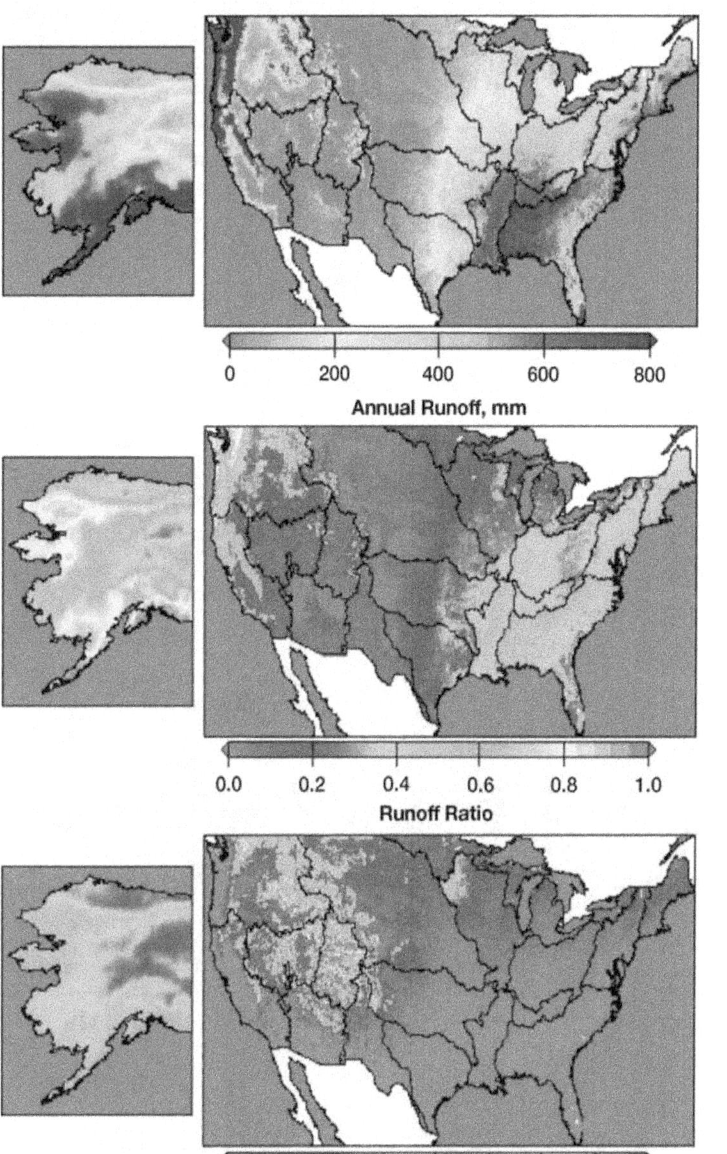

Figure 4.2 *Mean annual runoff, runoff ratio (annual mean runoff divided by annual mean precipitation), and ratio of maximum mean snow accumulation to mean annual runoff in the continental U.S. and Alaska. Data replotted from Maurer et al. (2002).*

Figure 4.3 shows two key aspects of runoff variability – the coefficient of variation of annual runoff, a measure of its variability, and its persistence in time (the latter expressed as the lag one correlation coefficient). The coefficient of variation of annual runoff generally follows that of precipitation; however, it is higher for the most part as the hydrologic system tends to amplify variability. Annual runoff persistence is generally low, but tends to be higher in the East

(and generally in more humid areas) than in the western United States. The differences between regions are, however, slight, and Vogel et al. (1998) argue that most of the United States can be considered to be a "homogeneous region" in terms of runoff persistence. It is nonetheless interesting that there is a general gradient downward in runoff persistence from east to west, which appears not to be entirely related to precipitation as the trend is not reversed in the generally more humid areas of the northwest and Pacific Coast regions.

4.1.2 Characteristics of Managed Water Resources in the United States

The water resources of the continental United States are heavily managed, mostly by surface water reservoirs. During the period from about 1930 through 1980, dams were constructed at most technically feasible locations, with the result that aside from headwater regions, the flow of most rivers, especially in the western United States, has been heavily altered by reservoir management. Figure 4.4 (modified from Graf 1999) shows the extent of reservoir storage across the continental United States. From the standpoint of water management, the lower panel in Figure 4.4, which shows variations in the ratio of reservoir storage to mean annual flow, is most relevant. Although the figure scale is in terms of quartiles, the lowest quartile has storage divided by mean annual runoff ratios in the range 0.25-0.36, and the upper quartiles 2.18-3.83 (see Graf 1999; Table 4.1). A storage to runoff ratio of one is usually taken as the threshold between reservoirs that are primarily used to shape within-year variations in runoff (small storage to runoff ratios; orange colors in Figure 4.4, lower panel) and those that are primarily used to smooth interannual variations in runoff (large storage to runoff ratios; dark blue in Figure 4.4, lower panel). In subsequent sections, these differences in storage capacity, coupled with the characteristics of the hydrologic systems, are defined as critical to the sensitivity of water resources to climate change.

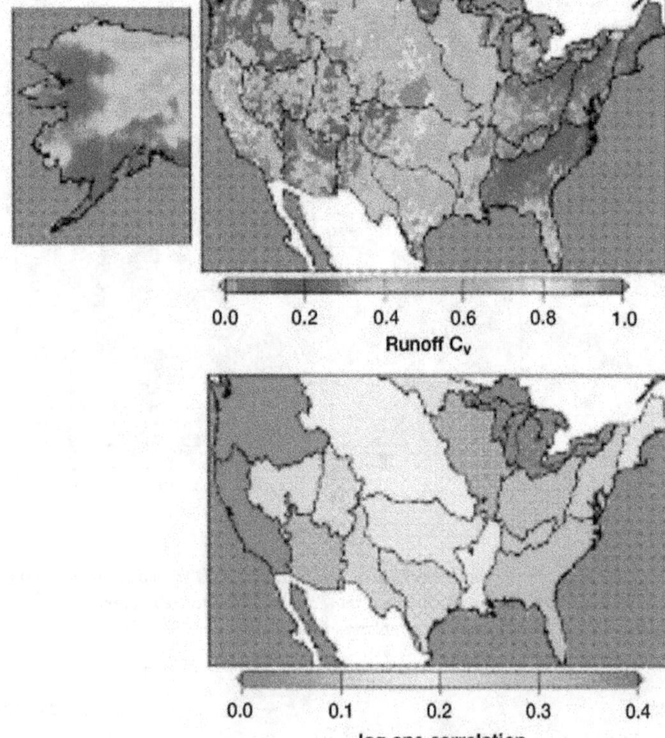

Figure 4.3 *Coefficient of variation of annual runoff (upper panel) and lag one correlation of annual runoff (lower panel). Upper panel replotted from Maurer et al. (2002); lower panel from Vogel et al. (1998).*

4.1.3 U.S. Water Use and Water Use Trends

The USGS compiles, at five-year intervals, information about the use of water in the United States. The most recent publication (Hutson et al. 2004) is for the period through 2000. The update to this publication, through 2005, unfortunately was not available as of the time of this writing. The data compiled by the USGS are somewhat limited in that they are for water withdrawals, rather than consumptive use. The distinction is important, as one of the largest uses of water is for cooling of thermoelectric power plants, and much of that water is returned to the streams from which it is withdrawn (use of water for hydroelectric power generation, virtually none of which is consumptively used, is not included in this category). On the other hand, a much higher fraction of the water withdrawn for irrigation is consumptively used.

Despite these limitations, the two key figures in the 2004 USGS publication, reproduced here as Figure 4.4, are instructive in that they further define the trends noted by Gleick et al. (2000) that U.S. water withdrawals have decreased slightly over the last 20 years in virtually all categories, and appear to have stabilized since about 1985. This is despite substantial population growth during the same period (Figure 4.4, upper panel).

These changes, which follow a 30-year period of rapid growth in water withdrawals, have occurred for somewhat different, but related reasons. Water withdrawals from many streams are now limited, particularly during periods of low flow, by environmental regulations. Furthermore, economic considerations have driven more efficient use of water. In the case of irrigation, there has been a transition from flood to sprinkler irrigation, and (albeit in a much smaller number of cases) much more efficient drip irrigation. Irrigation water use has also been affected by economic considerations, such as the cost of electric power to pump irrigation water.

Industrial water use efficiency gains have been driven by pollution control regulations, which encourage reduction of wastewater discharge, and hence more recycling. Municipal water use reductions have been driven by improved efficiency of in-house appliances and plumbing fixtures, as well as trends to higher density housing, which reduces use of water for landscape irrigation. Economic considerations have also had an effect on municipal water use, especially in municipalities where the cost of wastewater treatment is linked to water use. The combined result, as shown in Figure 4.5, is that total U.S. water withdrawals have been stable, which implies that per capita water use has declined.
Comparison of U.S. per capita water use (see Gleick 1996) globally shows that U.S. water use is much higher than elsewhere, even compared to other industrialized parts of the world such as Europe. It seems reasonable then to assume that this overall trend toward reduced per capita use of water will continue, at least over the next decade or two – notwithstanding that the Hutson et al (2004) trends are for the continental U.S. (including Hawaii in some cases) and are not disaggregated spatially, hence regional trends, past and future, may well differ.

4.2 OBSERVED CHANGES IN U.S. WATER RESOURCES

In this section observed trends in U.S. water resources – both physical aspects, and water quality – are reviewed. In general, much more work has been done evaluating trends in physical aspects of the land surface hydrologic cycle than for water quality, and more attention has been focused on the western United States than elsewhere. For this reason, studies of physical aspects are reviewed by region, but water quality is in aggregate.

4.2.1 Observed Streamflow Trends

The most comprehensive study of trends in U.S. streamflow to date is reported by Lins and

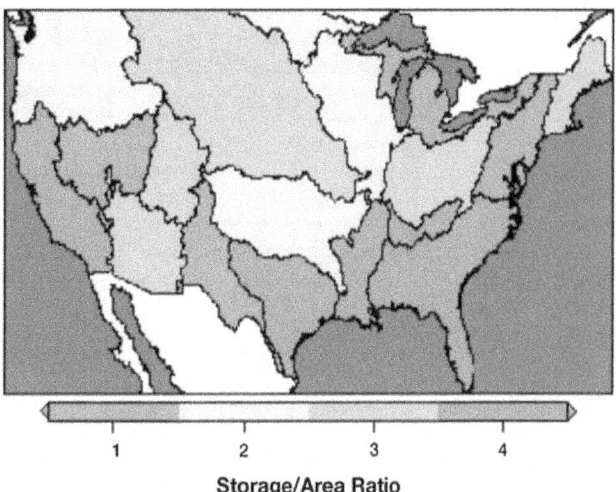

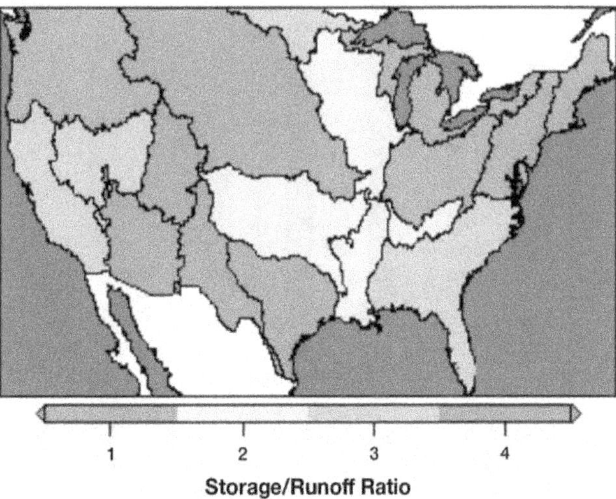

Figure 4.4 *Reservoir storage in the continental U.S. per unit area (upper panel) and storage/runoff ratio (lower panel). Colors are for four quartiles of cumulative probability distribution. Replotted from Graf (1999).*

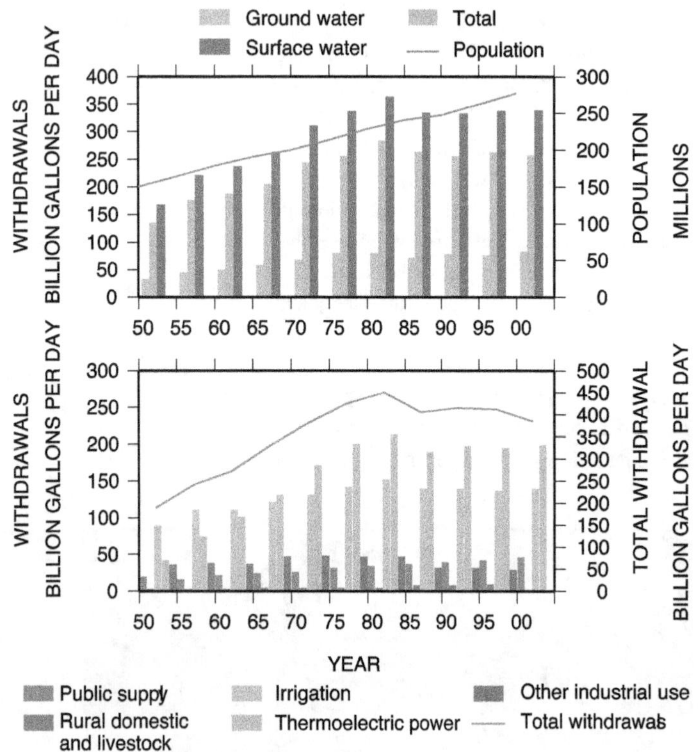

Figure 4.5 *Trends in U.S. water withdrawals, 1950-2000. Upper panel: trends in population, groundwater, and surface water withdrawals. Lower panel: withdrawals by sector. Figure from Hutson et al., 2004.*

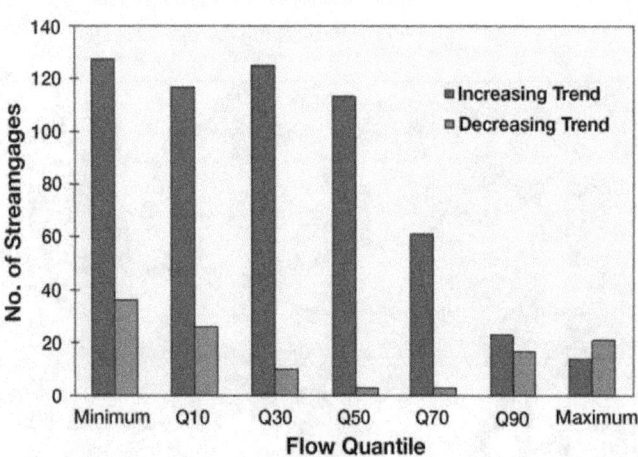

Figure 4.6 *Number of increasing and decreasing trends in continental U.S. streamflow records for a range of flow quartiles. From Lins and Slack (1999).*

Slack (1999; 2005). It follows an earlier study by Lettenmaier et al. (1994) that dealt also with precipitation and temperature, but in less detail with streamflow. Given that the Lins and Slack study concentrates more directly on streamflow, and is somewhat more current, it is the focus of the chapter. Although the methodologies, record lengths, and locations differ somewhat for the two studies, to the extent that the results can be compared, they are generally consistent.

Lins and Slack (1999) analyzed long-term streamflow records for a set of 395 stations across the continental United States for which upstream effects of water management were minimal, and which had continuous (daily) records for the period 1944-93 (updated to 435 stations for the period 1940-99 by Lins and Slack (2005)). For each station, they formed time series of minimum and maximum flows, as well as flows at the 10th, 30th, 50th, 70th, and 90th percentiles of the flow duration curve. They found, consistent with Lettenmaier et al. (1994), that there was a preponderance of upward streamflow trends (many more than would be expected due to chance) in all but the highest flow categories (see Figure 4.6), for which the number of upward and downward trends was about the same. In addition to the 50-year period of 1944-93, similar analyses were conducted for the smaller number of stations having 60, 70, and 80 years of record (all ending in 1993), and the fraction of upward and downward trends was about the same as for the analysis of the larger number of stations with at least 50 years of record.

Lins and Slack (2005) update the analysis to a "standard" 60-year period, 1940-99, but unlike their earlier paper, do not consider longer periods with smaller numbers of stations. Neither the 1999 nor the 2005 papers attempt to attribute the observed trends to climatic warming, although the spatial coherence in the trends suggest that non-climatic causes (e.g., land cover change), are not likely the cause. However, as noted in Cohn and Lins (2006), hydroclimatic records by nature reflect long term persistence associated with climate variability over a range of temporal scales, as well as low frequency effects associated with land processes, so the mere existence of trends in and of itself does not necessarily imply a causal link with climate change. Summaries of the Lins and Slack results are shown in Figure 4.7a-c, which plots the location and strength (as significance level) of trends at a subset of USGS Hydroclimatic Data Network (HCDN) stations with the longest records (note that in Figure 4.7 green indicates no significant trend at the 0.05 significance level).

Mauget (2003) used a method based on running time windows of length 6-30 years applied to streamflow records for the 1939-98 period extracted from the same USGS HCDN used by Lins and Slack (1999). The Mauget (2003) analysis was based only on the 167 stations for which data were available for the period 1939-98, and hence make up a somewhat different station set than was used by Lins and Slack. (It is worth noting that many of the stations used in the Mauget et al. study are likely the same as those used by Lins and Slack in their 60-year (1934-93) set of 193 stations. It should also be noted that the Mauget study is based on mean annual flow, while Lins and Slack use percentiles of the annual flow distribution, including the median.) The results of the Mauget et al. (2003) study are broadly similar to Lins and Slack (1999) to the extent that comparisons are possible. Mauget finds evidence of high streamflows being more likely toward the end of the record than the beginning in the eastern United States, especially in the 1970s, and "a coherent pattern of high-ranked annual flow … beginning during the later 1960s and early 1970s, and ending in either 1997 or 1998." By contrast, he found a more or less reverse pattern in the western United States, with an onset of dry conditions beginning in the 1980s.

4.2.2 Evaporation Trends

Several studies have been performed to assess changes in evapotranspiration (ET), another major influence on the land surface water balance. Unfortunately, there are no long-term ET observations. Methods that enable direct measurements (e.g., via eddy flux methods) have only been available for about 20 years, and are still used primarily in intensive research settings rather than for assessing long-term trends. Another source of evaporation data is records from evaporation pans, which are generally located in agricultural areas and have been used as an index to potential evaporation. These records are generally longer and a number (several

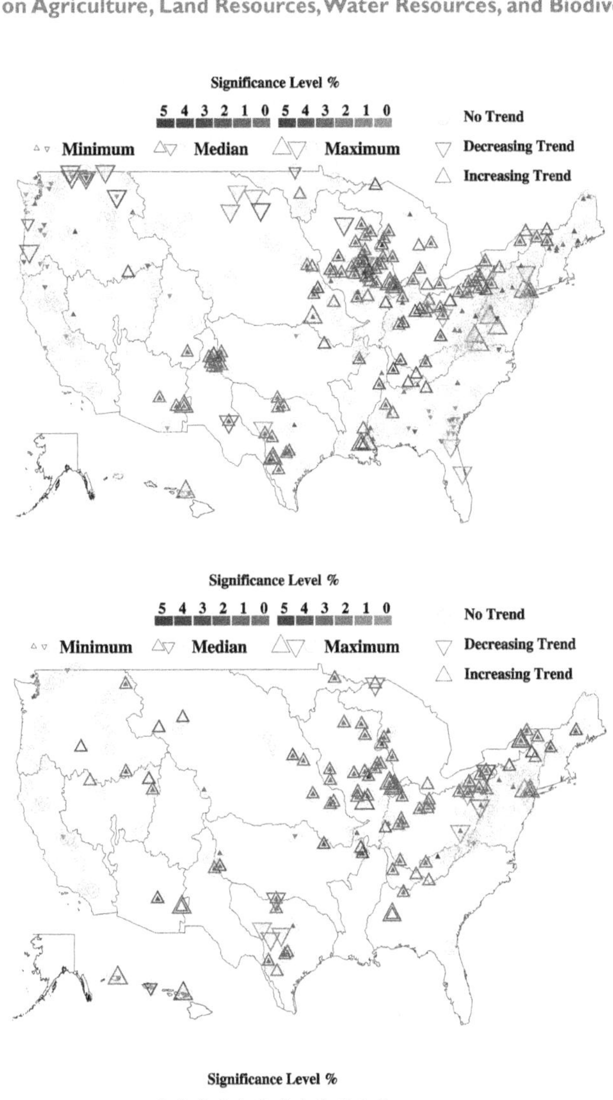

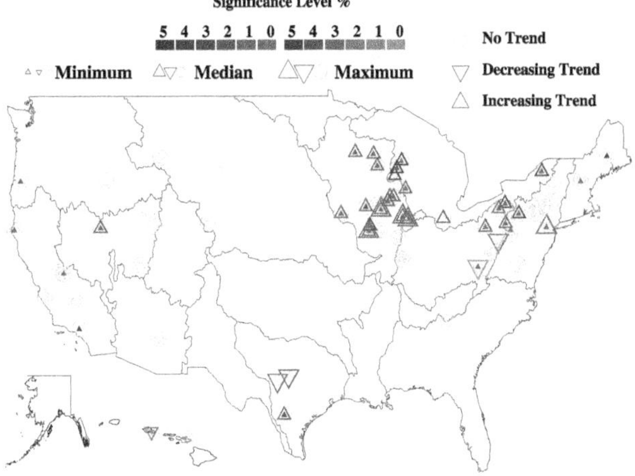

Figure 4.7 *Statistically significant trends in streamflow across the continental U.S. At each station location, direction of trend and significance level (if statistically significant at less than 0.05 level) are plotted for minimum, median, and maximum of the annual flows. Upper panel: 393 stations at which data were available from 1944-93; middle panel: same for 1934-93; lower panel: same for 1924-93. Data replotted from Lins and Slack (1999).*

hundred over the continental United States) have record lengths approaching 50 years. Several studies (e.g., Peterson et al. 1995; Golubev et al. 2001) have shown that pan evaporation records over the United States generally had downward trends over the second half of the 20th century. This is contrary to the expectation that a generally warming climate would increase evapotranspiration.

Two explanations have been advanced to account for this trend. The first is the so-called evaporation paradox (Brutsaert and Parlange, 1998), which holds that increasing evaporation alters the humidity regime surrounding evaporation pans, causing the air over the pan to be cooler and more humid. This "complementary hypothesis" suggests that trends in pan and actual evaporation should indeed be in the opposite direction. Observational evidence, using U.S. pan evaporation data and basin-scale actual evaporation, inferred by differencing annual precipitation and runoff, suggests that trends in U.S. pan and actual evaporation have in fact been in opposite directions (Hobbins et al. 2004).

The second hypothesis is that actual ET may also have declined due to reduced net radiation, resulting from increased cloud cover (Huntington et al. 2004). This hypothesis is consistent with observed downward trends in the daily temperature range (daily minimum temperatures have generally increased over the last 50 years, while daily maxima have increased more slowly, if at all); the temperature range is generally related to downward solar radiation, which would therefore have decreased. Unfortunately, as with actual evaporation, long-term records of surface solar radiation are virtually nonexistent, so indirect estimates (such as cloud cover, or daily temperature range) must be relied on. Roderick and Farquahr (2002) argue that decreasing net solar irradiance resulting from increased cloud cover and aerosol concentrations is a more likely cause for the observed changes, and that actual evaporation should generally have decreased, consistent with the pan evaporation trends.

Brutsaert (2006) argues that "the significance of this negative trend [in pan evaporation], as regards terrestrial evaporation, is still somewhat controversial, and its implications for the global hydrologic cycle remain unclear." The controversy stems from the apparently contradictory views that the observed changes result either from global radiative dimming, or from the complementary relationship between pan and terrestrial evaporation. Brutsaert (2006) argues that these factors are in fact not mutually exclusive, but act concurrently. He derives a theoretical relationship between trends in actual evaporation, net radiation, surface air temperature, and pan evaporation, and shows that the observed trends are generally consistent, accounting for the generally observed downward trend in net radiation ("global dimming") albeit from sparse observations.

4.2.3 U.S. Drought Trends

Andreadis and Lettenmaier (2006) investigated trends in droughts in the continental United States using a method that combined long-term observations with a land surface model. Their approach was to use gridded observations of precipitation and temperature that were adjusted to have essentially the same decadal variability as the Historical Climatology Network (HCN) stations, which have been carefully quality-controlled for changes in observing methods. These are used to force a land surface model, and then used it to evaluate trends in several drought characteristics in both model-derived soil moisture and runoff. Results show that the spatial character of trends in the model-derived runoff is in general consistent with the observed streamflow trends from Lins and Slack (1999). Andreadis and Lettenmaier also show that, generally, the continental United States became wetter over the period analyzed (1915-2003), which was reflected in upward trends in soil moisture and downward trends in drought severity and duration. However, there was some evidence of increased drought severity and duration in the western and southwestern United States. This was interpreted as increased actual evaporation dominating the trend toward increased soil wetness, which was evident through the rest of the United States.

There is evidence that much more severe droughts have occurred in North America prior to the instrumental record of roughly the last 100 years. For instance, Woodhouse and Overpeck (1998), using paleo indicators (primarily tree rings), find that many droughts over the last 2,000 years have eclipsed the major U.S.

droughts of the 1930s and 1950s, with much more severe droughts occurring as recently as the 1600s. Although the nature of future drought stress remains unclear, for those areas where climate models suggest drying, such as the Southwest (e.g., Seager et al. 2007), droughts more severe than those encountered in the instrumental record may become increasingly likely.

4.2.4 Regional Assessment of Changes in U.S. Water Resources

For purposes of this section, the continental United States is partitioned into four "super-regions" using aggregations of the USGS hydrologic regions chosen on the basis of hydro-climatic similarity (Figure 4.8) as follows: West (Pacific Northwest, California, Great Basin, Upper Colorado, Lower Colorado, Rio Grande, and upper Missouri); Central (Arkansas-Red, lower Missouri, Upper Mississippi, Souris-Red-Rainy, and Great Lakes); Northeast (New England, Mid Atlantic, Ohio, and northern half of South Atlantic-Gulf); and South and Southeast (Tennessee, Lower Mississippi, Texas-Gulf, and southern half of South Atlantic-Gulf), as well as Hawaii and Alaska. Hawaii and Alaska are each treated as separate regions. Observed changes over each of these parts of the country are summarized below.

4.2.4.1 West

The western United States has been more studied than any of the other regions in terms of both observed climate-related changes in hydrology and water resources, and the future implications. This is because: a) the western United States is, in general, more water-limited than is the rest of the United States, hence any changes in the availability of water have more immediate and widespread consequences, and b) much of the runoff in the western United States is derived from snowmelt, and therefore western U.S. streamflow is sensitive to ongoing and future climate change in ways that are more readily observable than elsewhere in the United States.

The western United States has been more studied than any of the other regions in terms of both observed climate-related changes in hydrology and water resources, and the future implications.

Much of the recent work on observed changes in the hydrology of the western United States has focused on changes in observed snowpack. Mote (2003) analyzed 230 time series of snow water equivalent (SWE) in the Pacific Northwest (defined as the states of Washington, Oregon, Idaho, and Montana west of the Continental Divide, and southern British Columbia) for the period 1950-2000 (in some cases longer). These records originate mostly from manual snow courses at which snow cores were taken at about the same time each year (in some cases, more than once, but at most locations around April 1), primarily for the purpose of predicting subsequent spring and summer runoff for water management purposes. Mote (2003) found that over this region, there was a strong preponderance of downward trends, especially in the Cascade Mountains, where winter temperatures generally are higher than elsewhere in the region. Also, the decreases in SWE were generally larger in

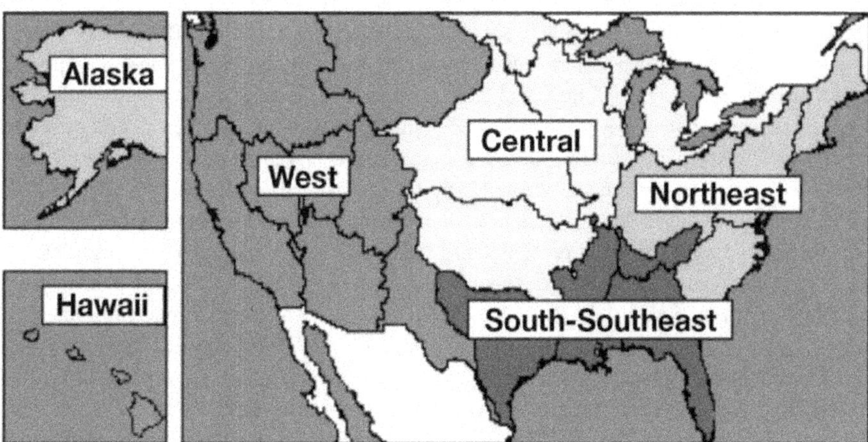

Figure 4.8 *Super-regions as aggregates of USGS hydrologic regions.*

absolute value at lower than at higher elevations. He noted that changes in precipitation, as well as decadal scale variability (especially the widely acknowledged shift in the Pacific Decadal Oscillation (PDO) in about 1977) may have contributed to the observed trends, but argued that the PDO shift alone could not explain changes in SWE over the period analyzed. He also concluded that while regional warming has played a role in the decline in SWE, "... regional warming at the spatial scale of the Northwest cannot be attributed statistically to increase in greenhouse gasses."

Mote et al. (2005) expanded the analysis of Mote (1999) to the western United States, and used a combination of modeling and data analysis, similar to the approach used by Andreadis and Lettenmaier in their continental United States drought analysis, to analyze changes in SWE over the western United States for the period 1915-2003. They used the snow accumulation and ablation model in the Variable Infiltration Capacity (VIC) macroscale hydrology model (Liang et al. 1994) to simulate SWE over the entire western United States for the period of interest, and then compared simulated trends and their dependence on elevation and average winter temperature with snow course observations. They found, notwithstanding considerable variability at the scale of individual snow courses, that the spatial and elevation patterns of trends agreed quite well over the region. They then analyzed reconstructed records for the entire period 1915-2003 and evaluated trends. The advantage of this approach is that the longer 1915-2003 period spans several phase changes in the PDO, and therefore effectively filters out its effect on long-term trends. They found that over the nearly 80-year period, there had been a general downward trend in SWE over most of the region. The exception was the southern Sierra Nevada, where an apparent upward trend in SWE, especially at higher elevations, appeared to have resulted from increased precipitation, which more than compensated for the generally warming over the period.

Hamlet et al. (2005) extended the work of Mote et al. (2005) and through sensitivity analysis determined that most of the observed SWE changes in the western United States can be attributed to temperature rather than precipitation changes. Hamlet et al. (2007) used a similar strategy of driving the Variable Infiltration Capacity (VIC) hydrological model with observed precipitation and temperature and found, over the 1916-2003 period, that trends in soil moisture, ET, and runoff generally can be traced to shifts in snowmelt timing associated with a general warming over the period. In a companion paper, Hamlet and Lettenmaier (2007) assessed changes in flood risk using a similar approach. Their analysis showed that in cold (high elevation and continental interior) river basins flood risk was reduced due to overall reductions in spring snowpack. In contrast, for relatively warm rain-dominant basins (mostly coastal and/or low elevation) where snow plays little role, little systematic change in flood risk was apparent. For intermediate basins, a range of competing factors such as the amount of snow prior to the onset of major storms, and the contributing basin area during storms (i.e., that fraction of the basin for which snowmelt was present) controlled flood risk changes, which were less easily categorized.

Stewart et al. (2005) analyzed changes in the timing of spring snowmelt runoff across the western United States. They computed several measures of spring runoff timing using 302 streamflow records across the western United States, western Canada, and Alaska for the period 1948-2002. The most useful was the center of mass timing (CT), which is the centroid of the time series of daily flows for a year. As shown in Figure 4.9, they found consistent shifts earlier in time of CT for snowmelt-dominated (mostly mountainous) river basins, but little change (or changes toward later runoff) for coastal basins without a substantial snowmelt component. Although they noted the existence of the PDO shift part way through their period of record, Stewart et al. (2005) argue that the variance in CT is explained both by temporal changes in the PDO and a general warming in the region, and that variations in PDO alone are insufficient to explain the observed trends. This finding is supported by the absence of coherent shifts in CT for non-snowmelt-dominated streams.

Pagano and Garen (2005) found that the variability of April-September streamflow at 141 unregulated sites across the western United States has generally increased from about 1980

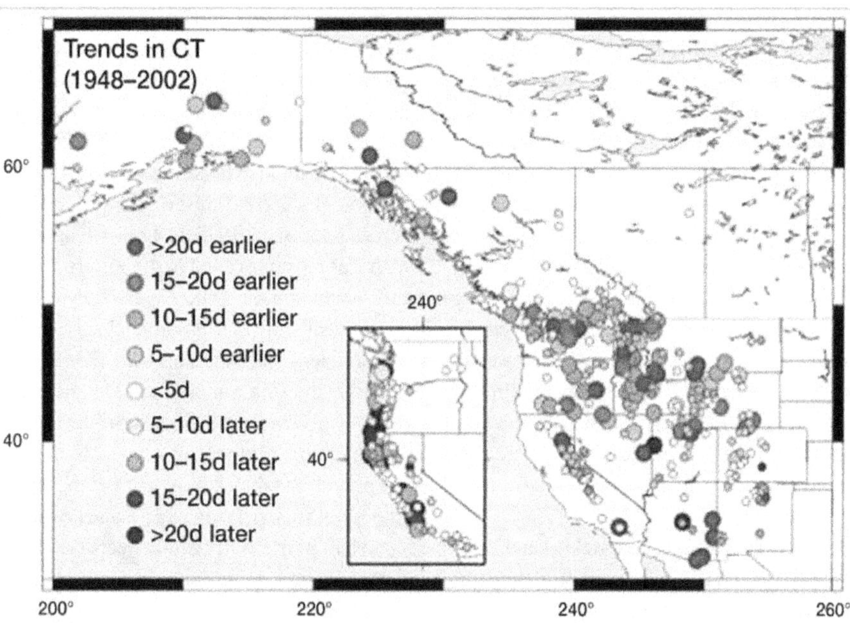

Figure 4.9 *Changes in western U.S. snowmelt runoff timing, 1948-2002. Source: Stewart et al. (2005).*

...increased streamflow variability is a major concern for water managers, as it tends to diminish the reliability with which water demands can be satisfied.

onward. This contrasts with a period of markedly low variability over much of the region from about 1930 through the 1970s. Although such shifts at decadal time scales have been observed before, and are even expected due to the nature of decadal scale variability, increased streamflow variability is a major concern for water managers, as it tends to diminish the reliability with which water demands can be satisfied.

4.2.4.2 Central

There has been relatively little work evaluating hydrologic trends in the central United States more specific than the U.S.-wide work of Lins and Slack (1999), and Mauget (2003). Garbrecht et al. (2004) analyzed trends in precipitation, streamflow, and evapotranspiration over the Great Plains. They found in an analysis of 10 watersheds in Nebraska, Kansas, and Oklahoma with streamflow records starting from 1922 to 1950 (median start year about 1940) and all ending in 2001, a common pattern of increasing annual streamflow in all watersheds. Most of this occurred in spring and winter (notwithstanding that most of the annual precipitation in these basins occurs in spring and summer). Garbrecht et al. also found that the relative changes in annual streamflow were much larger than in annual precipitation, with an average 12 percent increase in precipitation, leading to an average 64 percent increase in streamflow, but only a 5 percent increase in evapotranspiration. They also note that the large increases in streamflow had mostly occurred by about 1990 and in some (but not all) of the basins the trend appeared to have reversed in the last decade of the record. Mauget (2004) analyzed annual streamflow records at 42 USGS Hydro-Climatic Data Network stations in a large area of the central and southern United States (stations included were as far west as eastern Montana and Colorado, as far east as Ohio, as far north as North Dakota, and as far south as Texas). They used an approach similar to that of Mauget (2003). Although the patterns vary somewhat across the stations, in general higher flow periods tended to occur more toward the end of the period than the beginning, indicating general increases in streamflow over the period. A more detailed analysis of daily streamflows indicates negative changes in the incidence of drought events (defined as sequences of days with flows below a station-dependent threshold) and increases in the incidence of "surplus" days (days with flows above a station-dependent surplus threshold). These results are broadly consistent with those of Lins and Slack (1999), and Andreadis and Lettenmaier (2006).

4.2.4.3 NORTHEAST

The Northeast region is distinctive in that many records relating to hydrologic phenomena are relatively long. Burns et al. (2007) report that based on data from 1952 to 2005 in the Catskill region of New York State (the source of most of New York City's water supply), peak snowmelt generally shifted from early April at the beginning of the record to late March at the end of the record, "consistent with a decreasing trend in April runoff and an increasing trend in maximum March air temperature." Burns et al. (2007) also report increases in regional mean precipitation and regional mean potential evapotranspiration (PE), with generally increased regional runoff.

Hodgkins et al. (2003) and Hodgkins and Dudley (2006a) studied high flows in rural, unregulated rivers in New England, where snowmelt dominates the annual hydrological cycle. They showed significantly earlier snowmelt runoff (using methods similar to those applied in the western United States by Stewart et al. (2005)), with most of the change (advances of center of volume of runoff by one to two weeks) occurring in the last 30 years. Hodgkins et al. (2002) also noted reductions in ice cover in New England. Spring ice-out (when lake ice cover ends) records between 1850 and 2000 indicate an advancement of nine days for lakes in northern and mountainous regions, and 16 days for lakes in more southerly regions. These changes were generally found to be related to warmer air temperatures.

Huntington et al. (2004) analyzed the ratio of snow to precipitation (S/P) for Historical Climatology Network (HCN) sites in New England and found a general decrease in the ratio and decreasing snowfall amounts, which is consistent with warming trends. Hodgkins and Dudley (2006b) found that 18 of 23 snow course sites in and near Maine with records spanning at least 50 years had decreases in snowpack depth or increases in snowpack density, changes that are also consistent with a warming climate.

The Ohio Basin, also included within the defined northeast "super-region," is relatively understudied in terms of climate change (Liu et al. 2000) despite its economic and demographic importance and the significance of its flow (it contributes 49 percent of the total Mississippi River flow at Vicksburg). The Lins and Slack (1999) study of streamflow trends across the United States found increases in minimum and median flows at several locations in the Ohio basin, but no trend in maximum flows. McCabe and Wolock (2002a) describe a step change (increases) around 1970 in U.S. streamflow, which was most prevalent in the eastern United States, including the Ohio River. They related this apparent shift to a possible change in climate regime. Easterling and Karl (2001) note that during the 20th century there was a cooling of about 0.6°C in the Ohio basin, with warming in the northern Midwest of about 2°C for the same period. But they also report that the length of the snow season in the Ohio Valley over the second half of the 20th century decreased by as much as 16 days. In a study of evaporation and surface cooling in the Mississippi basin (including the Ohio River), Milly and Dunne (2001) suggest that high levels of precipitation were caused by an internal forcing, and that a return to normal precipitation could reveal warming in the basin.

Moog and Whiting (2002) studied the relationship of hydrologic variables (precipitation, streamflow, and snow cover) to nutrient exports in the Maumee and Sandusky river basins adjacent to the northern boundary of the Ohio. While not focused on climate-related changes directly, it allows inferences to be made of how climate change might impact water quality in the basin. Antecedent precipitation and streamflow were found to be negatively correlated to pollution loading, and snow cover tended to delay nutrient export. These results suggest how shifts in seasonal streamflow, and the increases in low and median flows observed by Lins and Slack (1999), might impact nutrient export from the basin.

4.2.4.4 SOUTH AND SOUTHEAST

No studies were found that dealt specifically with hydrologic trends in the South and Southeast, although the national study of Lins and Slack shows generally increasing streamflow over most of this region in the second half of the 20th century. This result is consistent with Mauget (2003) and the part of the domain studied by Mauget (2004) that lies in the South and Southeast super-region. A related study by Czikowsky and Fitzjarrald (2004) analyzed several aspects of seasonal and diurnal streamflow patterns at several hundred USGS stream gauge stations in the eastern and southeastern United States, as they might be related to evapotranspiration changes that occur at the onset of spring. They found a general shift in runoff patterns earlier in the spring in Virginia (as well as in New England and New York), but not in Pennsylvania and New Jersey.

4.2.4.5 Alaska

Hinzman et al. (2005) review evidence of changes in the hydrology and biogeochemistry of northern Alaska (primarily Arctic regions). They showed decreases in warm season surface water supply, defined as precipitation minus potential evapotranspiration, at several sites on the Arctic coastal plain over the last 50 years. Precipitation was observed and potential evapotranspiration was computed using observed air temperature. These downward trends were related primarily to increased air temperature, as precipitation trends generally were not statistically significant over the period. Permafrost temperatures from borehole measurements at 20-meter depth have increased over the last half-century, with the increases most marked over the last 20 years. The authors also found some evidence of increasing discharge of Alaskan Arctic rivers over recent decades, although short records precluded a rigorous trend analysis. Records of snow cover at Barrow indicate that the last day of snow cover has become progressively earlier, by about two weeks over 60 years. Stewart et al. (2005), in their study of seasonal streamflow timing, included stations in Alaska (mostly south and southeast), and found that the shifts toward earlier timing of spring runoff in the western United States extended into Alaska (see Figure 4.8). Lins and Slack (1999) included a handful of HCDN stations in southeast Alaska, for which there did not appear to be significant trends over the periods they analyzed.

4.2.4.6 Hawaii

Oki (2004) analyzed 16 long-term USGS streamflow records from the islands of Hawaii, Maui, Molokai, Oahu, and Kauai for the period 1913-2002. They found that for all stations, there were statistically significant downward trends in low flows, but that trends were generally not significant for annual or high flows. When segregated into baseflow and total flow, baseflow trends were significant across almost the entire distribution (mean as well has high and low percentiles). In general, low and base flows increased from 1913 to about the early 1940s, and decreased thereafter. Oki also found that streamflow was strongly linked to the El Niño-Southern Oscillation (ENSO), with winter flows tending to be low following El Niño events and high following La Niña events. The signal is modulated to some extent by the PDO, and is most apparent in the total flows, and to a lesser extent in baseflows. Oki (2004) noted that changes in ENSO patterns could be responsible for the observed long-term trends, but did not attempt to isolate the portion of the observed trends that could be attributed to interannual and interdecadal variability attributable to ENSO and the PDO.

A warming climate is, in general, expected to increase water temperatures and modify regional patterns of precipitation, and these changes can have direct effects on water quality.

4.2.5 Water Quality

Water quality reflects the chemical inputs from air and landscape and their biogeochemical transformation within the water (Murdoch et al. 2000). The inputs are determined by atmospheric processes and movement of chemicals via various hydrologic flowpaths of water through the watershed, as well as the chemical nature of the soils within the watershed. Water quality is also broadly defined to include indicators of ecological health (e.g. sensitive species). Regional scale variation in natural climatic conditions (precipitation patterns and temperature) and local variation in soils generates spatial variation in "baseline" water quality and specific potential response to a given scenario of climate change. A warming climate is, in general, expected to increase water temperatures and modify regional patterns of precipitation, and these changes can have direct effects on water quality. However, a major challenge in attributing altered water quality to climate change is the fact that water quality is very sensitive to other nonstationary human activities, particularly land use practices that alter landscapes and modify flux of water, as well as thermal and nutrient characteristics of water.

In general, water quality is sensitive to temperature and water quantity. Higher temperatures enhance rates of biogeochemical transformation and physiological processes of aquatic plants and animals. As temperature increases, the ability of water to hold dissolved oxygen declines, and as water becomes anoxic, animal species begin to experience suboptimal conditions. Nutrients in the water enhance biological productivity of algae and plants, which increases oxygen concentration by day, but at night these producers consume oxygen; oxygen sags can impose suboptimal anoxic conditions. Increased streamflow can dilute nutrient concentrations and thus diminish excessive biological production, however higher flows can flush excess nutrients from sources of origin in a stream. The overall balance of these competing effects in a changing climate is not yet known.

Most studies examining the responses of water quality over time have focused on nutrient loading. This factor has changed significantly over time, and there are specific U.S. laws (e.g., Clean Water Act) designed to reduce nutrient inputs into surface waters to increase their quality. For example, Alexander and Smith (2006) examined trends in concentrations of total phosphorus and total nitrogen and the related change in the probabilities of trophic conditions from 1975 to 1994 at 250 river sites in the United States with drainage areas greater than 1,000 km2. Concentrations in these nutrients generally declined over the period, and most improvements were seen in forested and shrub-grassland watersheds compared to agricultural and urban watersheds. Ramstack et al. (2004) reconstructed water chemistry before European settlement for 55 Minnesota lakes. They found that lakes in forested regions showed very little change in water quality since 1800. By contrast, about 30 percent of urban lakes and agricultural lakes showed significant increases in chloride (urban) or phosphorus (agricultural). These results indicate the strong influence of land use on water quality indicators. Detecting the effects of climate change requires the identification of reference sites that are not influenced by the very strong effects of human land use activities.

Recent historical assessments of changes in water quality due to temperature trends have largely focused on salmonid fishes in the western United States. For example, Bartholow (2005) used USGS temperature gauges to document a 0.5°C per decade increase in water temperatures in the lower Klamath River from the early 1960s to 2001, driven by basin-wide increase in air temperatures. Such changes may be related to PDO. Increases in water temperature can directly and indirectly influence salmon through negatively affecting different life stages. Crozier and Zabel (2006) reported that air temperatures have risen 1.2°C from 1992 to 2003 in the Salmon River basin in Idaho. Because water temperatures show a correlation with air temperature, smaller snowpacks that reduce autumn flows and cause higher water temperatures are expected to reduce salmon survival. Temperature effects can be indirect as well. For example, Petersen and Kitchell (2001) examined climate records for the Columbia River from 1933 to 1996 and observed variations of up to 2°C between "natural" warm periods and cold periods. Using a bioenergetics model, they showed that warmer water temperatures are associated with an expected higher mortality rate of young salmon due to fish predators.

4.3 ATTRIBUTION OF CHANGES

Trend attribution essentially amounts to determining the causes of trends. Among the various agents of hydrologic change, the most plausible are: a) changing climate, b) changing land cover and/or land use, c) water management, and d) instrumentation changes, or effects of other systematic errors. Among the causes of streamflow trends (the variable assessed by most studies reviewed in this chapter), water management changes are the easiest to quantify. With respect to changes in streamflow, the studies cited have all used streamflow records selected to be as free as possible of water management effects. For instance, USGS HCDN stations, used by Lins and Slack (1999; 2005), as well as several other studies reviewed, were selected specifically based on USGS metadata that indicate the effects of upstream water management. Certainly, it is not impossible for the metadata to be in error. An earlier study by Lettenmaier et al. (1994) that used a set of USGS records that pre-dated HCDN, selected using similar methods and identified some stations where there were obvious water management effects upstream, despite metadata entries to the contrary. However, the number of such stations was small, and the clear spatial structure in the Lins and Slack results shown in Figure 4.7, for instance, are unlikely to be the result of water management effects. If they were, it would require a corresponding spatial structure to errors in the metadata, which seems highly unlikely. In short, while it could be that some of the detected trends are attributable to undocumented water management effects; it is highly unlikely that the same could be said for the general patterns and conclusions.

Changes in instrumentation are always of concern in trend detection studies, as shifts in instrumentation often are implemented at a particular time, and hence can easily be confounded with other trend causes. This is a problem, for instance, with precipitation measurement, where changes in gauge types, wind shields, and other particulars complicate trend attribution (it should be noted that these problems are addressed in precipitation networks like the U.S. Historical Climatology Network, which has had adjustments made for observing system biases). In contrast, for streamflow observations, the methods are relatively straightforward; the measured variable is river stage, which is converted to discharge via a stage-discharge relationship, formed from periodic coincident measurements of discharge and stage. The USGS has well-established protocols for updating stage-discharge relationships, especially following major floods, which may affect the local hydraulic control. Therefore, while there almost certainly are cases where bias is introduced into discharge records following rating curve shifts, it is unlikely that such shifts would persist though a multi-decadal record, and even more unlikely that observed spatial patterns in trends could be caused by rating curve errors.

Distinguishing between the two remaining possible causes of trends – land cover and/or land use change and climate – requires more complicated analysis. Some land cover/land use change effects have striking effects on runoff. Urbanization is one such change agent, which typically decreases storm response time (the time between peak precipitation and peak runoff), increases peak runoff following storms, and decreases base flows (as a result of decreased infiltration). However, urban areas are generally avoided in selection of stations to be included in networks like HCDN, so urbanization is unlikely to be a major contributor. Other aspects of land cover change, however, such as conversion of land use to or from agriculture and forest harvest, tend to affect much larger areas. Conversions often occur over many decades. Hence, they have time constants that are similar to decadal and longer scale climate variability. Although many studies at catchment scale or smaller have attempted to quantify the effects on runoff of vegetation change such as forest harvest (Stednick et al. 1996), few studies have evaluated the larger scale effects. Matheussen et al. (2000) studied land cover change in the Columbia River basin from 1900 to 1990, and estimated that changes to annual runoff from forest harvest and fire suppression were at most 10 percent (in one of eight sub-basins analyzed, more typical changes were of order 5 percent) over this time period. Other studies have indicated larger changes (Brauman et al. 2007). On the other hand, studies of smaller basins, where a large fraction of the basin can be perturbed over relatively short periods of time, have projected or measured much larger changes (see Bowling and Lettenmaier (2001) for an example of modeled changes of forest harvest, and Jones and Grant (1996) for an observational study). However, over basins the size of which have been analyzed within networks like HCDN, more modest changes are likely, and over basins with drainage areas typical of HCDN (drainage areas hundreds to thousands of km^2 and up) efforts to isolate vegetation change from climate variability have been complicated by signal-to-noise ratios that are usually smaller for the vegetation than the climatic signal (see Bowling et al. 2000 for an example). It must be acknowledged, however, that some studies have reported changes in the hydrologic response of intermediate sized drainage basins, such as those included in the HCDN, that appear to be attributable to land cover rather than to climate change (see e.g. Potter, 1991). In summary, it is unlikely that the hydrologic trends detected in the various studies reviewed above can be attributed, at least in large part, to land cover and land use change, but sufficient questions remain that it cannot be definitively ruled out.

The final cause to which long-term hydrological trends might be attributed is climate change. Although it is essentially impossible to demonstrate cause and effect conclusively, streamflow (and other land surface hydrological variables) clearly are highly sensitive to climate, especially precipitation. Therefore, it is possible to compare trends in precipitation, for instance, with those in runoff, and most efforts in the continental United States that do so (some explicit, others more indirect), show a general correspondence. Certainly, this effect is clear in the Lins and Slack (1999; 2005) results, which show generally increased streamflow over most percentiles of the flow frequency distribution. These trends seem to correspond to generally upward trends in precipitation across much of the continental United States. For the annual maxima (floods), the correspondence to precipitation is less obvious. While various studies have shown increases in intense precipitation across the continental United States (e.g., Groisman et al., 1999), the absence of corresponding increases in flood incidence remains a somewhat open question. Groisman et al. (2001) used the same data as Lins and Slack (1999) and performed an analysis (updated by Groisman et al. (2004), who also used an area averaging approach rather than station-specific time series) to show that shifts in the probability distribution of extreme precipitation in general correspond to shifts in flood distributions. Possible reasons for the discrepancy between the two sets of studies include: a) the "floods" analyzed by Groisman et al. (2001) are not of the same general magnitude as the annual maxima series analyzed by Lins and Slack (1999); b) the shifts in intense precipitation observed by Groisman et al. (1999) and others occur mostly during periods of the year when extreme floods are uncommon; and/or c) the area-averaging approach used in the Groisman et al. studies filters out natural variability that obscures trends in the station data. Lins (2007), however, offers a more straightforward explanation. Groisman et al. (2001; 2004) test for trends in a variable that essentially is the fraction of the mean contributed by the highest 5th percentile of the flow distribution (which in turn is averaged spatially). Because the distribution of (e.g., daily) streamflow is positively skewed, a disproportionate fraction of the mean flow is accounted for by the upper percentiles, which tends to amplify changes. Lins (2007) concludes that "..the differences between the Groisman et al. findings and those of the [other studies] are apparent and interpretive rather than substantive." It is also noteworthy that Groisman et al. (2004) note that by extending their data record through 2003, several relatively dry years were included in the analysis, and the spatially averaged discharge change for the upper 5th percentile no longer had a statistically significant change.

Notwithstanding these difficulties related to the upper tail of the streamflow distribution, most streamflow trends do generally correspond to observed trends in precipitation. The question remains, though, whether these changes are evidence of climate change or decadal (or longer) scale variability? This question cannot be addressed through hydrologic analysis alone. For example, observed downward trends in streamflow in the Pacific Northwest are difficult to discriminate from changes associated with a mid-70s shift in the PDO, especially because this change occurred at about the mid-point of many streamflow records (many stations in the Pacific Northwest date to the late 1940s). One way to deal with this issue is through use of model reconstructions (e.g. Mote et al. 2005; Hamlet et al. 2007), which attempt to segregate decadal scale variability from longer-term (century or longer) shifts. An alternative approach reported by Barnett et al. (2008) involves use of a "climate fingerprinting" technique. Barnett et al. (2008) used a 1600-year control run in which a global climate model was used to force a regional hydrological model to characterize natural variability. Examination of the 1950-99 period of observations in the context of longer-term natural variability indicated that as much as 60 percent of the observed trends in streamflow, winter air temperature, and snow water equivalent (SWE) were human-induced.

Most of the studies reviewed in this chapter do not incorporate methods of trend attribution, and conclusions must be qualified to this effect (as the authors have done explicitly in many cases). Trend attribution for hydrologic applications is an evolving field and methods that are presently available are not nearly as refined as are trend estimation methods. This is an area to which research attention seems likely to turn in the future.

4.4 FUTURE CHANGES AND IMPACTS

This section examines recent work that assesses the potential impacts of climate change over the next several decades on the water resources and water quality of the United States. Numerous studies of the impacts of climate change on U.S. water resources have been performed, many of which are reviewed in, for instance, special issues of journals (see, for instance, Gleick 1999) and IPCC reports (e.g., Arnell and Liu 2001). An exhaustive review of this considerable body of research is beyond the scope of this chapter, and instead is limited to a review of the work that derives directly from climate scenarios archived for the 2007 IPCC assessment.

This recent work has several particular features. First, the global greenhouse gas emissions scenarios used in global model runs archived for use with the 2007 IPCC assessment are generally more consistent across models than in previous IPCC studies. Most models were run with transient scenarios where global greenhouse gases increased over time from an initial condition that typically is consistent with conditions as of about 2000, as specified in the IPCC (2000) Special Report on Emissions Scenarios (SRES). Although this report was issued prior to the 2001 IPCC Third Assessment Report, the full effect of the SRES report was not felt until the IPCC Fourth Assessment Report (2007) because of the lag time of several years that is required to run GCMs (often incorporating model improvements) and to archive their output. Second, the GCM physical parameterizations have improved with time, as has their spatial resolution, notwithstanding that the spatial resolution of most models is still coarse relative to the spatial scales required for regional impact assessments. Third, the length of GCM model runs has generally increased, with most modeling centers that have made runs available for IPCC analyses now producing simulations of length at least 100 years, and in many cases with multiple ensembles for each of several emissions scenarios. Finally, archiving model runs at the Lawrence Livermore National Laboratory's Program for Climate Model Diagnosis and Intercomparison (PCMDI) in common formats has greatly facilitated user access to the climate model scenarios.

Milly et al. (2005) evaluated global runoff from a set of 24 model runs archived for the IPCC AR4. They pre-screened model results by comparing model-estimated runoff from 20th century retrospective runs (GCM runs using estimated global emissions during the 20th century) with observations. The 12 models (total of 65 model runs, including multiple ensembles for some models) that had the lowest root mean square error (RMSE) of runoff per unit area over 165 large river basins globally, for which observations were available, were retained for evaluation of 21st century projections. The rationale for retaining only those models with plausible reproductions of 20th century runoff globally was that future projections for models that are unable to reproduce past runoff characteristics may be called into question. For the same 12 models, a set of 24 model runs was extracted from the PCMDI archive. Each of the model runs was performed by the parent global modeling center using the IPCC A1B global emissions scenario, which reflects modest reductions in current global greenhouse gas emissions trends over the 21st century. There were 24 runs for the 12 models because multiple ensembles were available for some models.

Milly et al. (2005) show projected changes in runoff globally for the 24 model runs, as both mean changes in fractional runoff for the future period 2041-2060 relative to the period 1900-1970 in the same model's 20th century run, and in the difference between the number of models showing increases less the number showing decreases. Figure 4.10 shows the same results replotted for the 18 USGS water resources regions in the continental United States, plus Alaska. In Figure 4.10, the shading identifies the median fractional change in runoff over the 24 model run pairs for 2041-2060 relative to 1901-1970 (using the median rather than the mean as in the original paper, which results in slightly improved statistical behavior). Figure 4.10 shows that, taken over all 24 of the model run pairs, the projections are for increased runoff over the eastern United States, gradually transitioning to little change in the Missouri and lower Mississippi, to substantial (median decreases in annual runoff approaching 20 percent) in the interior of the West (Colorado and Great Basin). Runoff changes along the West Coast (Pacific

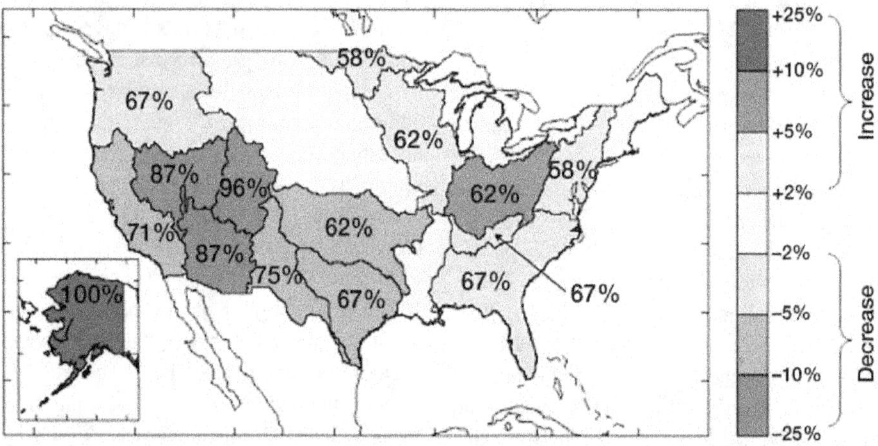

Figure 4.10 *Median changes in runoff interpolated to USGS water resources regions from Milly et al. (2005) from 24 pairs of GCM simulations for 2041-2060 relative to 1901-1970. Percentages are fraction of 24 runs for which differences had same sign as the 24-run median. Results replotted from Milly et al. (2005) by Dr. P.C.D. Milly, USGS.*

Northwest and California) are also negative, but smaller in absolute value than in the western interior basins.

Figure 4.10 also shows the consistency in the direction of changes across the 24 model pairs. In particular, the percentages given in the figure body are the fraction of model pairs for which the change was in the same direction as the indicated change in the model median. Hence, for Alaska, all 24 model pairs (100 percent) showed runoff increases, whereas for the Pacific Northwest, 16 pairs (67 percent) showed runoff decreases, while eight pairs (33 percent) showed runoff increases.

It is important to note several caveats and clarifications with respect to these results. First, the results for the various GCMs were interpolated to the USGS water resources regions, and some of the regions are small and are not well resolved by the GCMs (the highest resolution GCMs are less than three degrees latitude-longitude; others are coarser). Therefore, important spatial characteristics, such as mountain ranges in the western United States, are only very approximately accounted for in these results. Second, for some regions there is considerable variability across the models as indicated above. In some cases (for instance, see the example for the Pacific Northwest above), there may be a substantial number of models that do not agree with the median change direction. On the other hand, however, it is noteworthy that 23 of 24 model pairs showed runoff decreases for the upper Colorado, which is the source of most of the runoff for the entire Colorado basin.

Several other studies have used essentially the same model results pool, although not necessarily the same specific group of models, as in Milly et al. (2005). These studies use downscaling methods to produce forcings (usually precipitation and temperature, but occasionally other variables downscaled from the GCMs) for a land hydrology model. Downscaling results from a higher special resolution grid mesh and the lower resolution GCM grid being "trained" using historical observations. The advantage of these "off line" approaches is that the higher resolution land scheme is able to resolve spatial features, such as topography in the western United States, which may control runoff response. As an example, in mountainous areas there are strong seasonal differences in the period of maximum runoff generation and ET with elevation and these differences are not captured at the coarse spatial resolution of the GCM. However, the downside of the off-line approaches is that they do not generally preserve the water balance at the larger (GCM) scale. At this point, the nature of high-resolution feedbacks to the continental and global scale remains an area for research.

4.4.1 Hydrology and Water Resources

As in Section 4.2.4, the United States is partitioned into the same four super-regions, plus Alaska and Hawaii, (Figure 4.7) for review. For each of these super-regions, recent studies that have evaluated hydrologic and water resources implications of the IPCC AR4 archived model results were reviewed.

4.4.1.1 The West

Two recent studies have used IPCC AR4 multimodel ensembles to evaluate climate change effects on hydrology of the western United States. Maurer (2007) used statistical downscaling methods applied to 11 21st century AR4 simulations to produce one-eighth degree latitude-longitude forcings for the VIC macroscale hydrology model over the Sacramento and San Joaquin River basins of California. The GCM runs used reflected SRES A2 and B1 emissions scenarios. Maurer (2007) focused on four river basins draining to California's Central Valley from the Sierra Nevada, more or less along a transect from north to south: the Feather, American, Tolumne, and Kings rivers. Maurer's work primarily emphasized the variability across the ensembles relative to current conditions and the statistical significance of implied future changes given natural variability. All ensembles for both emissions scenarios are warmer than the current climate, whereas changes in precipitation are much more variable from model to model – although in the ensemble mean there are increases in winter precipitation and decreases in spring precipitation. These result in shifts in peak runoff earlier in the year, most evident in the higher elevation basins in the southern part of the domain. Notwithstanding variability across the ensembles, these runoff shifts are generally statistically significant, i.e., outside the bounds of natural variability, especially later in the 21st century (three periods were considered: 2011-2041, 2041-2070, and 2071-2100).

Although not considered explicitly in the paper, the results presented for 2041-2070 and emissions scenario A2 (which generally yields larger precipitation and temperature changes than B1) imply changes in ensemble mean runoff for the four basins as follows: +6.8 percent (increase) for the Feather; +3.1 percent for the American; +2.2 percent for the Tolumne; and -3.4 percent for the Kings River. By comparison, the Milly et al. (2005) results for emissions scenario A1B, which results in slightly less warming than the A2 scenario used by Maurer, indicate reductions in annual runoff of 5-10 percent for California.

Christensen and Lettenmaier (2007) used similar methods as Maurer (2007) for the Colorado River basin. The 11 GCM scenarios, two emissions scenarios, and the statistical downscaling methods used in the two studies were identical. Christensen and Lettenmaier (2007) found that in the multimodel ensemble average for emission scenario A2 for 2040-2069, discharge for the Colorado River at Lees Ferry was predicted to decrease by about 6 percent, with a larger decrease of 11 percent indicated for 2070-2099. By comparison, the Milly et al. (2005) results suggest approximately 20 percent reductions in Colorado River runoff by mid-century.

The differences in the two downscaled studies as compared with the global results raise the question of why the off-line simulations (that is, simulations in which a hydrology model is forced with GCM output, rather than extracting hydrologic variables directly from a coupled GCM run) imply less severe runoff reductions (or in the case of three of the four California basins, increases rather than decreases) than do the GCM results. The comparisons between Milly et al.'s (2005) global results and the off-line results from Maurer (2007) and Christensen and Lettenmaier (2007) should be interpreted with care. The emissions scenarios are slightly different, as are the models that make up the ensembles in the two studies. Furthermore, the statistical downscaling method used by Christensen and Lettenmaier (2007) and also Maurer (2007) does not necessarily preserve the GCM-level changes in precipitation. However, these factors do not seem likely to account entirely for the differences. First, as noted above, there is a negative feedback, reflected in the macroscale hydrology model results for snowmelt runoff under rising temperatures. Because this feedback is specific to the relatively high elevation headwaters portions of western U.S. watersheds, it is not well resolved at the GCM scale. However, while this feedback does appear to be present in the model

results, it remains to be evaluated whether the extent of the feedback in the model is consistent with observations.

Second, spatial resolution issues also imply that precipitation (and temperature) gradients are less in the GCM than in either the off-line simulations or the true system; for instance, the GCM resolution tends to "smear out" precipitation over a larger area, and hence nonlinear effects (such as much higher runoff generation efficiency at high elevations) are lost at the GCM scale. A third factor is the role of the seasonal shift (present in both the California and Colorado basins) from spring and summer to winter precipitation. Although this shift is present in the GCMs, the differential effect may well be amplified in the off-line, higher resolution runs, where increased winter precipitation leads to much larger increases in runoff than would the same amount of incremental precipitation spread uniformly over the entire basin. It should be emphasized, as indicated in Section 4.0, that these possible explanations should be cast as hypotheses, and not as definitive explanations.

4.4.1.2 CENTRAL

No studies based on IPCC AR4 were found that have examined water resources implications for this region specifically. However, a general idea of potential impacts of climate change on the Central super-region can be obtained from the global results from Milly et al. (2005) as plotted to the USGS regions in Figure 4.10. This figure shows a general gradation in the ensemble mean from increased runoff toward the eastern part of the Central super-region (e.g., Ohio, which has the largest ensemble mean runoff increases within the continental United States), to essentially neutral in the upper Mississippi, to moderately negative in the Arkansas-Red. The concurrence among models is generally modest (i.e., typically at most two-thirds of the models are in agreement as to the direction of runoff changes) so even in the Ohio basin where the ensemble mean shows increased annual runoff of 10-25 percent, about one-third of the models show downward annual runoff. This contrasts, for instance, to the higher preponderance of models showing drying in the southwestern United States. Also, the results shown in Figure 4.10 are for annual runoff, and seasonal patterns vary. Due to increased summer evaporative stress some, although certainly not all, models that predict increases in annual runoff may predict decreased summer runoff.

Jha et al. (2004) used a regional climate model to downscale a mid-21st century global simulation of the HADCM2 global climate model to the upper Mississippi River basin. This is a relatively old GCM simulation (not included in AR4), and as the authors note, is generally wetter and slightly cooler than other GCMs and relative to the AR4 ensemble means shown in Figure 4.10. Their simulations showed that a 21 percent increase in future precipitation leads to a 50 percent net increase in surface water yield in the upper Mississippi River basin. This contrasts with the much smaller 2-5 percent increase in the multimodel mean runoff in Figure 4.10. Takle et al. (2006), using an ensemble of seven IPCC AR4 models, showed results that are more consistent with Figure 4.10 for the Upper Mississippi basin, specifically a multimodel mean increase in runoff of about 3 percent for the end of the 21st century. They found that these hydrologic changes would likely decrease sediment loading to streams, but that the implications for stream nitrate loading were indeterminate.

Schwartz et al. (2004) analyzed projections of Great Lakes levels produced by three GCM runs in the late 1990s for the IPCC TAR. Two of the three GCMs projected declines in lake levels, and one projected a slight increase. Declining lake levels were associated with increased harbor dredging costs, and some loss in vessel capacity. However, low confidence must be ascribed to the projected declines in lake levels, as FAR model output shows runoff changes in the multimodel mean (see Figure 4.10) to be on the margin between slightly negative and slightly positive, with nearly as many models projecting increases as decreases.

4.4.1.3 NORTHEAST

Several studies have evaluated potential future climate changes and impacts in the Northeast using climate model simulations performed for the IPCC's AR4. Hayhoe et al. (2006) produced climate scenarios for the Northeast, which they defined as the 9-state area from Pennsylvania through Maine, using output from nine atmosphere-ocean general circulation models (AOGCMs) archived in the IPCC AR4 database. Three IPCC emissions scenarios were included: B1, A2, and A1F1, which represent low, moderately high, and high global greenhouse gas emissions over the next century. Results were presented as model ensemble averages for two time periods: 2035-2064 and 2070-2099. For the earlier period, the model ensemble averages for increases in temperature are from 2.1 to 2.9°C, and for increases in annual precipitation, 5 percent to 8 percent. The authors also used hydrologic modeling methods to evaluate the corresponding range of hydrologic variables for the period 2035-2064. They found increases in ET ranging from +0.10 to +0.16 mm/day; increases from 0.09 to 0.12 mm/day; advances in the timing of the peak spring flow centroid from 5 to 8 days; and decrease in the mean number of snow days/month ranging from 1.7 to 2.2. The authors conclude that "the model-simulated trends in temperature and precipitation-related indicators...are reasonably consistent with both observed historical trends as well as a broad range of future model simulations."

Rosenzweig et al. (2007) use a similar approach applied to a smaller geographic region to determine how a changing climate might impact the New York City watershed region, which feeds one of the largest municipal water systems in the United States. They used five models, also from the IPCC AR4 archive. Three emissions scenarios were considered: B2, A1B and A2, representing low, moderate and relatively high emissions, respectively (A2 is also used in Hayhoe et al. 2006). The scenarios were downscaled to the New York watershed region using a weighting procedure for adjacent AOGCM gridboxes, and were evaluated using observed data. For the 2050s, temperature increases in the range 1.1-3.1°C were indicated relative to the 1970-1999 baseline period, with a median range of 2-2.2°C. Precipitation changes ranged from -2.5 percent to +12.5 percent, compared to the baseline, with the median in the range 5-7.5 percent. This study also produced scenarios of local sea level rise, a factor that effects groundwater through salt water intrusion; river withdrawals for water use through the encroachment of the salt front; and sewer systems of coastal cites and wastewater treatment facilities through higher sea levels and storm surges.

Several studies have been performed on potential future climate change and impacts that are relevant to the Ohio River basin, but none are based on the most recent IPCC AR4 scenarios with multiple models and emissions scenarios. McCabe and Wolock (2002b) used prescribed future changes in climate, in this case an increase in monthly temperatures of 4°C, to examine changes in mean annual precipitation minus mean annual potential evapotranspiration (P-PE) and potential evapotranspiration (PE). In the Ohio basin, the drop in the first is relatively low, and the increase in the latter is moderate, reflecting the greater impact on PE (and thus P-PE) in warm regions as compared to cooler regions. Another study used a 4°C benchmark to examine land use effects relating to climate change. It found that land use conversion from commercial to low-density residential use decreased runoff (Liu et al. 2000). The early scenarios cited by Easterling and Karl (2001) suggest decreases of up to 50 percent in the snow cover season in the 21st century, and it is possible that by the end of the 21st century sustained snow cover (more than 30 continuous days of snow cover) could disappear from the entire southern half of the Midwest. However, these scenario results and others given by Easterling and Karl are based on earlier GCMs, and a comprehensive multimodel, multi-emissions AR4 scenario evaluation for the Ohio needs to be undertaken.

4.4.1.4 South and Southeast

No studies were identified that have assessed the implications of IPCC AR4 scenarios for the hydrology of the South and Southeast super-region. However, a general idea of potential impacts can be obtained from the global results of Milly et al. (2005) as plotted to the USGS regions in Figure 4.10. This figure shows a general gradation in the ensemble mean from east to west, with slightly increased runoff in the Southeast, near zero change in the lower Mississippi, and moderate decreases in the Texas drainages. As for the Central super-region, the concurrence among models is modest. For all regions within the South and Southeast super-region, two-thirds of the models are in agreement as to the direction of runoff changes, meaning that even for the Texas basins where moderate decreases in runoff are predicted in the ensemble mean, one-third of the models predicted increases. Furthermore, as for the Central sub-region, these results are for annual runoff and shifts in the seasonality of runoff. Generally higher summer evaporative stress will tend to decrease the fraction of runoff occurring in summer, and increase the fraction occurring at other times of the year, especially winter and spring, although this pattern certainly will not be present in all models.

4.4.1.5 Alaska

No studies were identified that have assessed hydrologic changes for Alaska associated with the AR4 scenarios. However, Figure 4.10 shows that relatively large runoff increases are suggested in the global model output for Alaska, a result that is consistent with the generally higher increases in temperature expected toward the poles. This, in turn, results in higher precipitation, in part because of increased moisture holding capacity of the atmosphere at higher temperatures, which generally results in increased precipitation. Large increases in runoff (10-25 percent, larger than anywhere in the continental United States) are predicted in the ensemble mean, and all models (100 percent) concur that runoff will increase over Alaska, a level of agreement not present anywhere in the continental United States. Nonetheless, Alaska covers a large area that encompasses several different climatic regions, so considerable subregional, as well as seasonal, variability in these results should be expected.

4.4.1.6 Hawaii

No studies were identified that have assessed hydrologic changes for Hawaii associated with the AR4 scenarios. Furthermore, the Hawaiian Islands are far too small to be represented explicitly within the GCMs, so any results that are geographically appropriate to Hawaii are essentially for the ocean and not the land mass. This is important as precipitation, and hence runoff, over this region is strongly affected by orography. The nature of broader shifts in precipitation, as well as evaporative demand over land, interacts in ways that can only be predicted accurately with regional scale modeling – an analysis that has not yet been undertaken.

4.4.2 Water Quality

The larger scale implications of increasing water temperature across the nation are illustrated by several modeling studies. Eaton and Scheller (1996) calculated that cool-water and cold-water fishes will shift their distributions nationwide, and streams and rivers currently supporting salmonids may become inhospitable as temperatures cross critical thresholds (Keleher and Rahel 1996). Stefan et al. (2001) simulated the warming effects of a doubling of $CO2$ on 27 lake types (defined by combinations of three categories of depth, area, and nutrient enrichment) across the continental United States, and examined the responses of fish species to projected changes in lake temperature and dissolved oxygen. They found that suitable habitat would be reduced by 45 percent for cold-water fish and 30 percent for cool-water fish, relative to historical conditions (before 1980). Shallow and medium-depth lakes (maximum depths of 4 meters and 13 meters, respectively) were most affected. Habitat for warm-water fish was projected to increase in all lake types investigated.

Warmer temperatures will also enhance algal production and most likely the growth of nuisance species, such as bluegreen algae. Modeling results suggest that increased temperatures associated with climate warming will increase the abundance of bluegreen algae and thus reduce water quality. This effect is exacerbated by nutrient loading, pointing to the importance of human response to climate change in affecting some aspects of water quality (Elliott et al. 2006). Increased temperatures, coupled with lower water volumes and increased nutrients, would further exacerbate the problem.

Because warmer waters support more production of algae, many lakes may become more eutrophic due to increased temperature alone, even if nutrient supply from the watershed remains unchanged. Warm, nutrient-rich waters tend to be dominated by nuisance algae, so water quality will decline in general under climate change (Murdoch et al. 2000; Poff et al. 2002). The possible increase in episodes of intense precipitation projected by some climate change models implies that nutrient loading to lakes from storm-related erosion could increase. Further, if freshwater inflows during the summer season also are reduced, the dissolved nutrients will be retained for a longer time in lakes, effectively resulting in an increase in productivity. These factors will independently and interactively contribute to a likely increase in algal productivity.

A warmer and drier climate will reduce streamflows and increase water temperatures. Expected consequences would be a decrease in the amount of dissolved oxygen in surface waters and an increase in the concentration of nutrients and toxic chemicals due to a reduced flushing rate (Murdoch et al. 2000). Reduced inputs of dissolved organic carbon from watershed runoff into lakes can increase the clarity of lake surface waters, allow biological productivity to increase at depth, and ultimately deplete oxygen levels and increase the hypolimnetic stress in deeper waters (Schindler et al. 1996).

A warmer-wetter climate could ameliorate poor water quality conditions in places where human-caused concentration of nutrients and pollutants currently degrades water quality (Murdoch et al. 2000). However, a wetter climate, characterized by greater storm intensity and long inter-storm duration, may act to episodically increase flushing of nutrients or toxins into freshwater habitats. For example, Curriero et al. (2001) reported that 68 percent of the 548 reported outbreaks of waterborne diseases during the period 1948-1994 were statistically associated with an 80 percent increase in precipitation intensity, implying that increased precipitation intensity in the future carries a health risk via polluted runoff into surface waters.

In general, an increase in extreme events will likely reduce water quality in substantial ways. More frequent floods and prolonged low flows would be expected to induce water quality problems through episodic flushing of accumulated nutrients/toxins on the landscape followed by their retention in water bodies (Murdoch et al. 2000, Senhorst and Zwolsman 2005). Clearly, human actions in response to climate change will influence the ultimate effect of climate on water quality. In a modeling example, Chang (2004) used the HadCM2 GCM scenario for five subbasins in southeastern Pennsylvania for projected changes in 2030 and found that climate change alone would slightly increase mean annual nitrogen and phosphorus loads, but concurrent urbanization would further increase nitrogen (N) loading by 50 percent. This example illustrates how human land use activity interacts with warming climate and altered precipitation patterns to induce synergistic water quality changes.

A warmer-wetter climate could ameliorate poor water quality conditions in places where human-caused concentration of nutrients and pollutants currently degrades water quality (Murdoch et al. 2000). However, a wetter climate, characterized by greater storm intensity and long inter-storm duration, may act to episodically increase flushing of nutrients or toxins into freshwater habitats.

4.4.3 Groundwater

In contrast to the many studies that have been conducted over the last 20 years of surface water vulnerability to climate change (see Section 4.2), few studies have examined the sensitivity of groundwater systems to a changing climate. For this reason, analysis was not restricted to the studies based on IPCC AR4 scenarios as no such studies of groundwater impacts have been performed to date. Instead, several studies are summarized that have evaluated groundwater sensitivity to climate change across the continental United States (no studies are known that are applicable to Alaska or Hawaii).

Among the first published papers in this area was a study by Vaccaro (1992) on the sensitivity of the Ellensburg (WA) basin to climate and land cover change. Vaccaro examined the sensitivity of groundwater recharge to both land cover change (over half of the 937 km2 basin whose native vegetation was a combination of grasslands and arid shrublands is now irrigated, mostly from surface water sources) and climate change. The climate change scenario considered was the average of CO_2 doubling scenarios from three GCMs. A physically based model of deep percolation that accounted for the effects of evapotranspiration on percolation to deep soil, and hence groundwater recharge, was used. For the native vegetation scenario, Vaccaro found that under the future climate scenario, groundwater recharge increased, whereas under current vegetation and future climate conditions, recharge was projected to decrease. The reason for the difference in signs of predicted recharge under the future land use and climate scenarios was that for native vegetation evapotranspiration peaks during spring, whereas for the irrigated condition, it peaks during summer. Therefore, total evapotranspiration, and hence recharge, is less sensitive to warming for native vegetation than for irrigated land use, and the balance of

increased precipitation and increased evaporative demand under future climate tips towards increased precipitation for native vegetation, but toward increased evaporative demand for current vegetation.

Loaiciga et al. (2000) studied the sensitivity of the Edwards Balcones fault zone (BFZ) aquifer of south central Texas to climate change, using results from several GCMs. They used an adaptation of a simple water balance model to estimate recharge, based on the estimated streamflow deficit between upstream and downstream gauges (accounting for local inflow) of the major stream crossing the aquifer. A simple pro-rating method was used to relate unmeasured lateral inflows to the channel in the reach between an upstream and a downstream gauge, and climate change effects on streamflow were scaled directly from GCM output. For the single GCM used (CO_2 doubled), projected future precipitation and runoff were considerably higher than for current climate, resulting in projections of increased recharge, and therefore increased discharge of a key spring in the region that was considered an index to groundwater conditions. Predicted spring discharge was, however, highly sensitive to assumptions about future groundwater pumping. Loaiciga et al. (2000) also considered a more physically based approach to estimating groundwater recharge, which accounted directly for evapotranspiration as it would change for future climate. In this case, six GCM CO_2 doubling scenarios were considered, all of which, aside from the single climate scenario used in the water balance approach, projected reduced precipitation. Coupled with higher evaporative demand under a warming climate, this resulted in projected recharge that was considerably reduced relative to current climate.

Scibek and Allen (2006) evaluated the sensitivity of two unconfined aquifers that straddle the U.S.-Canadian border between British Columbia and Washington State to climate change, as predicted by the Canadian Climate Centre GCM. The Abbotsford-Sumas aquifer lies in a humid area west of the Cascade Mountains, whereas the Grand Forks aquifer lies in a much drier climate east of the Cascades. Stream-aquifer interactions dominate the Grand Forks aquifer, but are less important in the case of the Abbotsford-Sumas aquifer. Recharge was assumed in the case of the Abbotsford-Sumas aquifer to be directly proportional to precipitation (scaled appropriately for different spatially varying recharge zones). For the Grand Forks aquifer, river discharge was related to downscaled climate variables. River discharge dominates aquifer variations, and hence aquifer changes are in turn dominated by changes in projected river flows, rather than recharge. Projected groundwater level change closely followed projected changes in river discharge, with higher levels in winter and early spring accompanying earlier snowmelt runoff, and lower levels in summer and fall, which result from lower streamflows during those periods. An apparent limitation of this study is that effects of evapotranspiration, and changes therein, on recharge were not accounted for directly. For the Abbotsford-Sumas aquifer, groundwater levels were predicted to decline slightly for future climate by mostly less than 1m. In this case of Abbotsford-Sumas, projected groundwater level declines are related entirely to projected GCM (downscaled) changes in precipitation, and effects of warming are not directly considered.

Other studies (e.g., Hansen and Dettinger 2005; Gurdak et al. 2007) have investigated effects of climate variability at interannual to decadal time scales on groundwater levels. In the case of Hansen and Dettinger's (2005) study of a southern California coastal aquifer, downscaled GCM output was used to evaluate the role of climate variations on groundwater levels. However, the groundwater model was driven primarily by downscaled GCM precipitation. The effects of evapotranspiration on recharge were calibrated to water levels, rather than being driven by computation based on surface variables (e.g., air temperature and/or solar radiation) from the GCM. Gurdak et al. (2007) investigated the influence of climate variability (primarily the decadal scale PDO) on groundwater levels in the deep High Plains aquifer system. They show that in this system the linkage between climate and groundwater levels is controlled by hydraulic head gradients in the vadose zone, which in turn is influenced by evapotranspiration. However, their study did not include a modeling element, so no attempt was made to predict recharge explicitly.

Taken together, these studies suggest that the ability to predict the effects of climate and climate change on groundwater systems is nowhere near as advanced as for surface water systems. A body of literature on the subject is, however, beginning to evolve (e.g. Green et al. 2007). The interaction of groundwater recharge with climate is an area that requires further research. The papers reviewed have used a variety of approaches, some of them physically based, but others have essentially "tuned" recharge in ways that do not represent the full range of mechanisms through which climate change might affect groundwater systems.

4.5 HYDROLOGY-LANDSCAPE INTERACTIONS

Across much of the continental United States, annual precipitation increased during the 20th century, and especially in the second half of the century. The average precipitation increase was estimated to be about 7 percent by Groisman (2004). As noted in Section 4.2.3, Andreadis and Lettenmaier (2006) found that as a result, droughts generally became shorter, less frequent, and covered a smaller part of the country toward the end of the 20th century than toward the beginning, although they noted that the West and Southwest were apparent exceptions. Dai et al. (2004) found that the fraction of the country under extreme either wet or dry conditions was increasing. Walter et al. (2004) found that ET has increased by an average of about 55 millimeters in the last 50 years in the conterminous United States, but that stream discharge in the Colorado and Columbia River basins has decreased since 1950 (also coincidentally a period of major reservoir construction).

These changes in physical climate and hydrology are strongly coupled with terrestrial ecosystems. Terrestrial ecosystems both respond to and modulate hydroclimatic fluxes and states. The most direct and observable connection between climate and terrestrial ecosystems is in life cycle timing of seasonal phenology, and in plant growth responses, annually in primary productivity and decadally over changes in biogeographical range. These impacts on seasonality and primary productivity then cascade down to secondary producers and wildlife populations. The vegetation growing season as defined by continuous frost-free air temperatures has increased by an average of two days per decade since 1948 in the conterminous United States, with the largest change in the West and with most of the increase related to earlier warming in the spring (Easterling 2002; Feng and Hu 2004). Global daily satellite data available since 1981 has detected similar changes in earlier onset of spring "greenness" of 10-14 days in 19 years, particularly over temperate latitudes of the Northern Hemisphere (Myeni et al. 1997; Lucht et al. 2002). For example, honeysuckle first bloom dates have advanced 3.8 days per decade at phenology observation sites across the western United States (Cayan et al. 2001) and apple and grape leaf onset have advanced two days/decade at 72 sites in the northeastern United States (Wolfe et al. 2004).

As a result of these climatic and hydrologic changes, forest growth appears to be slowly accelerating (<1 percent/decade) in regions of the United States where tree growth is limited by low temperatures and short growing seasons (McKenzie et al. 2001; Joos et al. 2002; Casperson et al. 2000). On the other hand, radial growth of white spruce in Alaska has decreased over the last 90 years due to increased drought stress on the dry, southern aspects they occupy (Barber et al. 2000). Semi-arid forests of the Southwest also showed a decreasing growth trend since 1895, which appears to be related to drought effects from warming temperatures (McKenzie 2001).

Terrestrial ecosystems both respond to and modulate hydroclimatic fluxes and states.

Climatic constraints on ecosystem activity can be generalized as variable limitations of temperature, water availability, and solar radiation, the relative impacts of which vary regionally and even locally (e.g., south vs. north aspects) (Nemani et al. 2003; Jolly et al. 2005). Where a single climatic limiting factor clearly dominates, such as low temperature constraints on the growing season at high latitudes or water limitations of deserts, ecosystem responses will be fairly predictable. However, where a seasonally changing mix of temperature and water constraints is possible, projection of ecosystem responses depends both on temperature trends and the land surface water balance. While temperature warming trends for North America are well documented, the land water balance trends over the past half century suggest that roughly the western half of the continent is getting drier and the eastern half wetter (see e.g. Andreadis and Lettenmaier 2006).

These changes have important implications for wildfires, especially in the western United States, but elsewhere as well. From 1920 to 1980, the area burned in wildfires in the continental United States averaged about 13,000 km2/yr. Since 1980, average annual burned area has almost doubled to 22,000 km2 /yr, and three major fire years have exceeded 30,000 km2 (Schoennagel et al. 2004). The forested area burned from 1987 to 2003 is 6.7 times the area burned for the period 1970-1986, with a higher fraction burning at higher elevations (Westerling et al. 2006). Warming climate encourages wildfires by drying of the land surface, allowing more fire ignitions and desiccated vegetation. The hot dry weather allow fires to grow exponentially more quickly, ultimately determining the area burned (Westerling et al. 2003). Relating climatic trends to fire activity is complicated by regional differences in seasonality of fire activity. Most fires occur in April to June in the Southwest and Southeast, and July to August in the Pacific Northwest and Alaska. Earlier snowmelt, longer growing seasons, and higher summer temperatures observed particularly in the western United States are synchronized with increase of wildfire activity, along with dead fuel buildup from previous decades of fire suppression activity (Westerling et al. 2006).

While temperature warming trends for North America are well documented, the land water balance trends over the past half century suggest that roughly the western half of the continent is getting drier and the eastern half wetter.

Insects and diseases are a natural part of all ecosystems. However, in forests periodic insect epidemics can erupt and kill millions of hectare of trees, providing dead, desiccated fuels for large wildfires. The dynamics of these epidemic outbreaks are related to insect life cycles that are tightly tied to climate fluctuations and trends (Williams and Liebhold 2002). Many of the northern insects have a two-year life cycle, and warmer winter temperatures now allow a higher percentage of overwintering larvae to survive. Recently, Volney and Flemming (2000) found that spruce budworm in Alaska have successfully completed their life cycle in one year, rather than two. Earlier warming spring temperatures allow a longer active growing season, and higher temperatures directly accelerate the physiology and biochemical kinetics of the life cycles of the insects (Logan et al. 2003). The mountain pine beetle has expanded its range in British Columbia into areas previously too cold to support its survival (Carroll et al. 2003). Multi-year droughts also reduce the available carbohydrate balance of trees, and their ability to generate defensive chemicals to repel insect attack (Logan et al. 2003).

4.6 OBSERVING SYSTEMS

Observations are critical to understanding the nature of past hydroclimatic changes and for interpreting the projections of potential effects of future changes reviewed in Sections 4.4. However, essentially no aspect of the current hydrologic observing system was designed specifically for purposes of detecting climate change or its effects on the hydrologic cycle – whether relatively slow, decadal or longer changes in mean quantities, or more rapid, "abrupt" climate change.

In the case of streamflow observations, the stream gauging network was first established in the late 1800s to provide basic information on water resource availability. More specifically, stream gauges were installed to help determine the natural variability of runoff from which decisions about how much water could be extracted from a reservoir or reservoirs of a given size could be made. Over time, as the era of dam construction waned in the 1960s and 1970s, the purpose of the stream gauge network shifted to focus more on water management than on design. Arguably, the network now is configured more to address accounting issues (i.e., stations are situated above and below major water management structures and/or diversions) than to address questions of long-term change, which requires location of stations where the confounding effects of water management and other anthropogenic influences are minimized. The HCDN is a subset of the USGS stream gauges first identified by Langbein and Slack (1982), with then record lengths of at least 20 years, which were considered "suitable for the study of variation of surface-water conditions in relation to climate variation" (see also Slack et al. 1993). The stations were selected to be mostly free of major anthropogenic influences, especially regulation by dams. Originally, more than 1,600 stations were included in this network. However the number of active stations is now substantially smaller (see Figure 4.11) due to discontinuation of stations over the years. In most cases, HCDN stations are not supported, at least in their entirety, by federal funds. The most common funding mechanism is the USGS Cooperative (Co-op) Program, in which states and local agencies share the cost of station operation. Although the Co-op program allows leveraging of federal funds and hence operation of a much larger stream gauging program than would be possible from federal funds alone, it makes the station network susceptible to short-term budget issues in the cooperating agencies, and the loss of stations indicated in Figure 4.11 is, in large part, the result of such issues. It is important to note that essentially all of the studies reviewed in this chapter that have analyzed long-term streamflow trends in the United States (e.g., Lettenmaier et al. 1994; Lins and Slack, 1999, 2005; Garbrecht et al. 2004; Mauget 2004; and McCabe and Wolock 2002a, among others) have been based on subsets of the HCDN network, hence the absence of a long-term strategy is of critical concern and needs to be addressed.

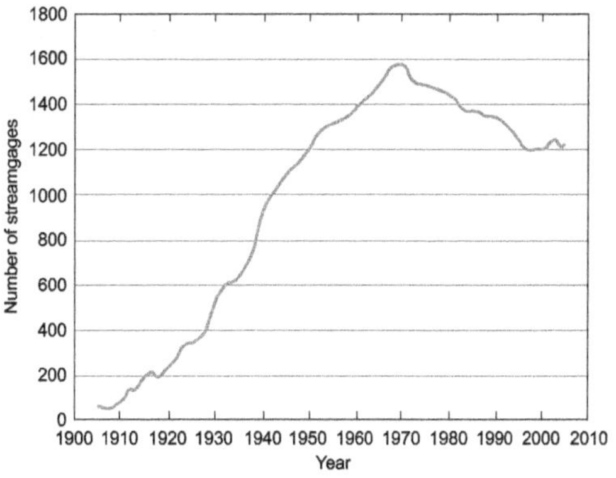

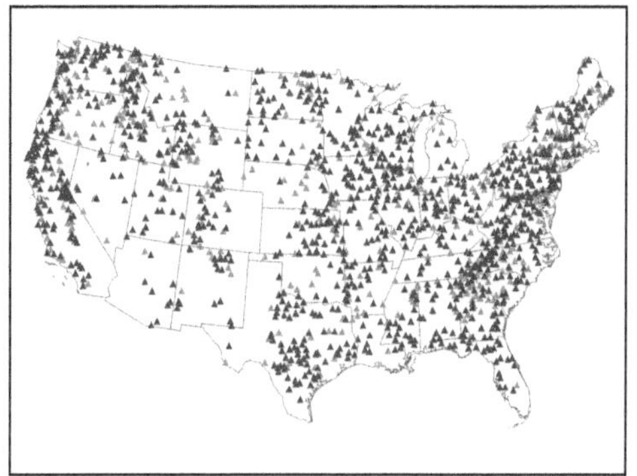

Figure 4.11 *Number of HCDN active stations 1905-2005 (upper panel), and location of discontinued stations as of 2005. Figure courtesy U.S. Geological Survey.*

Another key hydrologic variable that especially affects the western United States in addition to parts of the upper Midwest and Northeast is snow, specifically snow water equivalent or SWE. In the western United States, SWE was historically observed at manual snow courses, at which observations were mostly taken by Natural Resources Conservation Service (NRCS) (in California, observations have been taken by the Department of Water Resources). These observations are relatively costly to collect, as they involve travel to remote, mostly mountainous areas, and for this reason observations were collected only a few times per year (usually around April 1, at about the time of maximum snow accumulation). In the early 1980s, NRCS began to transition to an automated network of snow pillows, which essentially record the weight of snow on a pressure sensor and then convert to SWE. In California, there has been a similar transition from manual snow course to snow pillows, although California's Department of Water Resources continues to collect manual snow course data as well. The major advantage of the snow pillows is that data are essentially continuous, and the data transmission system provides additional channels that allow other variables such as temperature and precipitation to be transmitted as well. Analyses of long-term snow trends have faced the problem of merging the snow course and SNOTEL data. There are a variety of problems in doing so. For instance, thermodynamic properties of snow sensors are different from those of the surrounding natural landscape, and this can affect the rate of spring melt and statistics like "last date of snow." Furthermore, standard protocol for snow course measurements is to average a number (usually at least 10) of manual cores taken along or transects that cover a larger area than do the snow pillows,

so the representation of local spatial variability differs (see e.g. Dressler et al. 2006). Pagano et al. (2004) have shown how the transition from manual snow courses to the SNOTEL network has affected the accuracy of seasonal streamflow forecasts across the West.

Like HCDN, the purpose of the snow course and SNOTEL networks was not monitoring of climate change and variability, but rather support of water management through provision of basic data used in water supply forecasting. However, as demands for information related to long-term climate-related shifts in snow properties have grown, the networks have begun to be used increasingly for these other purposes. NRCS's National Water and Climate Center has initiated a study to evaluate effects of changes in SNOTEL instrumentation (e.g. metal or hypalon pillows), their comparison with manual snow courses, as well as systematic changes in snow courses and SNOTEL sites related to changes in vegetation and other site-specific characteristics, to provide better background information as to sources of systematic errors in long-term SWE records. A significant number of SNOTEL sites have been augmented with soil moisture and soil temperature sensors to improve spring runoff forecasts and basin-specific water management. The SNOTEL network also supports snow depth, relative humidity, wind speed/direction, and solar radiation measurements.

As noted in Section 4.2.2, evaporation pans do not provide a direct measurement of either actual or potential evaporation. Nonetheless, they provide a relatively uncomplicated measuring device, and the existing long-term records, taken together with the analyses discussed in Section 4.2.2, do provide a land surface data record that has some value. Pan evaporation data are most commonly collected at agricultural experiment stations, and are archived by the National Climatic Data Center.

Actual evaporation can be measured in several ways. One is weighing lysimeters, which generally are only practical for relatively short vegetation, such as crops, and are complicated by the disturbance to the surface inherent in their construction. The second is Bowen ratio sensors, which measure the gradient of humidity and air temperature close to the surface, the ratio of which is equal to the ratio of sensible to latent heat (the Bowen ratio). The Bowen ratio is used to partition the residual of net radiation and ground heat flux, both of which must be measured, into latent heat (equal to evapotranspiration, when adjusted by a proportionality factor) and sensible heat. Another method of estimating evapotranspiration (or more accurately, latent heat) directly is through eddy correlation, which measures high frequency variations in the vertical component of wind and humidity, the product of which, when averaged over time, is the latent heat flux. Both the Bowen ratio and eddy correlation methods require some assumptions (see Shuttleworth 1993). However, the eddy correlation method, which is somewhat more direct, seems to have gained favor recently. The AmeriFlux network consists of about 200 stations across the continental United States at which evapotranspiration is measured. The longest term records at these stations are somewhat longer than 10 years, not nearly long enough for meaningful trend analysis. Furthermore, instrumentation has evolved over time, and there is a need for careful calibration and maintenance, as well as quality control to assure, for instance, that the measured energy flux terms balance. In the long-term, however, the quality and reliability of the instrumentation will improve and this network appears to offer the best hope for direct, long-term measurements of evapotranspiration.

Soil moisture is a key indicator of the hydrologic state of the land system. However, until recently, there was no national soil moisture network, and the NRCS SCAN (Soil Climate and Analysis Network; Schaefer et al. 2007) dates only to 1998. At present it consists of fewer than 150 stations, although eventually, if fully funded, plans exist to create 1,000 stations. The most established soil moisture network is operated by the state of Illinois, and for about 25 years has produced data at about 20 stations statewide. More recently, the Oklahoma Mesonet network has observed soil moisture on a county-by-county basis in Oklahoma. A few other state networks have been initiated. These networks will become increasingly important as time passes, particularly given concerns over possible effects of climate change on drought. Steps are needed to assure the longevity of a core network of soil moisture stations with an appropriate national distribution. One shortcoming of most current in situ methods for soil moisture observation is that their "footprint" is quite small, typically considerably less than 1 meter, and hence the observations reflect the effects of local scale spatial variability that can only be reduced by replicate sampling (e.g., by clusters of instruments). This in turn substantially increases expense. Evolving technologies, such as cosmic ray probes (Zreda and Desilets 2005) have a footprint on the order of 100 meters, and hence are able to average out much of the local scale spatial variability that is inherent in current automated soil moisture observing systems.

4.7 FINDINGS AND CONCLUSIONS

Most of the United States has experienced increases in precipitation and streamflow and decreases in drought during the second half of the 20th century. It is likely these trends are due to a combination of decadal-scale climate variability, as well as long term change.

With respect to drought, consistent with streamflow and precipitation observations, most of the continental United States experienced reductions in drought severity and duration over the 20th century. However, there is some indication of increased drought severity and duration in the western and southwestern United States that may have resulted from increased actual evaporation dominating the trend toward increased soil wetness.

There is a trend toward reduced mountain snowpack, and earlier spring snowmelt runoff peaks across much of the western United States. This trend is very likely attributable, at least in part, to long-term warming, although some part may have been played by decadal scale variability, including shift in the Pacific Decadal Oscillation in the late 1970s. Where shifts to earlier snowmelt peaks and reduced summer and fall low flows have already been detected, continuing shifts in this direction are very likely and may have substantial impacts on the performance of reservoir systems.

Trends toward increased water use efficiency are likely to continue in the coming decades. Pressures for reallocation of water will be greatest in areas of highest population growth, such as the Southwest. Declining per capita (and for some water uses, total) water consumption will help mitigate the impacts of climate change on water resources.

Paleo reconstructions of droughts show that much more severe droughts have occurred over the last 2,000 years than those that have been observed in the instrumental record (notably, the Dust Bowl drought of the 1930s, and extensive drought in the 50s).

Water quality is sensitive both to increased water temperatures, and changes in patterns of precipitation, however most observed changes in water quality across the continental United States are likely attributable to causes other than climate change, primarily changes in pollutant loadings. There is some evidence, however, that temperatures have increased in some western U.S. streams, although a comprehensive analysis has yet to be conducted. Stream temperatures are likely to increase as the climate warms, and are very likely to have both direct and indirect effects on aquatic ecosystems. Changes in temperature will be most evident during low flow periods.

Stream temperatures are likely to increase as the climate warms, and are very likely to have both direct and indirect effects on aquatic ecosystems. Changes in temperature will be most evident during low flow periods, when they are of greatest concern. Stream temperature increases have already begun to be detected across some of the United States, although a comprehensive analysis similar to those reviewed for streamflow trends has yet to be conducted.

A suite of climate simulations conducted for the IPCC AR4 show that the United States may experience increased runoff in eastern regions, gradually transitioning to little change in the Missouri and lower Mississippi, to substantial decreases in annual runoff in the interior of the west (Colorado and Great Basin). Runoff changes along the West Coast are also negative, but smaller in absolute value than in the western interior basins. The projected drying in the interior of the West is quite consistent among models. The only projections that are more consistent among models are for runoff increases in Alaska. These changes are, very roughly, consistent with observed trends in the second half of the 20th century, which show increased streamflow over most of the United States, but sporadic decreases in the West.

Essentially no aspect of the current hydrologic observing system was designed specifically for purposes of detecting climate change or its effects on water resources. Many of the existing systems are technologically obsolete, are designed to achieve specific, often non-compatible management accounting goals, and/or their operational and maintenance structures allow for significant data collection gaps. As a result, many of the data are fragmented, poorly integrated, and in many cases unable to meet the predictive challenges of a rapidly changing climate.

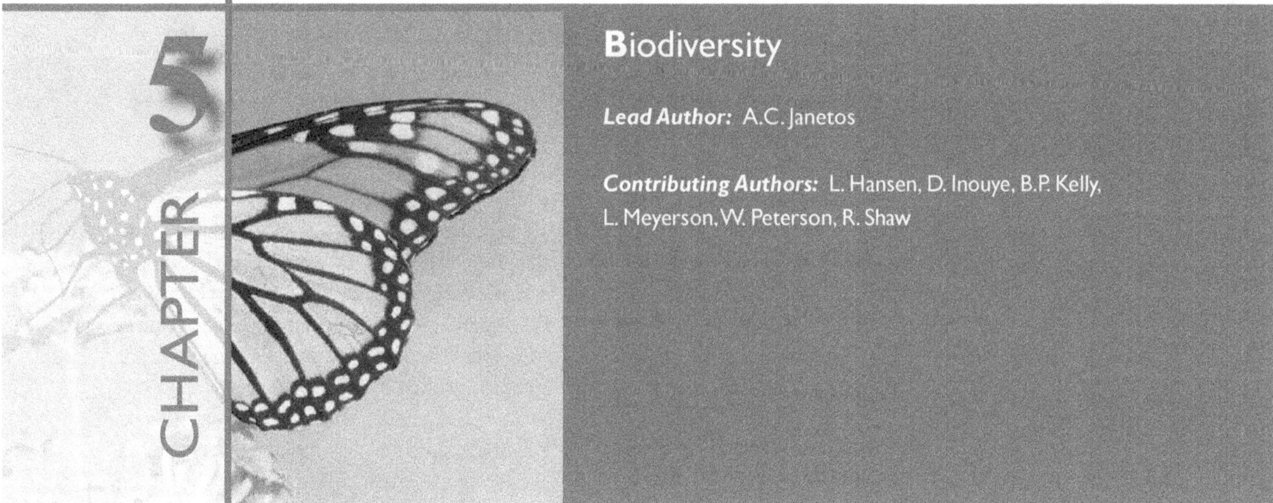

Biodiversity

Lead Author: A.C. Janetos

Contributing Authors: L. Hansen, D. Inouye, B.P. Kelly, L. Meyerson, W. Peterson, R. Shaw

5.1 INTRODUCTION AND FRAMEWORK

This synthesis and assessment report builds on an extensive scientific literature and series of recent assessments of the historical and potential impacts of climate change and climate variability on managed and unmanaged ecosystems and their constituent biota and processes. It identifies changes in resource conditions that are now being observed, and examines whether these changes can be attributed in whole or part to climate change. It also highlights changes in resource conditions that recent scientific studies suggest are most likely to occur in response to climate change, and when and where to look for these changes. As outlined in the Climate Change Science Program (CCSP) Synthesis and Assessment Product 4.3 (SAP 4.3) prospectus, this chapter will specifically address climate-related issues in species diversity and rare ecosystems.

In this chapter the focus is on the near-term future. In some cases, key results are reported out to 100 years to provide a larger context but the emphasis is on next 25–50 years. This nearer-term focus is chosen for two reasons. First, for many natural resources, planning and management activities already address these time scales through development of long-lived infrastructure, forest rotations, and other significant investments. Second, climate projections are relatively certain over the next few decades. Emission scenarios for the next few decades do not diverge from each other significantly because of the iinertiaî of the energy system. Most projections of greenhouse gas emissions assume that it will take decades to make major changes in the energy infrastructure, and only begin to diverge rapidly after several decades have passed (30–50 years).

The potential impacts of climate change on biological diversity at all levels of biological and ecological organization have been of concern to the scientific community for some time (Peters and Lovejoy 1992; IPCC 1990; Lovejoy and Hannah 2005). In recent years, the scientific literature has focused on a variety of observed changes in biodiversity and has continued to explore the potential for change due to variation in the physical climate system (IPCC 2001; IPCC 2007; Millennium Ecosystem Assessment (MEA) 2005). The focus of the chapter is mainly, although not exclusively, on ecosystems within the United States; in some areas, little work has been done here but analogs exist in other regions. Because there have been several recent comprehensive reviews of the overall topic (Lovejoy and Hannah 2005; Parmesan 2007; IPCC 2007), we will not attempt another encyclopedic review in this chapter. Instead, the chapter will focus on the particular issues of particular concern to U.S. decision-makers, as outlined in the governing prospectus. The chapter also explores the implications of

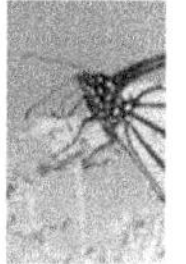

changes in biological diversity for the provision of ecosystem services (MEA 2005) and, finally, the implications of these findings for observation and monitoring systems. In each of the following sections, we provide a summary of current examples in the literature of the topics identified. There are inevitably some topics that have not been explored, although a growing literature exists (e.g., Poff et al. 2002). This is purely a function of the governing prospectus for the assessment.

This chapter thus summarizes and evaluates the current knowledge, based on both observed and potential impacts with respect to the following topics:

- Changes in Distributions and Phenologies in Terrestrial Ecosystems
- Changes in Coastal and Near-Shore Ecosystems
- Changes in Pests and Pathogens
- Changes in Marine Fisheries and Ecosystems
- Changes in Particularly Sensitive Ecosystems
- Ecosystem Services and Expectations for Future Change
- Adequacy of Monitoring Systems

5.2 CHANGES IN DISTRIBUTION AND PHENOLOGIES IN TERRESTRIAL ECOSYSTEMS

As previous chapters have demonstrated, terrestrial ecosystems are already being demonstrably impacted by climate change. Changes in the geographic distribution of species and timing of specific biological processes such as pollination or migration have long been expected because, as is widely known, over the long-term these are often controlled by large-scale patterns in climate. In this section, we examine some of those specific changes as they have been analyzed in the recent literature.

5.2.1 Growing Season Length and Net Primary Production Shifts

There is evidence indicating a significant lengthening of the growing season and higher net primary productivity (NPP) in the higher latitudes of North America, where temperature increases are relatively high. Over the last 19 years, global satellite data indicate an earlier onset of spring across the temperate latitudes by 10–14 days (Zhou at al. 2001; Lucht 2002), an increase in summer photosynthetic activity (Zhou et al. 2001), and an increase in the amplitude of the annual CO_2 cycle (Keeling 1996); climatological and field observations support these findings (Figure 5.1).

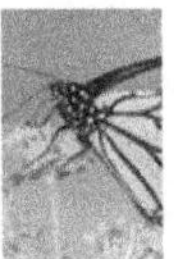

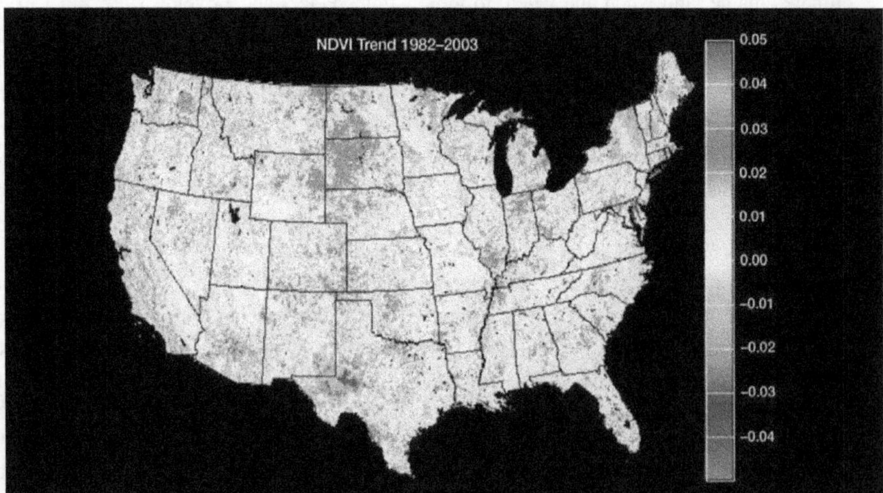

Figure 5.1 *Changes in U.S. vegetation observed by satellite (NDVI, or Normalized Difference Vegetation Index) between 1982 and 2003 (NDVI units per year). The NDVI reflects changes in vegetation activity related to climate variability, land-use change, and other influences, and shows substantial trends in much of the conterminous United States. Figure provided by J. Hicke, University of Idaho, based on data from C. Tucker, NASA Goddard Space Flight Center.*

In the higher latitudes in Europe, researchers detected a lengthening of the growing season of 1.1 to 4.9 days per decade since 1951, based on an analysis of climate variables (Menzel et al. 2003). Numerous field studies have documented consistent earlier leaf expansion (Wolfe et al. 2005; Beaubien and Freeland 2000) and earlier flowering (Schwartz and Reiter 2000; Cayan et al. 2001) across different species and ecosystem types. Accordingly, NPP in the continental United States increased nearly 10 percent between 1982–1998 (Boisvenue and Running 2006). The largest increases in productivity have been documented in croplands and grasslands of the central United States, as a consequence of favorable changes in water balance (Lobell et al. 2002; Nemani et al. 2002; Hicke and Lobell 2004).

Forest productivity, in contrast, generally limited by low temperature and short growing seasons in the higher latitudes and elevations, has been slowly increasing at less than 1 percent per decade (Boisvenue and Running 2006; Joos et al. 2002; McKenzie et al. 2001; Caspersen et al. 2000). The exception to this pattern is in forested regions that are subject to drought from climate warming, where growth rates have decreased since 1895 (McKenzie et al. 2001) and longer growing seasons have reduced productivity in forested subalpine regions (e.g., Monson et al. 2005; Sacks et al. 2007). Recently, widespread mortality over 12,000 km^2 of lower-elevation forest in the Southwest is consistent with the impacts of increased temperature and the associated multiyear drought (Breshears et al. 2005) even though previous studies had found productivity at treeline had increased (Swetnam and Betancourt 1998). Disturbances created from the interaction of drought, pests, diseases, and fire are projected to have increasing effects on forests and their future distributions (IPCC 2007). These changes in forests and other ecosystem types will cascade through the trophic structure with resulting impacts on other species.

5.2.2 Biogeographical and phenological shifts

Evidence from two meta-analyses (Root et al. 2003; Parmesan and Yohe 2003) and a synthesis (Parmesan 2006) on species from a broad array of taxa suggests that there is a significant impact from recent climatic warming in the form of long-term, large-scale alteration of animal and plant populations including changes in distribution (Root and Schneider 2006; Root et al. 2003; Parmesan 2003). If clear climatic and ecological signals are detectable above the background of climatic and ecological noise from a 0.6°C increase in global mean temperature over roughly the last century, by 2050 the impacts on ecosystems are very likely to be much larger (Root and Schneider 2006).

Movement of species in regions of North America in response to climate warming is expected to result in shifts of species ranges poleward, and upward along elevational gradients (Parmesan 2006). Species differ greatly in their life-history strategies, physiological tolerances, and dispersal abilities, which underlie the high variability in species responses to climate change. Many animals have evolved powerful mechanisms to regulate their physiology, thereby avoiding some of the direct influences of climate change, and instead interact with climate change through indirect pathways involving their food source, habitat, and predators (Schneider and Root 1996). Consequently, most distributional studies, which incorporate integrated measures of direct and indirect influences to changes in the climate environment, tend to focus on animals while phenological studies, which incorporate measures of direct influences, focus on plants and insects. Although most studies tend to separate distributional and phenological effects of climate change, it is important to keep in mind that the two are not independent and interact with other changing variables to determine impacts to species (Parmesan 2006). In addition, most of the observed species responses have described changes in species phenologies (Parmesan 2006). This section will cover both by major taxa type.

Parmesan (2006) describes three types of studies documenting shifts in species ranges: (1) those that measure an entire species' range, (2) those that infer large-scale range shifts from observations across small sections of the species' range, and (3) those that infer large-scale range shifts from small-scale change in species abundances within a local community. Although very few studies have been conducted at a scale that encompasses an entire species' range (amphibians (Pounds et al. 1999; Pounds et al. 2006), pikas

(Beever et al. 2003), birds (Dunn and Winkler 1999), and butterflies (Parmesan 2006, 1996)), there is a growing body of evidence that has inferred large shifts in species range across a very broad array of taxa. In an analysis of 866 peer-reviewed papers exploring the ecological consequences of climate change, nearly 60 percent of the 1,598 species studied exhibited shifts in their distributions and/or phenologies over the 20- and 140-year timeframe (Parmesan and Yohe 2003). Field-based analyses of phenological responses of a wide variety of different species have reported shifts as great as 5.1 days per decade (Root et al. 2003) with an average of 2.3 days per decade across all species (Parmesan and Yohe 2003).

5.2.2.1 Migratory Birds

For migratory birds, the timing of arrival to breeding territories and over-wintering grounds is an important determinant of reproductive success, survivorship, and fitness. Climate variability on interannual and longer time scales can alter phenology and range of migratory birds by influencing the time of arrival and/or the time of departure. The earlier onset of spring has consequences for the timing of migration and breeding in birds that evolved to match peak food availability (Visser et al. 2006). It should be expected that the timing of migration would track temporal shifts in food availability caused by changes in climate and the advancement of spring.

Climate variability on interannual and longer time scales change can alter phenology and range of migratory birds by influencing the time of arrival and/or the time of departure.

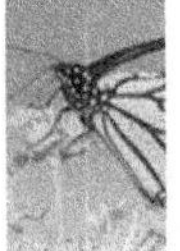

The phenology of migration to summer and wintering areas may be disrupted for long-distance, continental migrations as well regional local or elevational migrations. Since short-distance migrants respond to changes in meteorological cues whereas long-distance migrants often rely on photoperiod, it has been assumed that the climate signature on changes in phenological cycles would be stronger in short distance than in long-distance migrants (Lehikoinen et al. 2004). If true, this would lead to greater disruption in the timing of migration relative to food availability for long-distance, continental migrants relative to short-distance migrants. Recent studies of long-distance migration provide evidence to the contrary. In a continental-scale study of bird phenology that covered the entire United States and Canadian breeding range of a tree swallow (*Tachycineta biocolor*) from 1959 to 1991, Dunn and Winkler (1999) documented a 9-day advancement of laying date which correlated with the changes in May temperatures (Winkler et al. 2002; Dunn and Winkler 1999). In a study of the first arrival dates of 103 migrant bird species (long-distant, and very long-distant migrants) in the Northeast during the period 1951–1993 compared to 1903–1950, all migrating species arrived significantly earlier, but the birds wintering in the southern United States arrived on average 13 days earlier while birds wintering in South America arrived four days earlier (Butler 2003). MacMynowski and Root (2007) have found, in a study of 127 species over 20 years of migratory birds that use the migratory flyway through the central United States, that short-range migrants typically respond to temperature alone, which seems to correlate with food supply, while long-range migrants respond more to variation in the overall climate system.

Conversely, in a reversal of arrival order for short- and long-distance passerines, Jonzen et al. (2006) showed that long-distance migrants have advanced their spring arrival into Scandinavia more than short-distance migrants, based on data from 1980 to 2004. Similarly, in a 42-year analysis of 65 species of migratory birds through Western Europe, researchers found autumn migration of birds wintering south of the Sahara had advanced, while migrants wintering north of the Sahara delayed autumn migration (Jenni and Kéry 2003). Finally, a study that combined analysis of spring arrival and departure dates of 20 trans-Saharan migratory bird species to the United Kingdom found an 8-day advance in the arrival and the departure time to the breeding grounds, but with no change in the residence time. The timing of arrival advanced in relation to increasing winter temperatures in sub-Saharan Africa, whereas the timing of departure advanced in response to elevated summer temperatures in their breeding ground (Cotton 2003). But, without an understanding of how this change correlates with phenology of the food resource, it is difficult to discern what the long-term consequences might be (Visser and Both 2005).

As these studies suggest, when spring migration phenology changes, migrants may be showing a direct response to trends in weather or climatic patterns on the wintering ground and/or along the migration route, or there may be indirect microevolutionary responses to the selection pressures for earlier breeding (Jonzen et al. 2006). A climate change signature is apparent in the advancement of spring migration phenology (Root et al. 2003), but the indirect effects may be more important than the direct effects of climate in determining the impact on species persistence and diversity. Indeed, there is no *a priori* reason to expect migrants and their respective food sources to shift their phenologies at the same rate. A differential shift will lead to mistimed reproduction in many species, including seasonally breeding birds. There may be significant consequences of such mistiming if bird populations are unable to adapt (Visser et al. 2004). Phenological shifts in migration timing in response to climate change may lead to the failure of migratory birds to breed at the time of abundant food supply (Visser et al. 2006; Visser and Both 2005; Stenseth and Mystread 2002) and, therefore, may have implications for population success if the shift is not synchronous with food supply availability. Understanding where climate change-induced mistiming will occur and the underlying mechanisms will be critical in assessing the impact of global climate change on the success of migratory birds (Visser and Both 2005). The responses across species will not be uniform across their ranges, and are thus likely to be highly complex and dependent on species-specific traits, characteristics of local microhabitats, and aspects of local microclimates.

5.2.2.1.1 Mismatches and extinctions

Many migratory birds, especially short-range migrants, have adapted their timing of reproduction to the timing of the food resources. A careful examination of food resource availability relative to spring arrival and egg-laying dates will aid in the understanding of impacts of climate change. There is a suite of responses that facilitates an adaptive phenological shift; a shift in egg-laying date or a shift in the period between laying of the eggs and hatching of the chicks. In a long-term study of the migratory pied flycatcher (*Ficedula hypoleuca*), researchers found that the peak of abundance of their food resource (caterpillars) has advanced in the last two decades and, in response, the birds have advanced their laying date. In years with an early caterpillar peak, the hatching date was advanced and clutch sizes were larger. Populations of the flycatcher have declined by about 90 percent over the past two decades in areas where food for provisioning nestlings peaks early in the season, but not in areas with a late food peak (Both 2006).

In a long-term study of the migratory pied flycatcher (Ficedula hypoleuca), researchers found that the peak of abundance of their food resource (caterpillars) has advanced in the last two decades and, in response, the birds have advanced their laying date.

Climate change will lead to changing selection pressures on a wide complex of traits (Both and Visser 2005). It is the mistiming of the migration arrival, the provisioning of food resources and the lay dates that drive population declines. Predicting the long-term effects of ecological constraints and interpreting changes in life-history traits require a better understanding of both adaptive and demographic effects of climate change. Environmental stochasticity has the most immediate effect on the risk of population extinction because of its effects on parameters characterizing population dynamics, whereas the long-term persistence of populations is most strongly affected by the specific population growth rate (Saether et al. 2005). Research focused on both will aid in the understanding of the impacts of climate change.

5.2.2.2 BUTTERFLIES

Since temperature determines timing of migration and distribution, it is not surprising that many studies have documented changes in phenology of migration and significant shifts in latitudinal and elevational distribution of butterflies in response to current-day warming. The migration of butterflies in spring is highly correlated with spring temperatures and with early springs. Researchers have documented many instances of earlier arrivals (26 of 35 species in the United Kingdom (Roy and Sparks 2000); 17 of 17 species in Spain (Stefanescu et al. 2004); and 16 of 23 species in central California (Forister and Shapiro 2003)). An analysis of a 113-year record of nine migrating butterflies and 20 migrating moths found increasing numbers of migrants with increasing temperature along the migration route in response to fluctuation in the North Atlantic Oscillation (Sparks et al. 2005).

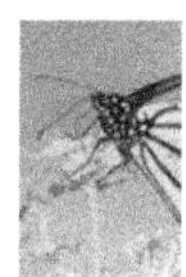

Butterflies are also exhibiting distributional and/or range shifts in response to warming. Across all studies included in her synthesis, Parmesan (2006) found 30–75 percent of species had expanded northward, less than 20 percent had contracted southward, and the remainder were stable (Parmesan 2006). In a sample of 35 non-migratory European butterflies, 63 percent have ranges that have shifted to the north by 35–240 km during this century, and 3 percent that have shifted to the south (Parmesan et al. 1999). In North America, butterflies are experiencing both distributional shifts northward, with a contraction at the southern end of their historical range, and to higher elevations, as climate changes.

In a 1993–1996 recensus of Edith's checkerspot butterfly (*Euphydryas editha*) populations, Parmesan et al. (1996) found that 40 percent of the populations below 730 meters had become extinct despite the availability of suitable physical habitat and food supply, compared to only 15 percent extinct above the same elevation (Parmesan 1996). Wilson et al. (2007) documented uphill shifts of 293 meters in butterfly species richness and composition in central Spain between 1967–1973 and 2004–2005, consistent with an upward shift of mean annual isotherms, resulting in a net decline in species richness in approximately 90% of the study region (Wilson et al 2007). In Britain, Franco et al. (2006) documented climate change as a driver of local extinction of three species of butterflies and found range boundaries retracted 70–100 km northward for *Aricia artaxerxes*, *Erebia aethiops* and 130–150 meters uphill for *Erebia epiphron;* these changes were consistent with estimated latitudinal and elevational temperature shifts of 88 km northward and 98 meters uphill over the 19-year study period.

In North America, butterflies are experiencing both distributional shifts northward, with a contraction at the southern end of their historical range, and to higher elevations as climate changes.

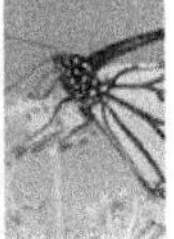

An investigation of a skipper butterfly (*Atalopedes campestris*) found that a 2–4°C warming had forced a northward range expansion over the past 50 years, driven by increases in winter temperatures (Crozier 2003, 2004). A study investigating the altitudinal and latitudinal movements of 51 British butterfly species related to climate warming found that species with northern and/or montane distributions have disappeared from low elevation sites, and colonized sites at higher elevations consistent with a climate warming, but found no evidence for a systematic shift northward across all species (Hill et al. 2002). A subsequent modeling exercise to forecast potential future distributions for the period 2070–2099 projects 65 and 24 percent declines in range sizes for northern and southern species, respectively (Hill et al. 2002).

5.2.2.2.1 *Mismatches and extinctions*

As is the case for birds, changes in timing of migrations and distributions are likely to present resource mismatches that will influence population success and alter the probability of extinction. Predictions of climate-induced population extinctions are supported by geographic range shifts that correspond to climatic warming, and a few studies have linked population extinctions directly to climate change (McLaughlin et al. 2002; Franco et al. 2006). As populations of butterfly species become isolated by habitat loss, climate change is likely to cause local population extinctions.

Modeling of butterfly distribution in the future under climate change found that while the potential existed to shift ranges northward in response to warming, lack of habitat availability caused significant population declines (Hill et al. 2002). Similarly, phenological asynchrony in butterfly-host interactions in California led to population extinctions of the checkerspot butterfly during extreme drought and low snowpack years (Singer and Harter 1996; Thomas et al. 1996; Ehrlich et al. 1980; Singer and Ehrlich 1979). A modeling experiment of two populations of checkerspot butterfly suggested that decline of the butterfly was hastened by increasing variability in precipitation associated with climate change. The changes in precipitation amplified population fluctuations leading to extinction in a region that allowed no distributional shifts because of persistent habitat fragmentation (McLaughlin et al. 2002).

Whether there is evidence of actual evolutionary change in insects in response to climate change is presently unclear. A study of the speckled wood butterfly (*Pararge aegeria*) in England found that evolutionary changes in dispersal were associated with reduced investment in reproduction, which affects the pattern and rate of expansion at range boundaries (Hughes 2003). But this result is only suggestive of a potential interaction of the factors that control the pattern and rate of expansion at range boundaries and the response to a changing climate system.

5.2.2.3 MAMMALS

Mammals are likely to interact with climate through indirect pathways involving their food source, habitat, and predators, perhaps more strongly than through direct effects on body temperature (Schneider and Root 2002), although Humphries et al. (2004) also demonstrate that overall bioenergetic considerations are important, especially in northern species. Over periods of geological time, mammals' geographic distributions have been demonstrated to respond to long-term changes in climatic conditions. Guralnick (2007) has shown that for mammal species of long duration in North America (i.e., those that have had good distributional records in both the Late Pleistocene and modern times), flatland species had large northward changes in the southern edge of their distributions as a response to the warming of the interglacial period. Montane species showed more upward and northward shifts during this time period, with the consequence that their overall ranges appeared to expand rather than to simply to track to new climatic conditions. Guralnick's results are not specific to the problems posed by recent changes in the physical climate system, or to projected changes, because these are happening much faster than interglacial warming. However, they are indicative of the direction of change that even mammal species are expected to undergo as the physical climate system changes.

Guralnick (2007) was not able to specify mechanisms by which such range adjustments occurred in his statistical analysis of existing data. It is likely, however, that climate change will alter the distribution and abundance of northern mammals through a combination of direct, abiotic effects (e.g., changes in temperature and precipitation) and indirect, biotic effects (e.g., changes in the abundance of resources, competitors, and predators). The similar results of Martinez-Meyer et al. (2004) suggest that the methods of modeling climate change response in mammals' geographic ranges as a function of changes in climate should provide robust results, at least over time periods that are long enough to allow the individual species to respond. In the United States, the General Accounting Office (2007) has identified several examples of mammals in the system of U.S. public lands for which the consequences of climate change are expected to be noticeable – among these are grizzly bears, bighorn sheep, pikas, mountain goats, and wolverines. In each case, the responses to climate-driven changes do not appear to be direct physiological responses to temperature and precipitation as much as they are responses to changes in the distribution of habitats, and in particular the compression and loss of habitats at higher elevations in mountainous areas.

The pika is a particularly interesting example, as several populations appeared to be extirpated in the United States when resampled during the 1990s (Beever et al. 2003). The pika lives in talus habitats at high elevations in mountainous areas and has a very short active season during the growing season, when it gathers grass for food for survival during the winter months. Seven out of 25 previously reported (early 20th century) populations appeared to have disappeared. Beever et al. (2003) concluded that local extirpations were best explained in a multifactorial way, and that changes in climatic factors that affected available habitat and food supply were one of the important factors. Similar phenomena have been reported for a different species of pika in Xinjiang Province in China (Li and Smith 2005). Climate effects are known to be important in both situations.

5.2.2.4 Amphibians

Many amphibian species are known to be undergoing rapid population declines, and there has been considerable discussion in the literature about the degree to which climate change might be involved (Stuart et al. 2004; Pounds, 2001; Carey et al. 2001). Carey et al. (2001) constructed a large database that included sites at which amphibian declines had been documented, and others at which they had not been. There were correlations of global environmental change in the climate system with evidence of decline, but their conclusion was that it was unlikely that the change in climate itself was the principal source of mortality in those populations. Rather, they hypothesized that changes in the global environment may have acted as an enabling factor, leading to other, more immediate causes of pathology and population declines.

There is some evidence that amphibian breeding is occurring earlier in some regions, and that global warming is likely the driving factor (Beebee 2002; Blaustein et al. 2001; Gibbs and Breisch 2001). Some temperate-zone frog and toad populations show a trend toward breeding earlier, whereas others do not (Blaustein et al. 2001). Statistical tests (Blaustein et al. 2002)

indicate that half of the 20 species examined by Beebee (1995), Reading (1998), Gibbs and Breisch (2001), and Blaustein et al. (2001) are breeding earlier. Of the half not exhibiting statistically significant earlier breeding, they are showing biologically important trends toward breeding earlier that, if continued, will likely become statistically significant (Blaustein et al. 2002). When taken together, these important data suggest that global warming is indeed affecting amphibian breeding patterns in many species. There is, however, marked unevenness of climate-change effects on amphibian breeding. For example, Fowler's Toad, *Bufo fowleri*, a late breeder, has bred progressively later in spring over the past 15 years on the north shore of Lake Erie (Blaustein 2001).

Kiesecker et al. (2001), in their study of amphibian populations in the U.S. Pacific Northwest, which are declining, point out that there are potential interactions among a number of environmental factors, including interannual climate variability, exposure to UV-B radiation causing egg and embryo mortality, and persistent climate change. It is very difficult to use field studies by themselves to sort out the relative contributions of each. However, two of the best-known examples of a climate-mediated rapid decline in amphibian populations are provided by the Golden Toad and Harlequin Frog, both of which are found in Costa Rica in the Monte Verde cloud forest. Pounds and Crump (1994) documented disappearances of the previously abundant populations of both animals as a consequence of climate-mediated stresses, in this case initially with the severe El Niño episode of 1987.

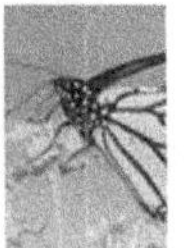

The discovery of a new disease caused by a previously unknown chytrid fungus has complicated the picture somewhat. But several studies, summarized recently by Wake (2007), conclude that even with the presence of the chytrid fungus, climate change has clearly had an impact in many of the well-documented amphibian declines and extinctions. Wake (2007) also points out that in at least one case, declines have also been found in nearby lizard species in the same habitats, although lizards are not known to be susceptible to the chytrid fungus.

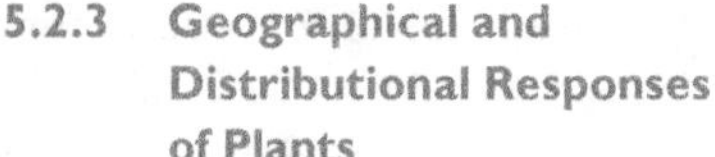

5.2.3 Geographical and Distributional Responses of Plants

In this assessment, the chapters on forests, arid lands and agriculture largely consider changes in either individual plant or ecosystem processes – e.g., photosynthesis and transpiration, soil respiration, allocation of carbon to above- and below-ground components of ecosystems, and overall carbon capture and sequestration. Those chapters, as well as a subsequent section in this chapter, also consider disturbances of different types as they affect ecosystem composition and processes, including fire, pests, and invasive species.

But a fundamental tenet of ecology is that the geographical distribution of plant species is determined in large part by climatic conditions. It is therefore natural to ask whether there is evidence of changes in plant distributions as a result of climate variability and change as well as in plant/ecosystem functional performance. It is also important to understand the degree to which changes might be expected to occur in the future in plant distributions, both at the functional level and at the individual species level.

Iverson and Prasad (2001) provide a comprehensive review of methods to determine both empirical and modeling approaches to understanding how vegetation responds to changes in climate. They point out that paleoecological observations demonstrate not only that tree species did respond to long-term changes in climate, but that they did so individually, leading to new combinations of species than previously existed. Iverson and Prasad (2001) show the results of statistical modeling for the potential future distribution of tree species in the eastern United States, using several different model-derived climate scenarios. Out of a pool of 80 common tree species, they conclude that some forest types (e.g., oak-hickory) are likely to expand, while others (maple-beech-birch) will likely contract, and still others (spruce-fir) are likely to be extirpated within the United States. Their results appear to be robust to different climate scenarios, and are consistent with what we know about these species in the paleo record.

Dirnbock et al. (2003) document both the existing relationships between the distribution of 85 alpine plant species in Europe and climate and land-use variables. They then use simple projections of both land-use and climate variables to assess the likely responses of these plants to changes in climate over the next several decades, concluding that climate forcings and land-use changes will interact substantially to determine future distributions.

Burkett et al. (2005) and the Government Accounting Office (2007) provide a number of current examples of vegetation changes that are clearly the result of responses to variability in climatic forcings, and supply mechanisms for those changes. Examples include changes in wetland vegetation in Michigan that occur as a result of the interaction of water withdrawals and drought occurrence, and extension of tree line in U.S. sub-Arctic and Arctic regions – the latter clearly responding to the observed large regional warming of the past several decades.

A growing community of ecosystem modelers using Dynamic Global Vegetation Models has developed a capability to simulate the changes in potential natural vegetation as a function of changes in the physical climate system (Cramer et al. 2001). These simulations can be used to investigate the potential for future changes in the distribution of plant functional types, and serve as a guide for assessing risk. Scholze et al. (2006) provide one such example, concluding that for an analysis that considered 16 different climate/atmospheric composition scenarios, there was a large risk of considerable change in forested ecosystems and freshwater supply in many regions around the world, including the eastern United States. However, such analyses do not include land management or land-use processes, and thus establish the potential for change, rather than serving as quantitative predictions of change.

5.3 CHANGES IN COASTAL AND NEAR-SHORE ECOSYSTEMS

Coastal and marine ecosystems are tightly coupled to both the adjacent land and open ocean ecosystems and are thus affected by climate in multiple ways. In the tropics, coral bleaching and disease events have increased, and in the Atlantic, hurricane intensity and destructive potential has increased. In temperate regions, there are demonstrated range shifts and possible alterations of ocean currents and upwelling strength. In the Arctic, there have been dramatic reductions in sea ice extent and thickness, as well as related coastal erosion. Marine species were the first to be listed as threatened species due to physical stresses that are clearly related to variability and change in the climate system (Federal Register 2006). Coastal and near-shore ecosystems are vulnerable to a host of climate change-related effects, including increasing air and water temperatures, ocean acidification, altered terrestrial run-off patterns, altered currents, sea level rise, and altered human pressures due to these and other related changes (such as development, shipping, pollution, and anthropogenic adaptation strategy implementation). This section will discuss some of the most prominent effects of climate change observed to date in the coastal and near-shore regions of the United States, with some consideration given to applicable examples from other parts of the world.

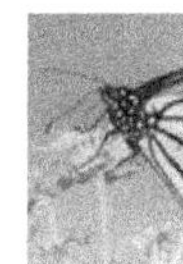

5.3.1 Coral Reefs

Tropical and subtropical coral reefs around the world have been known for some time to be under a wide variety of stresses, some of them related to changes in the climate system, and some not (Bryant et al. 1998). The United States has extensive coral reef ecosystems in both the Caribbean Sea and the Pacific Ocean. Coral reefs are very diverse ecosystems, home to a complex of species that support both local and global biodiversity and human societies. It has been estimated that coral reefs provide $30 billion in annual ecosystem service value (Cesar et al. 2003), including both direct market values of tourism, and estimates of the market value of other services, such as provision of habitat for fish breeding, and protection of coastline. A

Coastal and near-shore ecosystems are vulnerable to a host of climate change-related effects, including increasing air and water temperatures, ocean acidification, altered terrestrial run-off patterns, altered currents, sea level rise, and altered human pressures due to these and other related changes (such as development, shipping, pollution, and anthropogenic adaptation strategy implementation).

variety of regional estimates of economic value (Cesar 2000) have also been made that show substantial variation in their totals, depending in part on which services are taken into consideration. In some small developing countries, coral reefs may supply substantial fractions of total economic return through their contribution to tourism and as habitat for coastal fisheries; even in the United States and Australia where coral reefs provide small fractions of the total revenue, they generate many billions of dollars and can be very important in regional economies (Hoegh-Guldberg et al. 2007).

Corals and tropical regions where they live are experiencing increasing water temperatures, a reduction in surface water pH (Ravens et al. 2005), and there is evidence for increasing storm intensity (Emmanuel 2005), as well as a host of other ongoing challenges created as a result of development/tourism, fishing, and pollution. The effects of climate change in marine systems is highlighted by the 2006 proposed listing as Threatened under the Endangered Species Act of two species of corals in the Caribbean (Federal Register 2006). The major threats that motivated the proposed listings of Elkhorn (*Acropora palmata*) and Staghorn (*A. cervicornis*) corals were disease, elevated sea surface temperatures, and hurricanes – all of which relate to climate change and its effects (Muller et al. 2007; Mann and Emmanuel 2006).

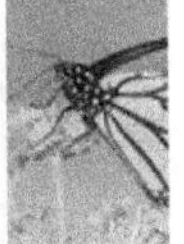

5.3.1.1 Increasing Temperature and Acidification of Ocean Waters

The El Niño-Southern Oscillation (ENSO) event of 1982–83 marked the first contemporary broad-scale coral reef bleaching and mortality event (Glynn 1984). Since then, there have been subsequent bleaching events including the 1997–98 ENSO. The rate of occurrence (annually in some cases), and almost global scale since the early 1980s is in stark contrast to the trend of the first half of the century in which bleaching events were localized and linked to local events (D'Elia et al. 1991; Glynn 1993). From 1876–1979 only three bleaching events were recorded, whereas 60 are on record between 1980 and 1993 (Glynn 1993). Bleaching is considered to be a stress response caused primarily by increased water temperature (Glynn 1993) and synergistically enhanced by increased irradiance levels (Fitt and Warner 1995; Jokiel and Coles 1990; Lesser et al. 1990). Corals become stressed if exposed to slight increases in water temperature – increases of only 1 to 2°C over the average annual thermal maxima for days to weeks can result in a bleaching event (Hoegh-Guldberg 1999). Field studies have correlated increased temperatures with mass bleaching events (Brown 1997; Hoegh-Guldberg et al. 1997; Glynn 1993). Additionally, the National Oceanic and Atmospheric Administration (NOAA) "Hotspot" program (Goreau and Hayes 1994) predicted bleaching for most geographic regions where bleaching occurred in 1998, adding further weight to the assessment that elevated temperature is the primary trigger for bleaching (Hoegh-Guldberg 1999). The final effect of the 1997–98 bleaching event has been assessed, with estimates indicating that 10–16 percent of world's living coral reefs died during this event. In the western Indian Ocean, coral reefs lost up to 46 percent of living, reef-building corals (Hoegh-Guldberg 2005).

In 2005, the Caribbean basin saw unprecedented water temperatures and some dramatic bleaching, followed by coral disease and mortality. The most dramatic monitored bleaching took place in the U.S. Virgin Islands, where National Park monitoring showed that at some sites 90 percent of the coral bleached. Afterward there appeared to be a period of recovery as water temperatures decreased. Unfortunately, this was short-lived as disease appeared in November of the same year on many of the previously bleached corals. To date there is an estimated 50 percent combined mortality from bleaching and disease in the Virgin Island National Park surveys. As of yet, there are no reports of recovery as amounts of mortality continue to increase (Eakin et al., *in press, accepted*). In the Florida Keys, equally massive bleaching was anticipated when temperatures exceeded 9-degree heating weeks in late August 2005 (NOAA Coral Reef Watch), and in fact some bleaching was observed. But the arrival of hurricanes Katrina and Rita reduced water temperatures and appear to have provided some respite for corals in the Keys. However, the same pattern of disease was seen in the Keys in those corals that did bleach, with bleaching setting in around mid-August, followed by disease in early September (Brandt, *in press, accepted*).

Both intensities and frequencies of bleaching events clearly driven by warming in surface waters have increased substantially over the past 30 years (Hughes et al. 2003). At least 30 percent of reefs globally have been severely damaged, and relatively simple projections based on temperature changes alone suggest that within the next several decades, as many as 60 percent of the world's reefs could be damaged or destroyed (Hughes et al. 2003). While there is some evidence of short-term recovery, in many locations the frequency of bleaching events could become nearly annual within several decades under a variety of reasonable climate scenarios (Donner et al. 2005). Such changes would be significantly more rapid and pose significant problems for coral reef management on a global scale (Hughes et al. 2003; Pandolfi et al. 2003; Hoegh-Guldberg et al. 2007).

Additionally, as CO_2 concentrations increase in the atmosphere, more CO_2 is hydrolyzed in the surface waters of the world's oceans, leading to their acidification (Orr et al. 2005; Hoegh-Guldberg et al. 2007) (Figure 5.2). The chemical reactions governing the dissolution of $CaCO_3$ in surface waters, and therefore the availability of material for building corals' calcium carbonate skeletons (as well as those of other calcifying organisms) are pH-dependent, and increases in acidity can lead to decreases in available $CaCO_3$ (Yates and Halley 2006). During the past 200 years, there has been a 30 percent increase in hydrogen-ion concentration in the oceans, and it is anticipated that this will increase by 300 percent by the end of this century (Ravens et al. 2005). There is evidence from site-specific studies (Pelejero et al. 2005) that in the Pacific Ocean there is natural decadal variability in the pH levels that individual reefs actually experience, and that the variability matches well with Interdecadal Pacific Oscillation variability.

However, even though some reef species may be more resistant to increases in acidity than others, the longer-term decreases in ocean pH due to increased atmospheric CO_2 concentrations may be occurring much more rapidly than in the recent history. And, when these long-term trends occur in phase with the IPO, even relatively resistant reefs would be exposed to extremely low pH levels that they have not experienced before. There are predictions that oceans could become too acidic over the long term for corals – as well as other species – to produce calcium carbonate skeletons (Caldeira and Wickett 2003;

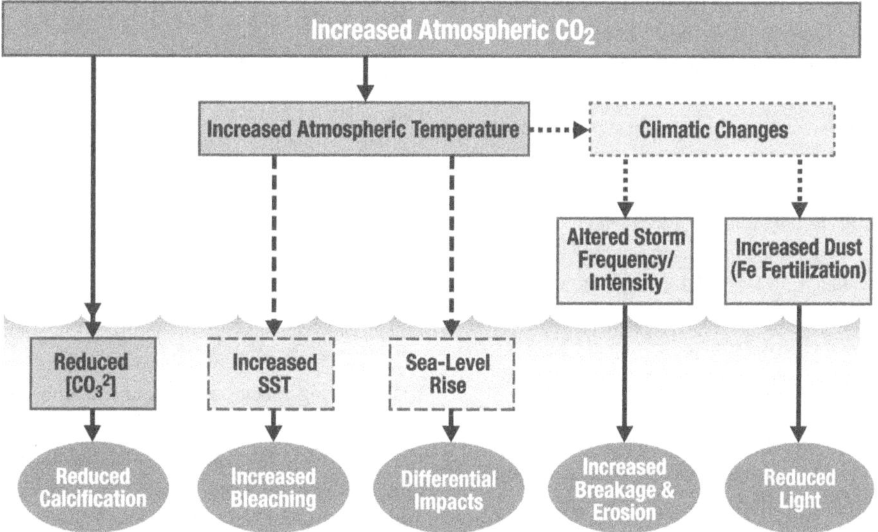

Figure 5.2 *The figure above depicts various direct and indirect effects of changes in atmospheric CO_2 concentrations on coral reef ecosystems. Solid lines indicate direct effects, dashed lines indicate indirect effects, and dotted lines indicate possible effects. Fe = iron; SST = sea surface temperature; CO_3^{2-} = carbonate ion.*

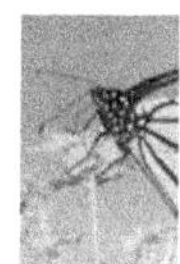

Hoegh-Guldberg 2005; Kleypas et al. 1999). More recent reviews of both experimental studies, modeling projections, and field observations suggest that the combination of changes in ocean surface temperatures, increasing ocean acidity, and a host of other stresses could bring coral reef ecosystems to critical ecological tipping points (Groffman et al. 2006) within decades rather than centuries, and that some regions of the ocean are already near that point from a biogeochemical perspective (Orr et al. 2005; Hoegh-Guldberg 2007).

Increasing sea surface temperatures are expected to continue as global temperatures rise. It is possible that these warmer waters are also increasing the intensity of the tropical storms in the region (Mann and Emmanuel 2006; Sriver and Huber 2006; Elsner 2006; Hoyos et al. 2006). As global temperatures rise, sea level will continue to rise providing additional challenges for corals. Increasing depths change light regimes, and inundated land will potentially liberate additional nutrients and contaminants from terrestrial sources, especially agricultural and municipal.

5.3.2 Coastal Communities and Ecosystems

5.3.2.1 Wetlands and Barrier Islands

The marine-terrestrial interface is vitally important for biodiversity as many species depend on it at some point in their life cycles, including many endangered species such as sea turtles and sea birds. In addition, coastal areas provide a wide variety of ecosystem services, including breeding habitat and buffering inland areas from the effects of wave action and storms (MEA 2005). There is a wide variety of different types of habitat in coastal margins, from coastal wetlands, to intertidal areas, to near-shore ecosystems, all of which are subject to a variety of environmental stresses from both the terrestrial, inland environments and from oceanic environments (Burkett et al. 2005). The additional proximity of large numbers of people makes coastal regions extremely important natural laboratories for global change.

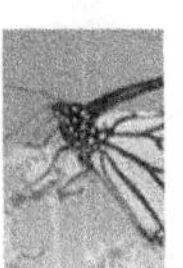

Mangroves and sea grasses protect coastlines from erosion, while also protecting near-shore environments from terrestrial run-off. Sea level rise, increased coastal storm-intensity and temperatures contribute to increased vulnerability of mangrove and sea grass communities (e.g., Alongi 2002). It has been suggested that the dominant sea grass species (*Zostera marina*) is approaching its thermal tolerance for survival in the Chesapeake Bay (Short and Neckles 1999). It has also been estimated that a 1-meter increase in sea level would lead to the potential inundation of 65 percent of the coastal marshlands and swamps in the contiguous United States (Park et al. 1989). In addition to overt loss of land, there will also be shifts in ìqualityî of habitat in these regions. Prior to being inundated, coastal watershed will become more saline due to saltwater intrusion into both surface and groundwater. Burkett et al. (2005) provide several excellent examples of documented and potential rapid, non-linear ecological responses in coastal wetlands to the combination of sea-level rise, local subsidence, salinity changes, drought, and sedimentation. Of particular concern in the United States are coastlines along the Gulf of Mexico and the Southeast Atlantic, where the combination of sea level rise and local subsidence has resulted in substantially higher relative, local rates of sea-level rise than farther north on the Atlantic Coast, or on the Pacific Coast (Burkett et al. 2005). In Louisiana alone, more than 1/3 of the deltaic plain that existed in the beginning of the 20th century has since been lost to this combination of factors. In the Gulf of Mexico and the South Atlantic, the ecological processes that lead to accretion of wetlands and continued productivity (Morris et al. 2002) have not been able to keep pace with the physical processes that lead to relative rising sea level (Burkett et al. 2005).

Barrier islands are particularly important in some regions where vulnerability to sea level rise is acute. In the northwest Hawaiian Islands, which were designated a National Monument in 2006, sea level rise is a threat to endangered beach nesting species and island endemics, including green sea turtles, Hawaiian monk seals, and the Laysan finch (Baker et al. 2006). Another example of an endangered island-locked species is the Key Deer, which is now limited to living on two islands in the Florida Keys. Their habitat is also at risk with most of the Keys at less than two meters above sea level. Median sea level rise coupled with storm surges would inundate most of the available habitat either permanently or episodically, further threatening this endangered species.

5.3.2.2 Rocky Intertidal Zones

Rocky intertidal habitats have been studied extensively with respect to their observed and potential responses to climate variability and change, both in Europe and in the United States (Helmuth et al. 2006; Mieszkowska et al. 2007; Mieszkowska et al. 2005; Bertness et al. 1999; Sagarin et al. 1999; Thompson et al. 2002; Mieszkowska et al. 2006; Barry et al. 1995). These systems react quite differently from wetlands because of the large differences in substrates. Nevertheless, the typical biota of gastropods, urchins, limpets, barnacles, mussels, etc., show reproductive, phenological, and distributional responses, similar in kind to responses of birds, butterflies, and mammals reported earlier in this chapter. However, Helmuth et al. (2006) point out that range shifts of up to 50 kilometers per decade have been recorded for intertidal organisms – far faster than documented for any terrestrial species to date.

Responses include reacting to changes in the thermal habitat, which results in heat stress, and subsequent low growth rates and early, stress-induced spawning of mussel species in New Zealand (Petes et al. 2007). Long time-series of observational data across several quite different taxonomic groups in the British Isles show consistent trends for species in response to strong regional warming trends observed since the 1980's, including: range extensions of northern species into previously colder waters; some range extension eastward of southern species into the English channel; a few species with southern range retractions; and several southern species showing earlier reproduction, greater survival rates, and faster growth rates than northern species (Mieszkowska et al. 2005). These responses are extremely similar to the biological responses shown by rocky intertidal species in the United States in several different locations (Bertness et al. 1999; Helmuth et al. 2006; Barry et al. 1995; Sagarin et al. 1999) on both the Pacific and Atlantic coasts. There is some suggestion in Europe that there could be food-web level effects on the supply of food for shore birds, but interactions among shore bird predators, gastropods and other rocky intertidal organisms, and algal cover are complex and extremely difficult to predict (Kendall et al. 2004).

Thompson et al. (2002), Helmuth et al. (2005) and Helmuth et al. (2006) all point out that the observational base of responses of intertidal organisms to changes in climate is well enough understood that reasonable projections of future change can be made. However, knowledge of the particular physiological mechanisms for the individual species' responses is especially important (Helmuth et al. 2005) in order to distinguish the reasons for the variation in responses, and in order to understand how climate changes operate in these systems in the presence of other physical and biological stresses.

Because of its importance as a contributing stress to coastal and intertidal habitats, projections of mean sea-level rise have been important to understand. Projections for sea level rise by 2100 vary from 0.18 to 0.59 m (±0.1-0.2) (IPCC 2007) to 0.5 to 1.4 m (Rahmstorf 2007). Some observational evidence suggests that recent IPCC estimates may be conservative and underestimate the rate of sea level rise (Meehl et al. 2007). The IPCC projection of 18–59 cm in this century assumes a negligible contribution to sea level rise by 2100 from loss of Greenland and Antarctic ice. Melting of the Greenland ice sheet has accelerated far beyond what scientists predicted even just a few years ago, with a more than doubling of the mass loss from Greenland due to melting observed in the past decade alone (Rignot and Kangaratnam 2006). The acceleration in the rate of melt is due in part to the creation of rivers of melt water, called ìmoulins,î that flow down several miles to the base of the ice sheet, where they lubricate the area between the ice sheet and the rock, speeding the movement of the ice toward the ocean. Paleoclimatic data also provide strong evidence that the rate of future melting and related sea-level rise could be faster than previously widely believed (Overpeck et al. 2006).

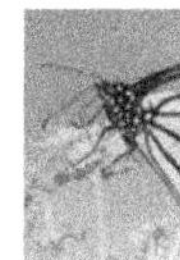

5.4 CLIMATE CHANGE, MARINE FISHERIES AND MARINE ECOSYSTEM CHANGE

The distribution of fish and planktonic species are also predominately determined by climatic variables (Hays et al. 2005; Roessig et al. 2004) and there is recent evidence that marine species are moving poleward, and that timing of plankton blooms is shifting (Beaugrand et al. 2002; Hays et al. 2005; Richardson and Schoeman 2004). Similar patterns have been observed in marine invertebrates and plant communities (Beaugrand et al. 2002; Sagarin et al 1999), Southward et al. (1995) document extensive movement of ranges and distributions of both warm and cold-water species of fish and other marine life around the British Isles and northern Europe over the past several decades, with long-time series of data from fish landings. They point out that much of the original research on fisheries biology in these regions took place from the 1930s–1970s, a period of relative constancy in the marine climate system in these regions. Changes in distributions since then appear to be much more pronounced.

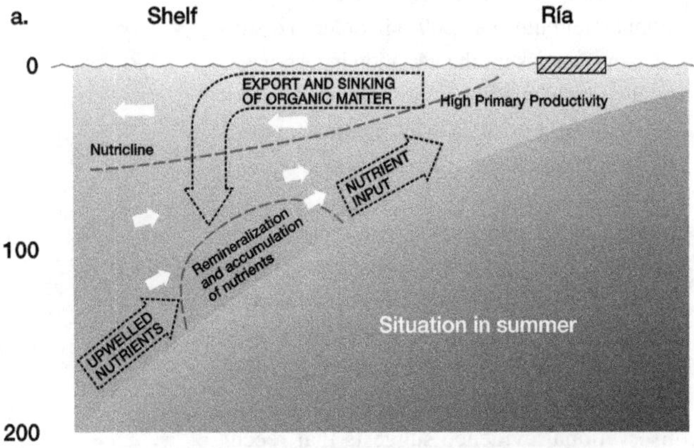

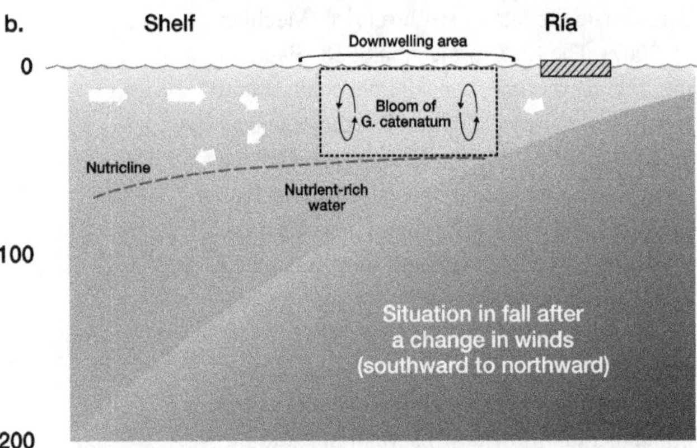

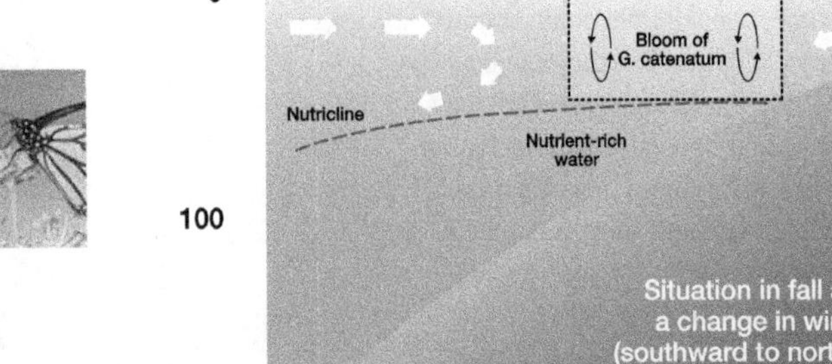

Figure 5.3 *Diagram of nutrient dynamics. a) Summer: a profile view of a ria and the adjacent continental shelf, illustrating the "loop" consisting of upwelling-enriched primary production, which leads to export, sinking, and accumulation on the bottom of particulate organic matter. This organic matter decays and remineralizes, enriching the waters beneath the nutricline. b) Fall: After fall relaxation of upwelling, lighter oceanic surface water collapses toward the coast, producing a zone of downwelling in the ria. This depresses the nutricline and cuts off upwelling-produced enrichment of the photic zone. Vertically migrating dinoflagellates may access the nutrient pool beneath the nutricline and transport them upward to levels of higher illumination, where they can use them to support photosynthesis. From Bakun 1996.*

Similar phenomena have been documented in Europe for Arctic and Norwegian cod in the Barents Sea (Dippner and Ottersen 2001), and Atlantic cod (Drinkwater 2005), where spawning, survival, and growth rates are affected in predictable ways by ocean temperature anomalies. In each case, the climate variability analyzed is tied to particular oscillations in the physical climate system (e.g., the North Atlantic Oscillation for cod), or to longer-term changes in climate. Fields et al. (1993) provide a general overview of the factors associated with the marine ecosystem responses to climate change. As in other systems examined in this report, the particular biological mechanisms of species responses are important in determining overall patterns. In addition, Hsieh et al. (2005) show that these large marine ecosystems are intrinsically non-linear, and thus subject to extremely rapid and large changes in response to small environmental forcings.

In coastal regions, decreased upwelling can decrease nutrient input to surface waters, reducing primary productivity (Soto 2002; Field et al. 2001). The food-web-level effects that such changes cause have been documented off the coast of Southern California after an abrupt, sustained increase in water temperature in the 1970s (Field et al. 1999). Conversely, climate change may alter wind patterns in ways that accelerate offshore winds and thus upwelling (Bakun 1990) (Figure 5.3).

Seven large marine ecosystems (LMEs) are recognized for U.S. waters: eastern Bering Sea, Gulf of Alaska, California Current, Gulf of Mexico, southeast U.S. continental shelf, northeast U.S. continental shelf and the greater Hawaiian Islands. Each is being studied to varying degrees with regard to the impacts of climate variability and change on ecosystem structure, biodiversity and marine fisheries. Much of the research in these systems has been carried out by U.S. and Canadian scientists associated with the International Geosphere-Biosphere Programme GLOBal Ocean ECosystem Dynamics (IGBP-GLOBEC), or by scientists following GLOBEC standards. The GLOBEC model focuses on study of the coupling of physical forcing and biological response in fisheries-rich ecosystems, and is detailed at www.globec.org. This approach has been taken due to the tight coupling between physics and biology in the oceans as compared to terrestrial ecosystems (Henderson and Steele 2001).

It has been well established that the large basin-scale atmospheric pressure systems that drive basin scale winds can suddenly shift location and intensity at interannual-to-decadal time scales, with dramatic impacts on winds and ocean circulation patterns. These low frequency oscillations are known as the North Atlantic Oscillation (NAO), the Pacific Decadal Oscillation (PDO), and the El Nino-Southern Oscillation (ENSO). Perhaps the greatest discovery of the past 10 years is that these shifts have dramatic impacts on marine ecosystems.

The NAO has been strongly positive since the 1980s. Increases in the strength of the winds have resulted in dramatic impacts on Northeast Atlantic ecosystems. For instance, increased flow of oceanic water into the English Channel and North Sea has contributed to a northward shift in the distribution of zooplankton such that the zooplankton communities are dominated by warm water species (Beaugrand, 2004) with concomitant changes in dominance in fish communities from whiting (hake) to sprat (similar to a herring). Similar ecosystem shifts in the Baltic Sea have occurred where drastic changes in both zooplankton and fish communities have been observed (Kenny and Mollman 2006). Linkages between the NAO, zooplankton and fisheries have also been described for the Northwest Atlantic waters off eastern Canada and the United States. The recovery of the codfish populations, which collapsed in the early 1990s (presumably as a result of overfishing), may be difficult due to changes in the structure of forage and food chains (Pershing and Green 2007).

In the North Pacific, the PDO refers to the east-west shifts in location and intensity of the Aleutian Low in winter (Mantua et al. 1997). Widespread ecological changes have been observed including increased productivity of the Gulf of Alaska when the PDO is in positive phase, resulting in dramatic increases in salmon production (Mantua et al. 1997), and a reversal of demersal fish community dominance from a community dominated by shrimps to one dominated by pollock (Anderson and Piatt 1991). Associated changes to the California Current ecosystem include dramatic decreases in zooplankton (McGowan et al. 1998) and salmon (Pearcy 1991) when the PDO changed to positive phase in 1977. There is also evidence that the large oscillations in sardine and anchovy populations are associated with PDO shifts, such that during positive (warm) phases, sardine stocks are favored but during negative (cool) phases, anchovy stocks dominate (e.g., Chavez et al. 2003).

In coastal regions, decreased upwelling can decrease nutrient input to surface waters, reducing primary productivity

ENSO is another major driver of climate variability. El Niño events negatively impact zooplankton and fish stocks resulting in a collapse of anchovy stocks in offshore ecosystems of Peru. Loss of anchovies, which are harvested for fish meal, affect global economies because fish meal is an important component of chicken feeds as well high-protein supplements in aquaculture feed. In waters off the west coast of the United States, plankton and fish stocks may collapse due to sudden warming (by 4–10°C) of the waters as well as through poleward advection of tropical species into temperate zones. Many of the countries most affected by ENSO events are developing countries in South America and Africa, with economies that are largely dependent upon agricultural and fishery sectors as a major source of food supply, employment, and foreign exchange.

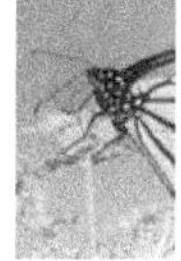

5.4.1 Other climate-driven physical forces that affect marine ecosystems

The California Current (CC) example represents an excellent case study for one Large Marine Ecosystem. The CC flows in the North

Pacific Ocean from the northern tip of Vancouver Island (Canada), along the coasts of Washington, Oregon and California, midway along the Baja Peninsula (Mexico) before turning west. For planktonic organisms and some fish species, the northern end of the Current is dominated by sub-arctic boreal fauna whereas the southern end is dominated by tropical and sub-tropical species. Faunal boundaries, i.e., regions where rapid changes in species composition are observed, are known for the waters between Cape Blanco, Oregon/Cape Mendocino, California, and in the vicinity of Point Conception, California. Higher trophic level organisms often take advantage of the strong seasonal cycles of production in the north by migrating to northern waters during the summer to feed. Animals that exhibit this behavior include pelagic seabirds such as black-footed albatross and sooty shearwaters, fishes such as Pacific whiting and sardines, and gray and humpback whales.

5.4.2 Observed and Projected Impacts

Based on long-term observation records, global climate models, regional climate models, and first principles, there is a general consensus on impacts of climate change for the United States with regard to climate modes, biophysical processes, community and trophic dynamics and human ecosystems. The type, frequency and intensity of extreme events are expected to increase in the 21st century, however Meehl et al. (2007) suggest that there is no consistent indication of discernable changes in either the amplitude or frequency of ENSO events over the 21st century (Meehl et al. 2007). Climate models from the fourth IPCC assessment project roughly the same timing and frequency of decadal variability in the North Pacific under the impacts of global warming. By about 2030, it is expected that the *minima* in decadal regimes will be *above* the historical mean of the 20th century (i.e., the greenhouse gas warming trend will be as large as natural variability). Regional analyses suggest that for California, temperatures will increase over the 20th century with variable precipitation changes by region (Bell et al. 2004), which is consistent with global projections (Tebaldi et al. 2006).

Among other findings, IPCC assessment results for the United States suggest there will be a general decline in winter snowpack with earlier snowmelt triggered by regional warming (Hayhoe et al. 2004; Salathé 2005).

Additionally, warmer temperatures on land surfaces, contributing to low atmospheric pressure combined with ocean heating may contribute to stronger and altered seasonality of upwelling in western coastal regions (Bakun 1990; Snyder et al. 2003). Migration patterns of animals within the California Current (e.g., whiting, sardines, shearwaters, loggerhead turtles, Grey Whales) may be altered to take advantage of feeding opportunities. Recent disruptions of seasonal breeding patterns of a marine seabird (Cassin's Auklet) by delayed upwelling have been reported by Sydeman et al. (2006).

Warmer ocean temperatures will contribute to changes in upwelling dynamics and decreased primary production along the California Current. Global declines in NPP (as estimated from the SeaWiFS satellite sensor) between 1997 and 2005 were attributed to reduced nutrient enhancement due to ocean surface warming (Behrenfeld et al. 2006; Carr et al. 2006). A recent example during the summer of 2005 was characterized by a three-month delay to the start of the upwelling season resulting in a lack of significant plankton production until August (rather than the usual April–May time period). Fish, birds and mammals that relied upon plankton production occurring at the normal time experienced massive recruitment failure (Schwing et al. 2006; Mackas et al. 2006; Sydeman et al. 2006). In contrast, the summer of 2006 had some of the strongest upwelling winds on record yet many species again experienced recruitment failure, in part because there was a one-month period of no winds (mid-May to mid-June).

Snyder et al. (2003) suggest that wind-driven upwelling in the California current is likely to continue its long, 30-year increase in the future, as a function of changes in the physical climate. Such a change could lead to enhanced productivity in the coastal marine environment, and subsequent changes throughout the ecosystem.

5.5 CHANGES IN PESTS AND PATHOGENS

5.5.1 Interactions of Climate Change with Pests, Pathogens, and Invasive Species

Increasing temperatures and other alterations in weather patterns (e.g., drought, storm events) resulting from climate change are likely to have significant effects on outbreaks of pests and pathogens in natural and managed systems, and are also expected to facilitate the establishment and spread of invasive alien species. For the purposes of this chapter, ìpests and pathogensî refers to undesirable outbreaks of either native or introduced insects or pathogens. Non-native species are those that are non-indigenous to a region, either historically or presently, while invasive species are those non-native species that harm the environment, the economy or human health. Initially, the most noticeable changes in plant and animal communities will most likely result from direct effects of climate change (for example, range expansions of pathogens, and invasive plants). The longer term consequences, however, may be the result of indirect effects such as disruptions of trophic relationships or a species decline due to the loss of a mutualistic relationship (Parmesan 2006).

Interactions between increasing global temperature and pests and pathogens are of particular concern because of the rapid and sweeping changes these taxa can render. While it is still difficult to predict specifically how climate change will interact with insect pests, or plant and animal diseases, some recent events have provided glimpses into the kinds of impacts that might unfold.

5.5.1.1 Mountain Pine Beetle Explosion

The mountain pine beetle (*Dendroctonus ponderosae*) is a native species that has co-existed with western conifers for thousands of years and plays an important role in the life cycle of North American western forests (Bentz et al. 2001; Powell and Logan 2001). However, the magnitude of recent outbreaks is above historical levels with historically unprecedented mortality (Logan et al. 2003). A recent outbreak in 2006 caused the death of nearly five million lodgepole pines (*Pinus contorta*) in Colorado, a four-fold increase from 2005. The infestation covers nearly half of all Colorado's forests. Such outbreaks are not confined to Colorado, but are also occurring in other parts of the United States and Canada, affecting tens of thousands of square miles of forest (Figure 5.4).

Figure 5.4 *Aerial view of the U.S. Forest Service Rocky Mountain Research Station's Fraser Experimental Forest near Winter Park, Colorado, May 2007 and a mountain pine beetle (inset). The green strips are areas of forest that had been harvested decades earlier, and so have younger faster growing trees. The red and brown areas show dead and dying trees caused by bark beetle infestation. A more recent photo would show less contrast because, due to drought and beetle epidemic, mortality rates of young trees have also risen. Photo courtesy USFS, Rocky Mountain Research Station.*

Multiple factors, including climate change, have been implicated in driving outbreaks in North America (e.g., Romme et al. 2006; Logan and Powell 2001; Logan et al. 2003). First, many North American conifer forests are primarily mature, even-aged stands due to widespread burning and heavy logging of the region during settlement 100 years ago. Mountain pine beetles prefer the mature trees resulting from these disturbances. Second, long-term drought stresses trees and makes them more vulnerable to the beetles because they cannot effectively defend themselves. Third, warmer summers also cause stress and increase growth rates of the insects, and, fourth, milder winters increase the chances of survival for the insect larvae (Romme et al. 2006; Powell and Logan 2005; Powell et al. 2000). While there is not yet definitive proof that climate change is behind the high levels of mountain pine beetle infestation, a recent study showed that over the last century Colorado's average temperatures have warmed (NRC 2007). It is therefore reasonable to expect that warmer temperatures in the future may lead to similar or more intensive events than those that are now occurring.

5.5.1.2 Poleward Migration of Plant Pests and Pathogens

Latitudinal gradients in plant defenses and herbivory are widely known but the basis for these defenses (i.e., genetic versus environment) are not fully understood. A potential outcome under warming global temperatures is a relatively rapid poleward migration of pests and pathogens, and a relatively slower rate of adaptation (e.g., increased defense against herbivory) for plants. Biogeographic theory predicts increased insect herbivory (i.e., greater loss of leaf area to herbivores) in the lower latitudes relative to higher latitudes (MacArthur 1972; Vermeij 1978; Jablonski 1993). As with the mountain pine beetle described above, higher population densities of other herbivorous insects and therefore herbivory occur because dormant season death (i.e., winter dieback) of herbivores is absent, or greatly reduced at warmer temperatures, and/or plant productivity is generally greater than at higher latitudes (Coley and Aide 1991; Coley and Barone 1996). Because of this greater herbivory, plants are thought to be better defended or otherwise less palatable at low latitudes as a result of natural selection (e.g., MacArthur 1972; Hay and Fenical 1988; Coley and Aide 1991; Coley and Barone 1996). Alternatively, plants at low latitudes could be better defended because high latitude populations have had fewer generations since the last glaciation to evolve such defenses (Fischer 1960).

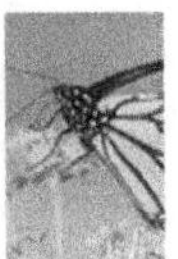

5.5.1.3 Climate Change and Pathogens

Evidence is beginning to accumulate that links the spread of pathogens to a warming climate. For example, the chytrid fungus (*Batrachochytrium dendrobatidis*) is a pathogen that is rapidly spreading worldwide and decimating amphibian populations. A recent study by Pounds et al. (2006) showed that widespread amphibian extinction in the mountains of Costa Rica is positively linked to global climate change. To date, geographic range expansion of pathogens related to warming temperatures has been the most easily detected (Harvell et al. 2002), perhaps most readily for arthropod-borne infectious disease (Daszak et al. 2000). However, a recent literature review found additional evidence gathered through field and laboratory studies that supports hypotheses that latitudinal shifts of vectors and diseases are occurring under warming temperatures. Based on their review, Harvell et al. (2002) gathered evidence that:

- Arthropod vectors and parasites die or fail to develop below threshold temperatures.
- Rates of vector reproduction, population growth, and biting increase (up to a limit) with increasing temperature.
- Parasite development rates and period of infectivity increase with temperature.

Furthermore, Ward and Lafferty (2004) conducted an analysis that revealed that disease for some groups of marine species is increasing while others are not. Turtles, corals, mammals, urchins, and mollusks all showed increasing trends of disease, while none were detected for sea grasses, decapods, or sharks/rays. The effects of increasing temperature on disease are complex, and can increase or decrease disease depending on the pathogen (Ward and Lafferty 2004).

Expansion of an invader may not always be simply explained by warming temperatures. For example, Roman 2006 suggests that the northern expansion of the invasive European green crab (*Carcinus maenas*) in North America was facilitated through the introduction of new lineages of *C. maenas* to Nova Scotia from the northern end of its native range in Europe. These northern populations may be better adapted to the colder temperatures found in northern Nova Scotia, relative to more southerly waters. Furthermore, the construction of a causeway and subsequent ìsuper portî in the Strait of Canso, Nova Scotia, appears to be at the epicenter of the high diversity of new *C. maenas* haplotypes (Roman 2006).

5.5.1.4 Climate Change and Invasive Plants

Projected increases in CO_2 are expected to stimulate the growth of most plants species, and some invasive plants are expected to respond with greater growth rates than non-invasive plants (Dukes 2000; Ziska and George 2004; Moore 2004; Mooney et al. 2006). Some invasive plants may have higher growth rates and greater maximal photosynthetic rates relative to native plants under increased CO_2, but definitive evidence of a general benefit of CO_2 enrichment to invasive plants over natives has not emerged (Dukes and Mooney 1999). Nonetheless, invasive plants in general may better tolerate a wider range of environmental conditions and may be more successful in a warming world because they can migrate and establish in new sites more rapidly than native plants, and they are not usually limited by pollinators or seed dispersers (Vila et al, *in press, accepted*).

Finally, it is critical to recognize that other elements of climate change (e.g., nitrogen deposition, land conversion) will play significant roles in the success of invasive plants in the future, either alone or under elevated CO_2 (Vila et. al., *in press, accepted*). For example, several studies have brought to light the role of increasing nitrogen availability and the success of invasive grass species (e.g., Huenneke et al. 1990; Brooks 2003). Disturbance at both global and local scales has been shown to be an important factor in facilitating species invasions (e.g., Sher and Hyatt 1999; Mooney and Hobbs 2001; D'Antonio and Meyerson 2002), and land conversion that occurred more than 100 years ago may play a role in current invasions (Von Holle and Motzkin 2007). Recent work by Hierro et al. (2006), which compared the effects of disturbance on *Centaurea solstitialis* in its native and introduced ranges, suggests that disturbance alone does not fully explain invasion success. Instead, it appears that, for *C. solstitialis*, it is the combination of disturbance and escape from soil pathogens in the native range that has encouraged invasion.

5.6 PARTICULARLY SENSITIVE SYSTEMS

5.6.1 Impacts of Climate Change on Montane Ecosystems

Temperate montane ecosystems are characterized by cooler temperatures and often increased precipitation compared to surrounding lowlands. Consequently, much of that precipitation falls in the form of snow, which serves to insulate the ground from freezing air temperatures, stores water that will be released as the snow melts during the following growing season, and triggers vertical migration by animal species that cannot survive in deep snow. Changes in historical patterns of snowfall and snowpack are predicted as a consequence of global climate change, in part due to changes in spatial patterns of precipitation, and in part due to the warming that will result in more precipitation falling as rain rather than snow (Beniston and Fox 1996; MacCracken et al. 2001). Areas that historically have most of their annual precipitation as snow are now seeing more of it as rain; documentation of this trend comes from the Sierra Nevada Mountains, where Johnson found from analysis of a 28-year dataset (Johnson 1998) that below 2400 meters, less snow is accumulating and it is melting earlier. Diaz et al. (2003) (Figure 5.5) also reported that all the major continental mountain chains exhibit upward shifts in the height of the freezing level surface over the past three to five decades.

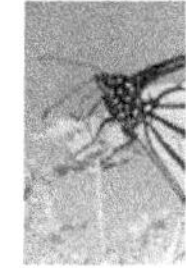

Increased variation in precipitation and temperatures is also predicted by climate change models, and Johnson (1998) also found that "Higher elevations exhibit greater variability, with most stations accumulating more snow and melting earlier. This could be the result of warmer air masses having higher moisture contents."

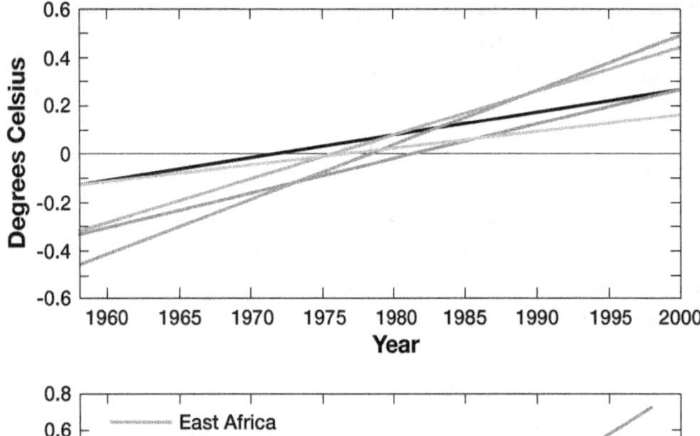

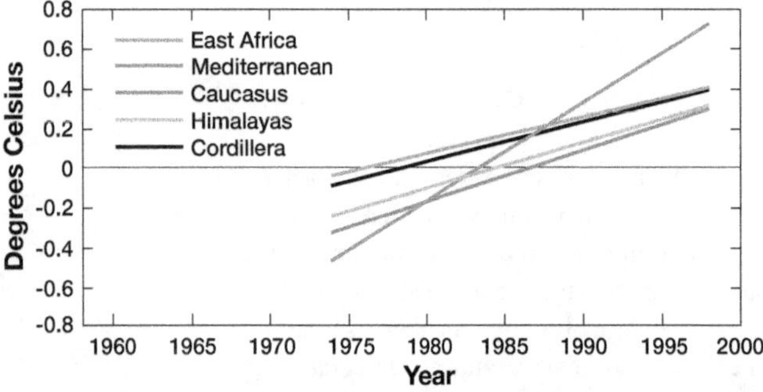

Figure 5.5 *Linear trends in near surface air temperatue for 5 different mountainous regions, based on the NCEP/NCAR Reanalysis data set. Top panel is for 1958–2000, lower panel for the period 1974–1998. From Diaz et al. 2003.*

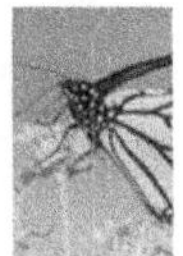

In addition to the influences of global climate change, which could affect both precipitation and temperature, regional effects can be important. For example, in the Colorado Rocky Mountains there are significant ENSO and PDO effects on winter precipitation. ENSO has also been shown to effect changes in freezing level in the American Cordillera (Diaz et al. 2003). Of course, all downstream water flows with headwaters in mountain areas are also affected by the variation in both timing and quantity of snowmelt (e.g., Karamouz and Zahraie 2004). These environmental changes are resulting in the disappearance of glaciers in most montane areas around the world. The changes in patterns and abundance of melt water from these glaciers have significant implications for the sixth of the world's population that is dependent upon glaciers and melting snowpack for water supplies (Barnett et al. 2005). Plant and animal communities are also affected as glaciers recede, exposing new terrain for colonization in an ongoing process of succession (e.g., for spider communities, see Gobbi et al. 2006).

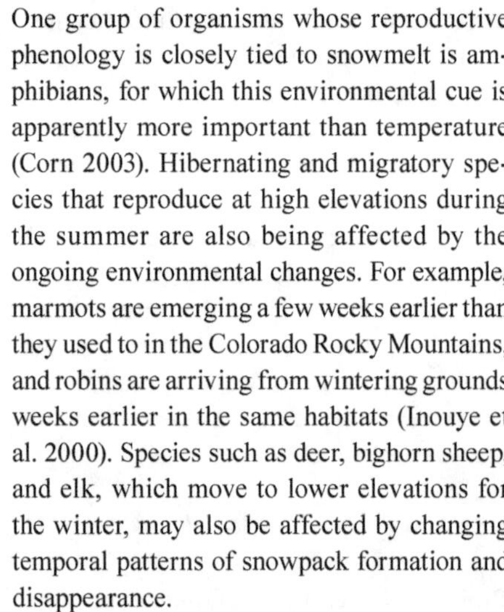

One group of organisms whose reproductive phenology is closely tied to snowmelt is amphibians, for which this environmental cue is apparently more important than temperature (Corn 2003). Hibernating and migratory species that reproduce at high elevations during the summer are also being affected by the ongoing environmental changes. For example, marmots are emerging a few weeks earlier than they used to in the Colorado Rocky Mountains, and robins are arriving from wintering grounds weeks earlier in the same habitats (Inouye et al. 2000). Species such as deer, bighorn sheep, and elk, which move to lower elevations for the winter, may also be affected by changing temporal patterns of snowpack formation and disappearance.

The annual disappearance of snowpack is the environmental cue that marks the beginning of the growing season in most montane environments. Flowering phenology has been advancing in these habitats (Inouye and Wielgolaski 2003) as well as in others at lower altitudes, mirroring what is going on at higher latitudes (Wielgolaski and Inouye 2003). There is a very strong correlation between the timing of snowmelt, which integrates snowpack depth and spring air temperatures, and the beginning of flowering by wildflowers in the Colorado Rocky Mountains (e.g., Inouye et al. 2002; Inouye et al. 2003). For some wildflowers there is also a strong correlation between the depth of snowpack during the previous winter and the abundance of flowers produced (Inouye et al. 2002; Saavedra et al. 2003). The abundance of flowers can have effects on a variety of consumers, including pollinators (Inouye et al. 1991), herbivores, seed predators, and parasitoids, all of which are dependent on flowers, fruits, or seeds.

An unexpected consequence of earlier snowmelt in the Rocky Mountains has been the increased frequency of frost damage to montane plants, including the loss of new growth on conifer trees, of fruits on some plants such as *Erythronium grandiflorum* (glacier lilies), and of flower buds of other wildflowers (e.g., *Delphinium* spp., *Helianthella quinquenervis*, etc.) (Inouye 2008). Although most of these species are long-lived perennials, as the number of years in which frost damage has negative consequences on recruitment increases, significant demographic consequences may result. These and other responses to the changing montane environment are predicted to result in loss of some species at lower elevations, and migration of others to higher elevations. Evidence that this is already happening comes from studies in both North America (at least on a latitudinal scale, Lesica and McCune 2004) and Europe (Grabherr et al. 1994). It is predicted that some animal species may also respond by moving up in elevation, and preliminary evidence suggests that some bumble bee (*Bombus*) species in Colorado have moved as much as a couple of thousand feet over the past 30 years (J. Thomson, personal communication).

5.6.2 Arctic Sea-Ice Ecosystems

Sea ice seasonally covered as much as 16,000,000 km^2 of the Arctic Ocean before it began declining in the 1970s (Johannessen et al. 1999; Serreze et al. 2007). For millennia, that ice has been integral to an ecosystem that provisions polar bears and the indigenous people. The ice also strongly influences the climate, oceanography, and biology of the Arctic Ocean and surrounding lands. Further, sea ice influences global climate in several ways, including via its high albedo and its role in atmospheric and oceanic circulation. In the past 10 years, the rate of decline in the areal extent of summer sea ice in the Arctic Ocean has accelerated, and evidence that the Arctic Ocean will be ice-free by 2050 is increasing (Stroeve et al. 2005, 2007; Overland and Wang 2007; Serreze et al. 2007; Comiso et al. 2008). Many organisms that depend on sea ice – ranging from ice algae to seals and polar bears – will diminish in number and may become extinct. Ecosystem changes already have been observed and are predicted to accelerate along with the rates of climate change. Many of the changes will not be readily obvious and may even counterintuitive. Here, we summarize expected changes and provide a few expected responses involving upper trophic levels that are thought to be illustrative.

At the base of the sea ice ecosystem are epontic algae adapted to very low light levels (Kühl et al. 2001; Thomas and Dieckmann 2002). Blooms of the those algae on the undersurface of the ice are the basis of a food web leading through zooplankton and fish to seals, whales, polar bears, and people. Sea ice also strongly influences winds and water temperature, both of which influence upwelling and other oceanographic phenomena whereby nutrient rich water is brought up to depths at which there is sufficient sunlight for phytoplankton to make use of those nutrients (Buckley et al. 1979; Alexander and Niebauer 1981; Legendre et al. 1992).

Among the more southerly and seasonally ice-covered seas, the Bering Sea produces our nation's largest commercial fish harvests as well as supports subsistence economies of Alaskan Natives. Ultimately, the fish populations depend on plankton blooms regulated by the extent and location of the ice edge in spring. Naturally, many other organisms, such as seabirds, seals, walruses, and whales, depend on primary production, mainly in the form of those plankton blooms. As Arctic sea ice continues to diminish, the location, timing, and species make-up of the blooms are changing in ways that appear to favor marked changes in community composition (Hunt et al. 2002; Grebmeier et al. 2006). The spring melt of sea ice in the Bering Sea has long favored the delivery of organic material to a benthic community of bivalve mollusks, crustaceans, and other organisms. Those benthic organisms, in turn, are important food for walruses, gray whales, bearded seals, eider ducks, and many fish species. The earlier ice melts resulting from a warming climate, however, lead to later phytoplankton blooms that are largely consumed by zooplankton near the sea surface, vastly decreasing the amount of organic material reaching the benthos. The likely result will be a radically altered community favoring a different suite of upper level consumers. The subsistence and commercial harvests of fish and other marine organisms would also be altered.

Walruses (*Odobenus rosmarus*) feed on clams and other bottom-dwelling organisms (Fay 1982). Over a nursing period of two or more years, the females alternate their time between attending a calf on the ice and diving to the bottom to feed themselves. The record ice retreats observed in recent summers extend northward of the continental shelf such that the ice is over

An unexpected consequence of earlier snowmelt in the Rocky Mountains has been the increased frequency of frost damage to montane plants, including the loss of new growth on conifer trees, of fruits on some plants such as Erythronium grandiflorum (glacier lilies), and of flower buds of other wildflowers (e.g., Delphinium spp., Helianthella quinquenervis, etc.).

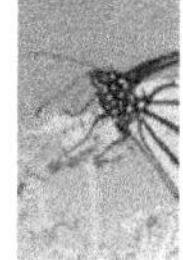

water too deep for the female walruses to feed (Kelly 2001). The increased distance between habitat suitable for adult feeding and that suitable for nursing young is likely to reduce population productivity (Kelly 2001; Grebmeier et al. 2006).

The major prey of polar bears and an important resource to Arctic Natives, ringed seals (*Pusa hispida*) are vulnerable to decreases in the snow and ice cover on the Arctic Ocean (Stirling and Derocher 1993; Tynan and DeMaster 1997; Kelly 2001). Ringed seals give birth in snow caves excavated above breathing holes they maintain in the sea ice. The snow caves protect the pups from extreme cold (Taugbøl 1984) and, to a large extent, from predators (Lydersen and Smith 1989). As the climate warms, however, snow melt comes increasingly early in the Arctic (Stone et al. 2002; Belchansky et al. 2004), and the seals' snow caves collapse before the pups are weaned (Kelly 2001; Kelly et al. 2006). The small pups are exposed without the snow cover and die of hypothermia in subsequent cold periods (Stirling and Smith 2004). The prematurely exposed pups also are more vulnerable to predation by Arctic foxes, polar bears, gulls, and ravens (Lydersen and Smith 1989). Gulls and ravens are arriving increasingly early in the Arctic as springs become warmer, further increasing their potential to prey on the seal pups.

Whether the changes underway today will be survived by walruses, seals, polar bears, and other elements of he ecosystem will depend critically on the pace of change.

Polar bears (Figure 5.6) are apex predators of the sea ice ecosystem, and their dependence on ice-associated seals makes them vulnerable to reductions in sea ice. While polar bears began diverging from brown bears (*Ursus arctos*) 150,000 to 250,000 years ago (Cronin et al. 1991; Talbot and Shields 1996; Waits et al. 1998), their specialization as seal predators in the sea ice ecosystem apparently is more recent, dating to 20,000 to 40,000 years ago (Stanley 1979; Talbot and Shields 1996). The bears' invasion of this novel environment was stimulated by an abundance of seals, which had colonized the region earlier in the Pleistocene (Deméré et al. 2003; Lister 2004). Adapting to the sea ice environment and a dependence on seals – especially ringed seals – exerted strong selection on the morphology, physiology, and behavior of polar bears.

The polar bear's morphological adaptations to the sea ice environment include dense, white fur over most of the body (including between foot pads), with hollow guard hairs; short, highly curved claws; and dentition specialized for carnivory. Physiologically, polar bears are extremely well adapted to feed on a diet high in fat; store fat for later future energy needs; and enter and sustain periods of reduced metabolism whenever food is in short supply (Derocher et al. 1990; Atkinson and Ramsay 1995). Feeding success is strongly related to ice conditions; when stable ice is over productive shelf waters, polar bears can feed throughout the year on their primary prey, ringed seals (Stirling and McEwan 1975; Stirling and Smith 1975; Stirling and Archibald 1977; Amstrup and DeMaster 1988; Amstrup et al. 2000). Less frequently, they feed on other marine mammals (Smith 1980, 1985; Calvert and Stirling 1990) and even more rarely on terrestrial foods (Lunn and Stirling 1985; Derocher et al. 1993). Polar bears exhibit the behavioral plasticity typical of top-level predators, and they are adept at capturing seals from the ice (Stirling 1974; Stirling and Derocher 1993).

Today, an estimated 20,000 to 25,000 polar bears live in 19 apparently discrete populations distributed around the circumpolar Arctic (Polar Bear Specialists Group 2006). Their overall distribution largely matches that of ringed seals, which inhabit all seasonally ice-covered seas in the Northern Hemisphere (Scheffer 1958; King 1983). Polar bears are not regularly found, however, in some of the marginal seas (e.g., the Okhotsk Sea) inhabited by ringed seals. The broad distribution of their seal prey is reflected in the home ranges of polar bears that, averaging over 125,000 km^2, are more than 200 times larger than the averages for terrestrial carnivores of similar size (Durner and Amstrup 1995; Ferguson et al. 1999). Most polar bear populations expand and contract their range seasonally with the distribution of sea ice, and they spend most of year on the ice (Stirling and Smith 1975; Garner et al. 1994). Most populations, however, retain their ancestral tie to the terrestrial environment for denning, although denning on the sea ice is common among the bears of the Beaufort and Chukchi seas (Harrington 1968; Stirling and Andriashek 1992; Amstrup and

Gardner 1994; Messeir et al. 1994; Durner et al. 2003). Dens on land and on ice are excavated in snow drifts, the stability and predictability of which are essential to cub survival (Blix and Lentfer 1979; Ramsay and Stirling 1988, 1990; Clarkson and Irish 1991).

The rapid rates of warming in the Arctic observed in recent decades and projected for at least the next century are dramatically reducing the snow and ice covers that provide denning and foraging habitat for polar bears (Overpeck et al. 1997; Serreze et al. 2000; Holland et al. 2006; Stroeve et al. 2007). These changes to their environment will exert new, strong selection pressures on polar bears. Adaptive traits reflect selection by past environments, and the time needed to adapt to new environments depends on genetic diversity in populations, the intensity of selection, and the pace of change. Genetic diversity among polar bears is evident in the 19 putative populations, suggesting some scope for adaptation within the species as a whole even if some populations will be at greater risk than others. On the other hand, the nature of the environmental change affecting critical features of polar bears' breeding and foraging habitats, and the rapid pace of change relative to the bears' long generation time (circa 15 years) do not favor successful adaptation.
The most obvious change to breeding habitats is the reduction in the snow cover on which successful denning depends (Blix and Lenter 1979; Amstrup and Gardner 1994; Messier et al. 1994; Durner et al. 2003). Female polar bears hibernate for four to five months per year in snow dens in which they give birth to cubs, typically twins, each weighing just over 0.5 kg (Blix and Lentfer 1979). The small cubs depend on snow cover to maintain thermal neutrality. Whether it remains within the genetic scope of polar bears to revert to the ancestral habit of rearing in earthen dens is unknown.

Changes in the foraging habitat that will entail new selection pressures include seasonal mismatches between the energetic demands of reproduction and prey availability; changes in prey abundance; changes in access to prey; and changes in community structure. Emergence of female and young polar bears from dens in the spring coincides with the ringed seal's birthing season, and the newly emerged bears depend on catching and consuming young seals to recover from months of fasting (Stirling and Øritsland 1995). That coincidence may be disrupted by changes in timing and duration of snow and ice cover. Such mismatches between reproductive cycles and food availability are increasingly recognized as a means by which animal populations are impacted by climate change (Stenseth and Mysterud 2002; Stenseth et al. 2002; Walther et al. 2002).

Figure 5.6 *Polar bear lounges near the Beaufort Sea, along Alaska's coastline. Image by Susanne Miller, from the U.S. Fish & Wildlife Service's digital library collection.*

The polar bear's ability to capture seals depends on the presence of ice (Stirling et al. 1999; Derocher et al. 2004). In that habitat, bears take advantage of the fact that seals must surface to breathe in limited openings in the ice cover. In the open ocean, however, bears lack a hunting platform, seals are not restricted in where they can surface, and successful predation is exceedingly rare (Furnell and Oolooyuk 1980). Only in ice-covered waters are bears regularly successful at hunting seals. When restricted to shorelines, bears feed little if at all, and terrestrial foods are thought to be of little significance to polar bears (Lunn and Stirling 1985; Ramsay and Hobson 1991; Stirling et al. 1999). Predation on reindeer observed in Svalbard, however, indicates that polar bears have some capacity to switch to alternate prey (Derocher et al. 2000).

Seal and other prey populations also will be impacted by fundamental changes in the fate of primary production. For example, in the Bering and Chukchi seas, the reduction in sea ice cover alters the physical oceanography in ways that diminish carbon flow to the benthos and increase carbon recycling in pelagic communities (Grebmeier et al. 2006). The resultant shift in community structure will include higher trophic levels. The exact composition of future communities is not known, nor is it known how effectively polar bears might exploit those communities.

Recent modeling of reductions in sea ice cover and polar bear population dynamics yielded predictions of declines within the coming century that varied by population but overall totaled 66% of all polar bears (Amstrup et al. 2007). Some populations were predicted to be extinct by the middle of the current century. While population reductions seem inevitable given the polar bear's adaptations to the sea ice environment (Derocher et al. 2004), quantitative predictions of declines are less certain as they necessarily depend on interpretations of data and professional judgments.

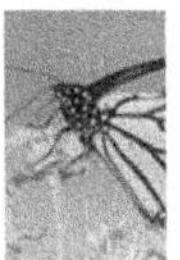

During previous climate warmings, polar bears apparently survived in unknown refuges that likely included some sea ice cover and access to seals. Within the coming century, however, the Arctic Ocean may be ice-free during summer (Overpeck et al. 2005), and the polar bear's access to seals will be diminished (Stirling and Derocher 1993; Lunn and Stirling 2001; Derocher et al. 2004). As snow and ice covers decline, polar bears may respond adaptively to the new selection pressures or they may become extinct. Extinction could result from mortality outpacing production, competition in terrestrial habitats with brown bears, and/or from re-absorption into the brown bear genome. Crosses between polar bears and brown bears produce fertile offspring (Kowalska 1965), and a hybrid was recently documented in the wild. Extinction through hybridization has been documented in other mammals (Rhymer and Simberloff 1996).

Predicted further warming inevitably will entail major changes to the sea ice ecosystem. Some ice-adapted species will become extinct; others will adapt to new habitats. Whether the changes underway today will be survived by walruses, seals, polar bears, and other elements of the ecosystem will depend critically on the pace of change. Ecosystems have changed before; species have become extinct before. Critically important in our changing climate is the rapid rate of change. Biological adaptation occurs over multiple generations varying from minutes to many years depending on the species. The current rates of change in the sea ice ecosystem, however, are very steep relative to the long generation times of long-lived organisms such as seals, walruses, and polar bears.

5.7 ECOSYSTEM SERVICES AND EXPECTATIONS FOR FUTURE CHANGE

The Millennium Ecosystem Assessment (2005) is the most comprehensive scientific review of the status, trends, conditions, and potential futures for ecosystem services. It is international in coverage, although individual sections focus on regions, ecosystem types, and particular ecosystem services. The MEA categorized services as supporting, provisioning, regulating, and cultural (Figure 5.7). Some of these services are already traded in markets, e.g. the provision of food, wood, and fiber from both managed and unmanaged ecosystems, or the cultural services of providing recreational activities, which generate substantial revenue both within the United States and globally. The United States, for example, has a $112 billion international tourism market and domestic outdoor recreation market (World Trade Organization 2002; Southwick Associates 2006).

Other services, in particular many cultural services, regulating services, and supporting services are not priced, and therefore not traded in markets. A few, like provision of fresh water or carbon sequestration potential, are mostly not traded in markets, but could be, and especially for carbon, there are many developing markets. In all cases, the recognition of a service provided by ecosystems is the recognition that they are producing or providing something of value to humans, and thus its value is shaped by the social dimensions and values of our societies as well as by physical and ecological factors (MEA 2005).

An example of an ecosystem service that has an increasingly recognized value is that provided by pollinators. Part of this increased recognition is a consequence of the recent declines in abundance that have been observed for some pollinators, particularly the introduced honey bee (*Apis mellifera*) (National Research Council 2006). The economic significance of pollination is underscored by the fact that about three-quarters of the world's flowering plants depend on pollinators, and that almost a third of the food that we consume results from their activity. The majority of pollinators are insects, whose distributions, phenology, and resources are all being affected by climate change (Inouye 2007). For example, an ongoing study at the Rocky Mountain Biological Laboratory (Pyke, Thomson, Inouye, unpublished) has found evidence that some bumble bee species have moved up as much as a few thousand feet in elevation over the past 30 years. Unfortunately, with the exception of honey bees and butterflies, there are very few data available on the abundance and distribution of pollinators, so it has been difficult to assess their status and the changes that they may be undergoing (National Research Council 2006).

Biological diversity is recognized as providing an underpinning for all these services in a fundamental way. A major finding of the MEA from a global perspective was that 16 out of 24 different ecosystems services that were analyzed were being used in ultimately unsustainable ways. While this finding was not specific to U.S. ecosystems, it does set a context for considering the consequences of documented ecosystem changes for services.

A subsequent question is whether any such changes in services can be reasonably attributed to climate change. The MEA evaluated the relative magnitudes and importance of a number of different direct drivers (Nelson et al. 2006) for changes in ecosystems, and whether the importance of those drivers was likely to increase, decrease, or stay about the same over the next several decades. The conclusion was that although climate change was not currently the most important driver of change in many ecosystems, it was one of the only drivers whose importance was likely to continue to increase in all ecosystems over the next several decades (Figure 5.8).

5.8 ADEQUACY OF OBSERVING SYSTEMS

One of the challenges of understanding changes in biological diversity related to variability and change in the physical climate system is the adequacy of the variety of monitoring programs that exist for documenting those changes.

It is useful to think about such programs as falling into three general categories. The first is the collection of operational monitoring systems that are sponsored by federal agencies, conservation groups, state agencies, or groups of private citizens that are focused on particular taxa (e.g., the Breeding Bird Survey) or particular ecosystems (e.g., Coral Reef Watch). These tend to have been established for very particular purposes, such as for tracking the abundance of migratory songbirds, or the status and abundance of game populations within individual states, or the status and abundance of threatened and endangered species.

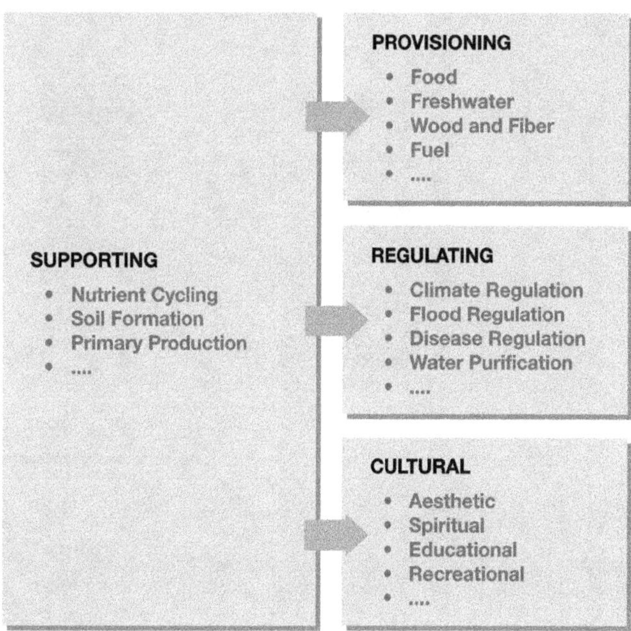

Figure 5.7 *Categorization of ecosystem services, from MEA 2005.*

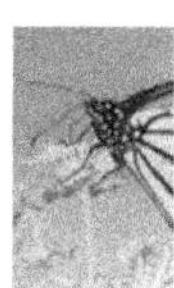

The second category of monitoring programs is those whose initial justification has been to investigate particular research problems, whether or not those are primarily oriented around biodiversity. For example, the existing Long-Term Ecological Research Sites (LTERs) are important for monitoring and understanding trends in biodiversity in representative U.S. biomes, although their original justification was oriented around understanding ecosystem functioning. The yet-to-be established National Ecological Observatory Network (NEON) would also fall into this category. NEON's design for site locations samples both climate variability and ecological variability within the United States in a much more systematic way than ever before done for a long-term research network, so there are likely to be very powerful results that can come from network-wide analyses.

The third category of monitoring systems is those that offer the extensive spatial and variable temporal resolution of remotely sensed information from Earth-orbiting satellites. These are not always thought of as being part of the nation's system for monitoring biological diversity, but in fact, they are an essential component of it. Remotely sensed data are the primary source of information on a national scale for documenting land-cover and land-cover change across the United States, for example, and thus they are essential for tracking

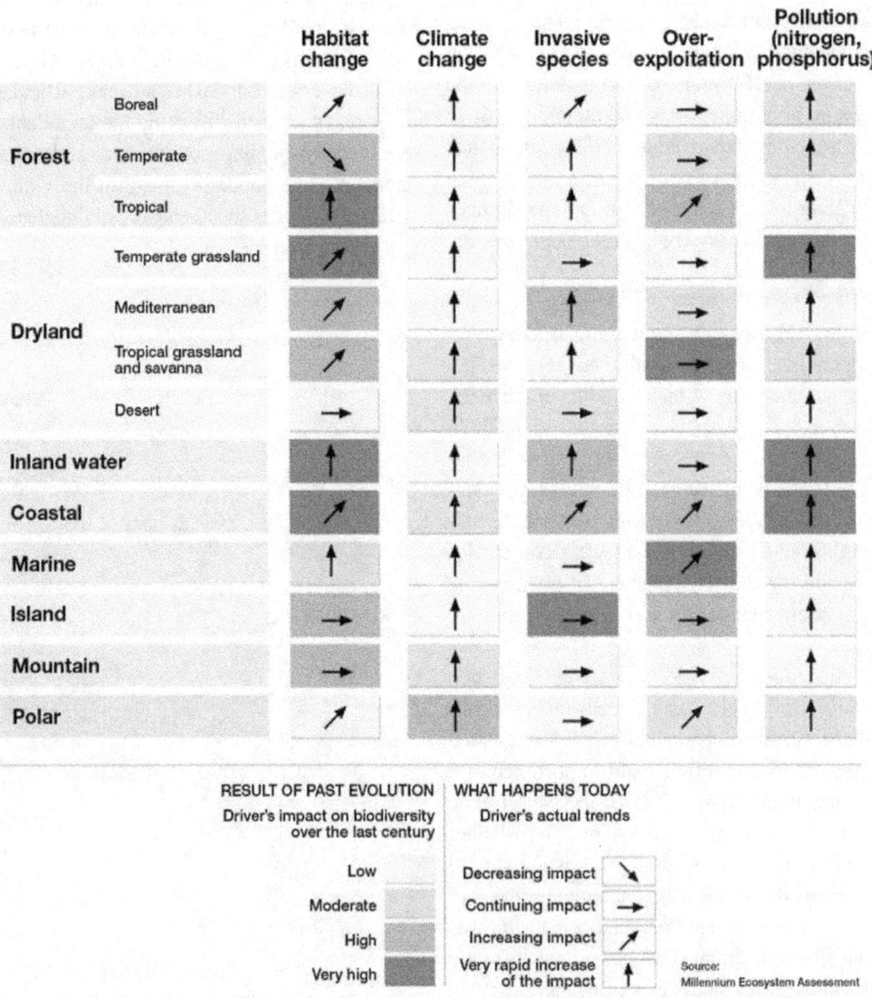

Figure 5.8 *Relative changes in magnitude to ecosystem services caused by changes in habitat, climate, species invasion, over-exploitation, and pollution.*

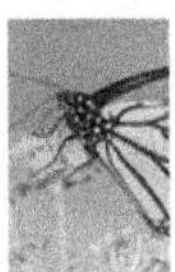

changes in perhaps the biggest single driver of changes in biodiversity, i.e. changes in habitat. Over the decades of the 1990s and 2000s, the remarkable profusion of Earth observation satellites has provided global coverage of many critical environmental parameters, from variability and trends in the length of growing season, to net primary productivity monitoring, to the occurrence of fires, to the collection of global imagery on 30-meter spatial resolution for more than a decade. Observational needs for biodiversity monitoring and research were recently reviewed by the International Global Observations of Land Panel, in a special report from a conference (IGOL 2006).

The National Research Council has recently released the first-ever Decadal Survey for Earth Science and Observations (NRC, 2007), which makes a comprehensive set of recommendations for future measurements and missions that would simultaneously enhance scientific progress, preserve essential data sets, and benefit a wide variety of potential applications. The report found that "the extraordinary U.S. foundation of global observations is at great risk. Between 2006 and the end of the decade, the number of operating missions will decrease dramatically, and the number of operating sensors and instruments on NASA spacecraft, most of which are well past their nominal lifetimes, will decrease by some 40 percent."

Although there are lists of specifications for monitoring systems that would be relevant and important for recording changes in biodiversity associated with climate variability and change (e.g., IGOL 2007), at present there is no analysis in the literature that directly addresses the question of whether existing monitoring systems are adequate. For the moment, there is no viable option but to use existing systems for recording biodiversity changes, even if it means that the scientific community is attempting to use systems originally designed for alternate purposes.

The Government Accountability Office (2007) has documented extensively that one of the greatest perceived needs of federal land management agencies is for targeted monitoring systems that can aid them in responding to climate change. These agencies (e.g., U.S. Forest Service, National Park Service, Bureau of Land Management, NOAA, the U.S. Geological Survey) each face situations in which they recognize that they are already beginning to see the biological and ecological impacts of climate change on resources that they manage. GAO identified the improvement of monitoring capabilities to formulate effective adaptation and management responses as a priority. Remedying this situation was identified as a critical priority.

The Government Accountability Office (2007) has documented extensively that one of the greatest perceived needs of federal land management agencies is for targeted monitoring systems that can aid them in responding to climate change.

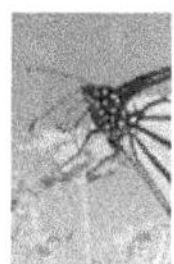

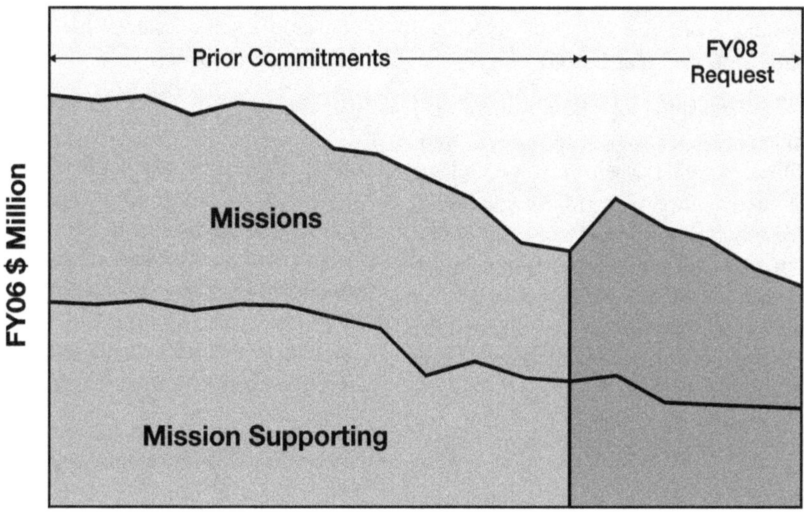

Figure 5.9 *NASA budget for earth science research and applications demonstrations (1996–2012, fixed 2006 dollars).*

5.9 MAJOR FINDINGS

In this section, we list the major findings from each section of the chapter, by topic heading. We then draw some general conclusions about the observed and potential impacts of climate change on biological diversity, the relationships to ecosystem services, and the adequacy of existing monitoring systems to document continuing change.

5.9.1 Growing Season and Phenology

- There is evidence indicating a significant lengthening of the growing season and higher NPP in the higher latitudes of North America where temperature increases are relatively high.
- Over the last 19 years, global satellite data indicate earlier onset of spring across the temperate latitudes by 10–14 days (Myneni, 2001; Lucht, 2002}, an increase in summer photosynthetic activity (normalized difference vegetation index satellite estimates) (Myneni 2001) and an increase in the amplitude of annual CO2 cycle (Keeling 1996), all of which are supported by climatological and field observations.
- Forest productivity, in contrast, which is generally limited by low temperature and short growing seasons in the higher latitudes and elevations, has been slowly increasing at less than 1 percent per decade (Boisvenue and Running 2006; Joos et al.2002; McKenzie 2001; Caspersen et al. 2000).
- The exception to this pattern is in forested regions that are subject to drought from climate warming, where growth rates have decreased since 1895 (McKenzie 2001). Recently, widespread mortality over 12,000 km^2 of lower elevational forest in the southwest United States demonstrates the impacts of increased temperature and the associated multiyear drought (Breshears et al. 2005) even as productivity at tree line had increased previously (Swetnam and Betancourt 1998).
- Disturbances created from the interaction of drought, pests, diseases, and fire are projected to have increasing impacts on forests and their future distributions (IPPC 2007).

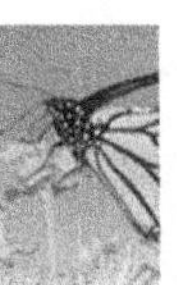

5.9.2 Biogeographical and Phenological Shifts

- Evidence from two meta-analyses (Root et al. 2003; Parmesan 2003) and a synthesis (Parmesan 2006) on species from a broad array of taxa suggests that there is a significant impact of recent climatic warming in the form of long-term, large-scale alteration of animal and plant populations.
- Movement of species in regions of North America in response to climate warming is expected to result in shifts of species ranges poleward and upward along elevational gradients (Parmesan, 2006).
- In an analysis of 866 peer-reviewed papers exploring the ecological consequences of climate change, nearly 60 percent of the 1598 species studied exhibited shifts in their distributions and/or phenologies over a 20- and 140-year time frame {Parmesan 2003).
- Analyses of field-based phenological responses have reported shifts as great as 5.1 days per decade (Root et al. 2003), with an average of 2.3 days per decade across all species (Parmesan and Yohe 2003).

5.9.3 Migratory Birds

- A climate change signature is apparent in the advancement of spring migration phenology (Root et al. 2003) but the indirect effects may be more important than the direct effects of climate in determining the impact on species persistence and diversity.

5.9.4 Butterflies

- The migration of butterflies in the spring is highly correlated with spring temperatures and with early springs. Researchers have documented many instances of earlier arrivals (26 of 35 species in the United Kingdom (Roy and Sparks 2000); 17 of 17 species in Spain, (Stefanescu 2004); and 16 of 23 species in central California (Forister and Shapiro 2003).
- Butterflies are also exhibiting distributional and/or range shifts in response to warming. Across all studies included in her synthesis, Parmesan (2006) found 30–75 percent of

species had expanded northward, less than 20 percent had contracted southward, and the remainder were stable (Parmesan 2006).

5.9.5 Coastal and Near Shore Systems

- In the tropics there have been increasing coral bleaching and disease events and increasing storm intensity.
- In temperate regions there are demonstrated range shifts in rocky intertidal organisms, coastal fisheries, and in marine fisheries as well, and possible alterations of ocean currents and upwelling sites.

5.9.6 Corals

- Corals and tropical regions where they live are experiencing increasing water temperatures, increasing storm intensity (Emmanuel 2005), and a reduction in pH (Ravens et al. 2005), all while experiencing a host of other on-going challenges from development/tourism, unsustainable fishing and pollution. Acidification presents a persistent threat that is increasing in magnitude for shallow water corals and free-swimming calcifying organisms.
- Corals in many tropical regions are experiencing substantial mortality from increasing water temperatures and intense storms, both of which could be exacerbated by a reduction in pH. Increases in ocean acidity are a direct consequence of increases in atmospheric carbon dioxide.

5.9.7 Marine Fisheries

- Large, basin-scale atmospheric pressure systems that drive basin-scale winds can suddenly shift their location and intensity at decadal time scales, with dramatic impacts on winds and ocean circulation patterns. Perhaps the greatest discovery of the past 10 years is that these shifts also have powerful impacts on marine ecosystems.
- Examples of ecosystems impacts include increased flow of oceanic water into the English Channel and North Sea resulting in a northward shift in the distribution of zooplankton. As a result, the zooplankton community became dominated by warm water species (Beaugrand, 2004) with concomitant changes in fish communities from one dominated by whiting (hake) to one dominated by sprat (similar to a herring).
- Similar (and drastic) ecosystem changes are known for the Baltic Sea (Kenny and Mollman 2006), where drastic changes in both zooplankton and fish communities were observed. Cod were replaced by sprat and dominance in zooplankton switched from lipid-rich (and high bioenergetic content) species to lipid-poor species.
- Linkages between the NAO, zooplankton and fisheries have also been described for the Northwest Atlantic, including waters off eastern Canada and the United States; Pershing and Green (2007) report a decrease in salinity, and an increase in biomass of small copepods (zooplankton).

5.9.8 Pests and Pathogens

- Evidence is beginning to accumulate that links the spread of pathogens to a warming climate. For example, the chytrid fungus (Batrachochytrium dendrobatidis) is a pathogen that is rapidly spreading worldwide, and decimating amphibian populations. A recent study by Pounds and colleagues (2006) showed that widespread amphibian extinction in the mountains of Costa Rica is positively linked to global climate change.
- To date, geographic range expansion of pathogens related to warming temperatures have been the most easily detected (Harvell et al. 2002), perhaps most readily for arthropod-borne infectious disease (Daszak et al. 2000). However, a recent literature review found additional evidence gathered through field and laboratory studies that support hypotheses that latitudinal shifts of vectors and diseases are occurring under warming temperatures.

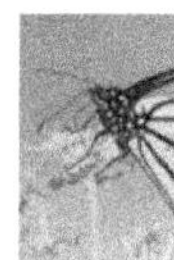

5.9.9 Invasive Plants

- Projected increases in CO_2 are expected to stimulate the growth of most plants species, and some invasive plants are expected to respond with greater growth rates than non-invasive plants. Some invasive plants may have higher growth rates and greater maximal photosynthetic rates relative to native plants under increased CO_2, but definitive evidence of a general benefit of CO_2 enrichment to invasive plants over natives has not emerged (Dukes and Mooney 1999).

- Nonetheless, invasive plants in general may better tolerate a wider range of environmental conditions and may be more successful in a warming world because they can migrate and establish in new sites more rapidly than native plants and they are not usually limited by pollinators or seed dispersers (Vila et al., in press, accepted).
- Finally, it is critical to recognize that other elements of climate change (e.g., nitrogen deposition, land conversion) will play a significant role in the success of invasive plants in the future, either alone or under elevated CO_2 (Vila et. al., in press, accepted).

5.9.10 Particularly Sensitive Systems

- One group of organisms whose reproductive phenology is closely tied to snowmelt is amphibians, for which this environmental cue is apparently more important than temperature.
- Hibernating and migratory species that reproduce at high elevations during the summer are also being affected by the ongoing environmental changes. For example, marmots are emerging a few weeks earlier than they used to in the Colorado Rocky Mountains, and robins are arriving from wintering grounds weeks earlier in the same habitats. Species such as deer, bighorn sheep, and elk, which move to lower elevations for the winter, may also be affected by changing temporal patterns of snowpack formation and disappearance.

- There is a very strong correlation between the timing of snowmelt, which integrates snowpack depth and spring air temperatures, and the beginning of flowering by wildflowers in the Colorado Rocky Mountains.

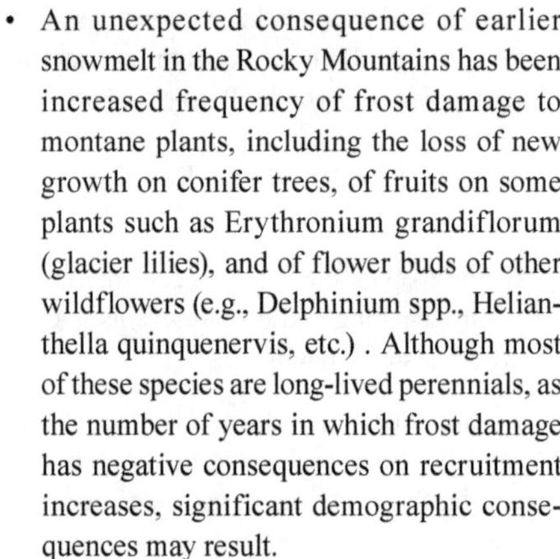

- An unexpected consequence of earlier snowmelt in the Rocky Mountains has been increased frequency of frost damage to montane plants, including the loss of new growth on conifer trees, of fruits on some plants such as Erythronium grandiflorum (glacier lilies), and of flower buds of other wildflowers (e.g., Delphinium spp., Helianthella quinquenervis, etc.) . Although most of these species are long-lived perennials, as the number of years in which frost damage has negative consequences on recruitment increases, significant demographic consequences may result.

5.9.11 Arctic Sea-Ice Ecosystems

- Today, an estimated 20,000–25,000 polar bears live in 19 apparently discrete populations distributed around the circumpolar Arctic (IUCN Polar Bear Specialists Group 2005). Their overall distribution largely matches that of ringed seals, which inhabit all seasonally ice-covered seas in the northern hemisphere (Scheffer 1958; King 1983), an area extending to approximately 15,000,000 km^2.
- Most polar bear populations expand and contract their range seasonally with the distribution of sea ice, and they spend most of year on the ice (Stirling and Smith 1975; Garner et al. 1994).
- The rapid rates of warming in the Arctic observed in recent decades and projected for at least the next century are dramatically reducing the snow and ice covers that provide denning and foraging habitat for polar bears (Roots 1989; Overpeck et al. 1997; Serreze et al. 2000; Stroeve et al. 2007).
- During previous climate warmings, polar bears apparently survived in some unknown refuges. Whether they can withstand the more extreme warming ahead is doubtful (Stirling and Derocher 1993; Lunn and Stirling 2001).

5.10 CONCLUSIONS

Terrestrial and marine systems are already being demonstrably affected by climate change. This conclusion can be made with very high confidence. There are observable impacts of climate change on terrestrial ecosystems in North America including changes in the timing of growing season length, phenology, primary production, and species distributions and diversity. Some important effects on components of biological diversity have already been observed and have been increasingly well documented over the past several decades. This statement is true both for U.S. ecosystems, and ecosystems and biological resources around the world (IPCC 2007).

There is a family of other impacts and changes in biodiversity that are theoretically possible, and even probable (e.g., mismatches in phenologies between pollinators and flowering plants), but for which we do not yet have a substantial observational database. However, we cannot conclude that the lack of a complete observational database in these cases is evidence that they are not occurring – it is just as likely that it is simply a matter of insufficient numbers or lengths of observations.

It is difficult to pinpoint changes in ecosystem services that are specifically related to changes in biological diversity in the United States. The Millennium Ecosystem Assessment (2005) is the most recent, and most comprehensive, scientific assessment of the state of ecosystem services around the world, the drivers of changes in both ecosystems and services, the inherent tradeoffs among different types of ecosystem services, and what the prospects are for sustainable use of ecological resources. The MEA concludes that climate change is very likely to increase in importance as a driver for changes in biodiversity over the next several decades, although for most ecosystems it is not currently the largest driver of change. But a specific assessment of changes in ecosystem services for the United States as a consequence of changes in climate or other drivers of change has not been done.

We can think of the monitoring systems that have been used to evaluate the relationship between changes in the physical climate system and biological diversity as having three components.

- There is a plethora of species-specific or ecosystem-specific monitoring systems, variously sponsored by the U.S. federal agencies, state agencies, conservation organizations, and other private organizations. However, in very few cases were these monitoring systems established with climate variability and climate change in mind.
- Augmenting the monitoring systems that make routine measurements is a set of more specific research activities that have been designed to create time-series of population data and associated climatic and environmental data.
- The third component is spatially extensive observations derived from remotely sensed data. Some of these satellite data are primarily focused on land-cover, and thus are a good indicator of the major single driver of changes in biodiversity patterns. Others produce estimates of NPP and changes in the growing season, and thus reflect functional changes in ecosystems. However, similarly to the in situ monitoring networks, the future of space-based observations is not assured. The NRC (2007) has recently released a major survey of data and mission needs for the Earth sciences to address this issue, so we will not pursue it further here.

Synthesis

Lead Author: D. Schimel, A.C. Janetos, P. Backlund

Contributing Authors: J. Hatfield, D.P. Lettenmaier, M.G. Ryan

6.1 INTRODUCTION

The preceding chapters have focused on the observed and potential impacts of climate variability and change on U.S. agriculture, land resources, water resources, and biodiversity. This section synthesizes information from those sectoral chapters to address a series of questions that were posed by the CCSP agencies in the prospectus for this report and formulate a set of overarching conclusions.

6.2 KEY QUESTIONS AND ANSWERS

CCSP Question: What factors influencing agriculture, land resources, water resources, and biodiversity in the United States are sensitive to climate and climate change?

Climate change affects average temperatures and temperature extremes, timing and geographical patterns of precipitation; snowmelt, runoff, evaporation and soil moisture; the frequency of disturbances, such as drought, insect and disease outbreaks, severe storms, and forest fires; atmospheric composition and air quality, and patterns of human settlement and land use change. Thus, climate change leads to myriad direct and indirect effects on U.S. ecosystems. Warming temperatures have led to effects as diverse as altered timing of bird migrations, increased evaporation and longer growing seasons for wild and domestic plant species. Increased temperatures often lead to a complex mix of effects. Warmer summer temperatures in the western U.S. have led to longer forest growing seasons, but have also increased summer drought stress, increased vulnerability to insect pests and increased fire hazard. Changes to precipitation and the size of storm events affect plant-available moisture, snowpack and snowmelt, streamflow, flood hazard, and water quality.

Further Details: The direct effects of changes to air temperature and precipitation are relatively well understood, though some uncertainties remain. This report emphasizes that a second class of climate changes are also very important. Changes to growing season length are now documented across most of the country and affect crops, snowmelt and runoff, productivity, and vulnerability to insect pests. Earlier warming has profound effects, ranging from changes to horticultural systems to expansion of the mountain pine beetleís range. Changes to humidity, cloudiness, and radiation may reflect both the effect of anthropogenic aerosols and the global hydrological systemís response to warming at the surface, humidity, and, hence, evaporation. Since plants and, in some cases, disease organisms are very sensitive to the near-surface humidity and radiation environment, this is emerging as an important ìhiddenî global change. Finally, changes to temperature

... climate change leads to myriad direct and indirect effects on U.S. ecosystems.

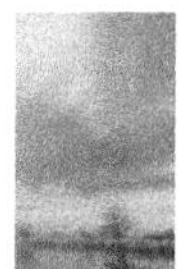

and water are hard to separate. Increasing temperatures can increase evapotranspiration and reduce the growing season by depleting soil moisture sooner, reduce streamflow and degrade water quality, and even change boundary layer humidity.

Disturbance (such as drought, storms, insect outbreaks, grazing, and fire) is part of the ecological history of most ecosystems and influences ecological communities and landscapes. Climate affects the timing, magnitude, and frequency of many of these disturbances, and a changing climate will bring changes in disturbance regimes to forests and arid lands. Ecosystems can take from decades to centuries to re-establish after a disturbance. Both human-induced and natural disturbances shape ecosystems by influencing species composition, structure, and function (productivity, water yield, erosion, carbon storage, and susceptibility to future disturbance). Disturbances and changes to the frequency or type of disturbance present challenges to resource managers. Many disturbances command quick action, public attention, and resources.

Climate change may also ameliorate stress. Carbon dioxide "fertilization," increased growing-season length, and increased rainfall may increase productivity of some crops and forests, increase carbon storage in forests, and reduce water stress in arid land and grazing land ecosystems.

Climate and air quality –chemical climate–also interact. Nitrogen deposition has major chemical effects in ecosystems. It can act as a fertilizer increasing productivity, but can also contribute to eutrophication. High levels of deposition have been associated with loss of species diversity and increased vulnerability to invasion, and there is some evidence that climate change exacerbates these effects. On the other side of the ledger, increases in atmospheric CO_2 and nitrogen availability can increase crop yields if soil water is available.

Climate change can also interact with socioeconomic factors. For example, how crop responses to changing climate are managed can depend on the relative demand and price of different commodities. Climate change mitigation practices, such as the promotion of biofuel crops, can also have a major impact on the agricultural system by increasing the demand and prices for some crops.

CCSP Question: How could changes in climate exacerbate or ameliorate stresses on agriculture, land resources, water resources, and biodiversity? What are the indicators of these stresses?

Ecosystems and their services (land and water resources, agriculture, biodiversity) experience a wide range of stresses, including effects of pests and pathogens, invasive species, air pollution, extreme events and natural disturbances such as wildfire and flood. Climate change can cause or exacerbate direct stress through high temperatures, reduced water availability, and altered frequency of extreme events and severe storms. Climate change can also modify the frequency and severity of other stresses. For example, increased minimum temperatures and warmer springs extend the range and lifetime of many pests that stress trees and cops. Higher temperatures and/or decreased precipitation increase drought stress on wild and crop plants, fruit and nut trees, animals and humans. Reduced water availability can lead to increased withdrawals from rivers, reservoirs, and groundwater, with consequent effects on water quality, stream ecosystems, and human health.

Further Details: Changes to precipitation frequency and intensity can have major effects. More intense storms lead to increased soil erosion, flooding, and decreased water quality (by transporting more pollutants into water bodies through runoff or leaching through soil layers), with major consequences for life and property. Changing timing, intensity and amount of precipitation can reduce water availability or the timing of water availability, potentially increasing competition between biological and consumptive water use at critical times. Flushing of pollutants into water bodies or concentration of contaminants during low-flow intervals can increase the negative consequences of effects of other stresses, such as those resulting from development, land use intensification, and fertilization.

Climate change may also ameliorate stress. Carbon dioxide ìfertilization,î increased growing-season length, and increased rainfall may increase productivity of some crops and forests, increase carbon storage in forests, and reduce water stress in arid land and grazing land ecosystems. Increased minimum temperatures during winter can reduce winter mortality in crops and wild plants and reduce low-temperature stresses on livestock. Increased rainfall can increase groundwater recharge, increase water levels in lakes and reservoirs, and flow levels in rivers. Increased river levels tend to reduce water temperatures and, other things being equal, can help limit the warming of water that might otherwise occur.

Indicators of climate change-related stress are incredibly diverse. Even a short list includes symptoms of temperature and water stress, such as plant and animal mortality, reduced productivity, reduced soil moisture and streamflow, increased eutrophication and reduced water quality, and human heat stress. Indicators of stress can also include changes in species ranges, occurrence and abundance of temperature- or moisture-sensitive invasive species and pest/pathogen organisms, and altered mortality and morbidity from climate-sensitive pests and pathogens. Many stresses are tied to changes in seasonality. Early warning indicators include timing of snowmelt and runoff, as early snowmelt has been related to increased summer-time water stress, leading to reduced plant growth, and increased wildfire and insect damage in the western U.S. Phenology can provide warning of stresses in many ways. Changes to crop phenology may presage later problems in yield or vulnerability to damage, changes to animal phenology (for example, timing of breeding) may come in advance of reduced breeding success and long-term population declines. Changes in the abundance of certain species, which may be invasive, rare, or merely indicative of changes, can provide warning of stress. For example, some C4 plants may be indicative of temperature or water stress, while other species reflect changes to nitrogen availability. Changes to the timing of animal migration may indicate certain types of stress, although some migration behavior also responds to opportunity (e.g., food supply or habitat availability).

CCSP Question: What current and potential observation systems could be used to monitor these indicators?

A wide range of observing systems within the United States provides information on environmental stress and ecological responses. Key systems include National Aeronautics and Space Administration (NASA) research satellites, operational satellites and ground based observing networks from the National Oceanic and Atmospheric Administration (NOAA) in the Department of Commerce, USDA forest and agricultural survey and inventory systems, Department of Interior/U.S. Geological Survey (USGS) stream gauge networks, Environmental Protection Agency (EPA) and state-supported water quality observing systems, the Department of Energy (DOE) Ameriflux network, and the LTER network and the proposed National Ecological Observing Network (NEON) sponsored by the National Science Foundation (NSF). However, many key biological and physical indicators are not currently monitored, are monitored haphazardly or with incomplete spatial coverage, or are monitored only in some regions. In addition, the information from these disparate networks is not well integrated. Almost all of the networks were originally instituted for specific purposes unrelated to climate change, and are challenged by adapting to new questions.

Climate change presents new challenges for operational management. Understanding climate impacts requires both monitoring many aspects of climate and a wide range of biological and physical responses. Understanding climate change impacts in the context of multiple stresses and forecasting future services requires an integrated analysis approach. Beyond the problems of integrating the data sets, the nation has limited operational capability for integrated ecological monitoring, analyses and forecasting. A few centers exist, aimed at specific questions and/or regions, but no coordinating agency or center has the mission to conduct integrated environmental analysis and assessment by pulling this information together. Operational weather and climate forecasting provides an analogy. Weather-relevant observations are collected in many ways, ranging from surface observations through radiosondes to operational and research satellites. These data are used as

the basis for analysis, understanding, and forecasting of weather through highly integrative analyses blending data and models in a handful of university, federal and private centers. This activity requires substantial infrastructure to carry out operationally and depends on multi-agency federal, university and private sector research for continual improvement. By contrast, no such integrative analysis of comprehensive ecological information is carried out, although the scientific understanding and societal needs have probably reached the level where an integrative and operational approach is both feasible and desirable.

Further Details: Operational and research satellite remote sensing provides a critical capability. Satellite observations have been used to detect a huge range of stresses, including water stress (directly and via changes to productivity), invasive species, effects of air pollution, changing land use, wildfire, spread of insect pests, and changes to seasonality. The latter is crucial, as much of what we know about changing growing season length comes from satellite observations. Changing growing seasons and phenology are crucial indicators of climate and climate stress on ecosystems. Aircraft remote sensing complements satellite remote sensing, and provides higher resolution and, in some cases, additional sensor types that are useful in monitoring ecosystems. Remote sensing also provides essential spatial context for site-based measurements, that, when used in the appropriate analysis framework, allows the results of local studies to be applied over regions large enough to be useful in management.

Broad purpose climate and weather networks are complemented by more specialized networks.

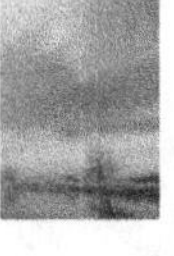

Ground-based measurements, such as USDA forest and agricultural surveys ,provide regular information on productivity of forest, rangeland, and crop ecosystems, stratified by region and crop type. Somewhat parallel information is reported on diseases, pathogens, and other disturbances, such as wind and wildfire damage. Current systems for monitoring productivity are generally more comprehensive and detailed than surveys of disturbance and damage. Agricultural systems are monitored much more frequently than are forest ecosystems, due to differences in both the ecological and economic aspects of the two systems.

Climate stress itself is monitored in a number of ways. NOAA operates several types of observing networks for weather and climate, providing detailed information on temperature and precipitation, somewhat less highly resolved information on humidity and incoming solar resolution, and additional key data products, such as drought indices and forecasts, and flood forecasts and analyses. DOEís ARM network provides key process information on some atmospheric processes affecting surface radiation balance, but at a limited number of sites. The USDA SNOTEL network provides partial coverage of snowfall and snowmelt in high elevation areas, though many of the highest and snowiest mountain ranges have sparse coverage. Several even more detailed meteorological networks have been developed, such as the Oklahoma Mesonet, which provide dense spatial coverage, and some additional variables.

Broad purpose climate and weather networks are complemented by more specialized networks. For example, the Ameriflux network focuses on measuring carbon uptake by ecosystems using micrometeorological techniques, and also provides very detailed measurements of the local microclimate. The National Atmospheric Deposition Network monitors deposition of nitrogen and other compounds in rainwater across the continent, while several sparser networks monitor dry deposition. Ozone is extensively monitored by the Environmental Protection Agency, though rural sites are sparse compared to urban because of the health impacts of ozone. The impact of ozone on vegetation, though believed to be significant, is less well observed. Water resources are monitored as well. Streamflow is best observed through the USGS networks of stream gauges. The number of watersheds, of widely varying scale, and the intensity of water use in the United States make monitoring in-stream water extremely complicated. Establishing basic trends thus requires very careful analysis. Lake and reservoir levels are fairly well measured. Groundwater, though critical for agricultural and urban water use in many areas, remains poorly observed and understood, and very few observations of soil moisture exist.

In addition to observing networks developed for operational decision making, several important research networks have been established. The Ameriflux network was described above. The LTER network spans the United States, and includes polar and oceanic sites as well. LTER provides understanding of critical processes, including processes that play out over many years, at sites in a huge range of environments, including urban sites. While the LTER network does not emphasize standardized measurements (but rather addresses a core set of issues, using site-adapted methods), the proposed new NEON program will implement a set of standardized ecological sensors and protocols across the county.

Because climate change is just one of the multiple stresses affecting ecosystems, the fact that the existing observing systems are at best loosely coordinated is a major limitation. Interoperability of data remains an issue, despite efforts in standardization and metadata, and co-location of observations of drivers and responses to change occurs only haphazardly. This contributes to the difficulties discussed below of quantifying the relative contributions of climate change and other stressors.

CCSP Question: Can observation systems detect changes in agriculture, land resources, water resources, and biodiversity that are caused by climate change, as opposed to being driven by other causal activities?

In general, the current suite of U.S. observing systems provides a reasonable overall ability to monitor ecosystem change and health in the United States, but neither the observing systems or the current state of scientific understanding are adequate to rigorously quantify climate contributions to ecological change and separate these from other influences. It is very difficult, and in most cases, not practically feasible, to quantify the relative influences of individual stresses, including climate change, through observations alone.

In the case of **agriculture**, monitoring systems for measuring long-term response of agricultural lands are numerous, but integration across these systems is limited. In addition, at present, there are no easy and reliable means to accurately ascertain the mineral and carbon state of agricultural lands, particularly over large areas.

For **land resources**, current observing systems are very likely inadequate to separate the effects of changes in climate from other effects. There is no coordinated national network for monitoring changes associated with disturbance (except for forest fires) and alteration of land cover and land use. Attempts to date lack spatial or temporal resolution, or the necessary supporting ground truth measurements, to themselves adequately distinguish climate change influences. Separating the effects of climate change from other impacts would require a broad network of indicators, coupled with a network of controlled experimental manipulations.

Essentially no aspect of the current hydrologic observing system was designed specifically for purposes of detecting climate change or its effects on **water resources**. Many of the existing systems are technologically obsolete, are designed to achieve specific, often non-compatible management accounting goals, and/or their operational and maintenance structures allow for significant data collection gaps. As a result, the data is fragmented, poorly integrated, and very likely unable to meet the predictive challenges of a rapidly changing climate.

In the case of **biodiversity**, there is a collection of operational monitoring systems that are sponsored by federal agencies, conservation groups, state agencies, or groups of private citizens that are focused on particular taxa (e.g. the Breeding Bird Survey), or on particular ecosystems (e.g. Coral Reef Watch), or even particular phenomena (e.g. the National Phenology Network). These tend to have been established for very particular purposes, e.g. for tracking the abundance of migratory songbirds, or the status and abundance of game populations within individual states, or the status and abundance of threatened and endangered species. There is a second category of monitoring programs whose initial justification has been to investigate particular research problems, whether or not those are primarily oriented around biodiversity. The third category of monitoring systems is those that offer the extensive spatial and variable temporal resolution of remotely sensed information from Earth-orbiting satellites. None of these existing monitoring systems are likely to be completely adequate for monitoring changes in biodiversity associated with climate variability and change.

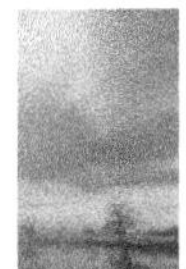

Although there are lists of specifications for monitoring systems that would be relevant and important for this purpose (e.g. IGOL 2007), there is at present no analysis in the literature that has addressed this question directly.

So for the moment, there is no viable alternative to using the existing systems for identifying climate change and its impacts on U.S. agriculture, land resources, water resources, and biodiversity, even though these systems were not originally designed for this purpose. There has obviously been some considerable success so far in doing so, but there is limited confidence that the existing systems provide a true early warning system capable of identifying potential impacts in advance. The authors of this report also have very limited confidence in the ability of current observation and monitoring systems to provide the information needed to evaluate the effectiveness of actions that are taken to mitigate or adapt to climate change impacts. Furthermore, we emphasize that improvements in observations and monitoring of ecosystems, while desirable, are not sufficient by themselves for increasing our understanding of climate change impacts. Experiments that directly manipulate climate and observe impacts are critical for developing more detailed information on the interactions of climate and ecosystems, attributing impacts to climate, differentiating climate impacts from other stresses, and designing and evaluating response strategies. Institutional support for such experiments is a concern.

So for the moment, there is no viable alternative to using the existing systems for identifying climate change and its impacts on U.S. agriculture, land resources, water resources, and biodiversity, even though these systems were not originally designed for this purpose.

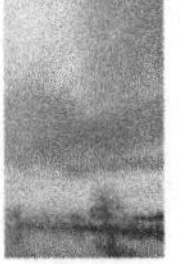

Further Details: One of the great challenges of understanding climate change impacts is that these changes are superimposed on an already rapidly changing world. Ecosystems across the United States are subject to a wide variety of stresses, most of which inevitably act on those systems simultaneously. It is rare in these cases for particular responses of ecosystems to be diagnostic of any individual stress – ecosystem-level phenomena, such as reductions in net primary productivity, for example, occur in response to many different stresses. Changes in migration patterns, timing, and abundances of bird and/or butterfly species interact with changes in habitat and food supplies.

In some cases, effects due to climate variability and change can be quite different from those expected from other causes. For example, the upward or northward movements of treeline in montane and Arctic environments are almost certainly driven by climate, as no other driver of change is implicated. Other changes, such as those in wildfire behavior, are influenced by climate, patterns of historical land management, and current management and suppression efforts. Disentangling these influences is difficult. Some changes are so synergistic that our current scientific understanding cannot separate them solely by observations. For example, photosynthesis is strongly and interactively controlled by levels of nitrogen, water availability, temperature, and humidity. In areas where these are all changing, estimating quantitatively the effects of, say, temperature alone is all but impossible. In regions of changing climate, separating effects of climate trends from other influencing factors with regard to biodiversity and species invasions is very challenging, and requires detailed biological knowledge, as well as climate, land use, and species data.

Separating climate effects from other environmental stresses is difficult but in some cases feasible. For example, when detailed water budgets exist, the effects of land use, climate change and consumptive use on water levels can be calculated. While climate effects can be difficult to quantify on small scales, sometimes, regional effects can be separated. For example, regional trends in productivity, estimated using satellite methods, can often be assigned to regional trends in climate versus land use, although on any individual small-scale plot, climate may be primary or secondary. In other cases, scientific understanding is sufficiently robust that models in conjunction with observations can be used to estimate climate effects. This approach has been used to identify climate effects on water resources and crop productivity, and could be extended to forests and other ecological systems as well.

While it is not yet possible to precisely determine and separate the effects of individual stresses, it is feasible to quantify the actual changes in ecosystems and their individual species, in many cases through observations. There are many monitoring systems and reporting efforts set up specifically to do this, and while each may individually have gaps and weaknesses, there are many opportunities for improvement. This report identifies a number of opportunities, and many other documents have addressed the nationís need for enhanced ecological observations as well. Many networks exist, but for the integrative challenges of understanding and quantifying climate change impacts, they provide limited capability. Most existing networks are fairly specialized, and at any given measurement site, only one or a few variables may be measured. The ongoing trend of more co-location of sensors, and development of new, much more integrative networks (such as NEON and the NOAA Climate Reference Network) is positive. By measuring drivers of change and ecological responses, the processes of change can be understood and quantified, and the ability to separate and ultimately forecast climate change enhanced.

6.3 DESIGNING SYSTEMS TO MONITOR CLIMATE CHANGE IMPACTS

This assessment makes clear that there are many changes and impacts in many US ecosystems that are being driven by changes in the physical climate system, including both long-term changes and climate variability. Documentation of such changes has largely been a function of assessing results from individual studies that have been creative in their use of information from existing monitoring systems, all of which were originally designed for other purposes. But because the observed changes are proving to be both large and rapid, and because management agencies and organizations are ill-prepared to cope with such changes (GAO 2007), there is a growing need to develop strategies for adapting to ecological changes, and for managing ecosystems to ameliorate climate impacts. In addition, because changes in climate and subsequent impacts in ecosystems are very likely to continue to occur, adaptive management strategies for adapting to change are going to be quite important (GAO 2007). Observation and monitoring systems, therefore, must be able to support analyses that would contribute to this management challenge, i.e. adapting to change, documenting the rapidity of ecological changes so that management strategies can be adjusted, and most importantly, forecasting when potential thresholds of change might occur, and how rapidly changes will occur. Ecological forecasting is one of the specific goals of international programs such as Global Earth Observation System of Systems (GEOSS), but exactly how such programs will fulfill these goals is still being developed.

In order to fulfill the goal of providing observations for responding the climate change, there are at least five issues that such systems must be able to address.

Monitoring changes in overall status, regardless of cause: This need has been most cogently articulated by the work of the NRC (NRC 2000 [Orians report]), and the Heinz Center on indicators of status of US ecosystems (Heinz Center 2004). The argument is straightforward; there is both scientific and societal value to the US to know the extent, status, and condition of its own natural resources and ecosystems. Both the NRC and the Heinz Center present recommendations for specific indicators that either derived from scientific concerns (NRC) or from a broader, stake-holder driven process (Heinz Center). In either case, no attempt is made to attribute changes in the indicators to particular stresses strictly through use of the monitoring data. Both recognize, however, that such analyses are necessary for both scientific and policy purposes. The system of indicators is ultimately dependent on existing monitoring systems, most of which have been put in place for other reasons. In addition, the degree to which the ecosystem indicators identified either by the NRC or the Heinz Center process are sensitive to expected changes due to climate variability and change is as yet unknown.

In order to fulfill the goal of providing observations for responding the climate change, there are at least five issues that such systems must be able to address.

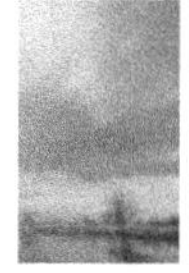

Early warning of changes due to climate: As of now, there are no routine monitoring systems established specifically for early warning of changes due to climate change. The impacts documented in this report and elsewhere are the results of analyses of existing monitoring systems and research projects, but those systems have not been optimized for early warning purposes. Without changing the configuration of existing in situ monitoring systems, or initiating new systems, it will be difficult to be sure that we have constructed an adequate early warning system and the ability to determine overall consequences of climate change may be limited.

Fortunately, enough is now known about the existing responses of ecosystems and species to changes in climate and climate variability to define monitoring systems that are optimized for early warning of subsequent changes. For example, one could set up systematic monitoring of ocean pH and alkalinity along with coral observations to track whether or not there were early indications of difficulties in calcification due to increasing pCO_2 in surface waters. One might also systematically sample vegetation along montane transects to detect early changes in flowering phenology and/or change in establishment patterns of seedlings that would result in species range changes to higher elevations as a result of warming temperatures.

...in principle it is possible to evaluate both damages and benefits from climate change for any region and/or ecosystem, but such studies will need to be very carefully designed and implemented in order to yield defensible quantification.

In the near-term, stratification of existing systems holds promise for providing reasonable information about early responses. Monitoring of snow pack and streamflow is being used in just this way, as are long time series of ice-out dates in northern lakes and national phenology data. At a minimum, identifying systems known to be at risk of early change (e.g. high latitude ecosystems, high elevation systems, coastal wetlands, migratory bird species), either because similar systems have already exhibited change or because they are in locations that are likely to experience rapid change, and investigating existing monitoring data from them would be more likely to reveal early evidence of expected changes than broad-based monitoring. Over the longer term, studies of existing monitoring data that are stratified with respect to either observation-based or model-based expectations of change would probably lead to better designs for future monitoring, but such studies have not yet been done.

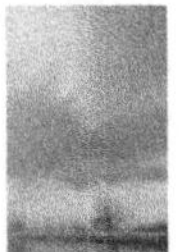

Monitoring programs that are optimized for early warning would not be appropriate for other purposes, such as calculating average damages in ecosystems or average changes in ecosystem services, precisely because they would be more likely to detect changes than the overall ecosystem average. This is not a drawback to early warning systems, but it is a caution that information from them cannot simply be used to calculate overall expected damages.

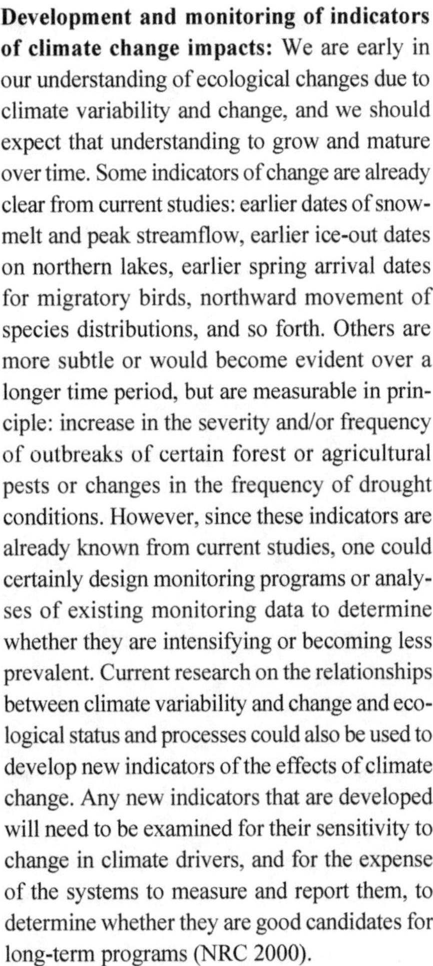

Development and monitoring of indicators of climate change impacts: We are early in our understanding of ecological changes due to climate variability and change, and we should expect that understanding to grow and mature over time. Some indicators of change are already clear from current studies: earlier dates of snowmelt and peak streamflow, earlier ice-out dates on northern lakes, earlier spring arrival dates for migratory birds, northward movement of species distributions, and so forth. Others are more subtle or would become evident over a longer time period, but are measurable in principle: increase in the severity and/or frequency of outbreaks of certain forest or agricultural pests or changes in the frequency of drought conditions. However, since these indicators are already known from current studies, one could certainly design monitoring programs or analyses of existing monitoring data to determine whether they are intensifying or becoming less prevalent. Current research on the relationships between climate variability and change and ecological status and processes could also be used to develop new indicators of the effects of climate change. Any new indicators that are developed will need to be examined for their sensitivity to change in climate drivers, and for the expense of the systems to measure and report them, to determine whether they are good candidates for long-term programs (NRC 2000).

Experiments to isolate the impacts of climate change from other impacts: Experiments that directly manipulate climate variables and observe impacts are a critical component in understanding climate change impacts and in separating the effects of climate from those caused by other factors. Direct manipulations of precipitation, CO_2, temperature, and nitrogen deposition have yielded much useful information and many surprises (such as the increased growth and toxicity of poison ivy when exposed to higher CO_2). Because many factors change in concert under ambient conditions, manipulations are especially useful at isolating the effect of specific factors. In fact, manipulative experiments that reveal information about underlying ecological processes are crucial to ensuring that a true forecasting capacity is developed.

Evaluating damages and benefits from climate change: Over the long term, we need to understand the extent to which climate change is damaging or enhancing the goods and services that ecosystems provide and how additional climate change would affect the future delivery of such goods and services. This information cannot currently be provided for any ecosystem for several reasons. In some cases we lack sufficient understanding to identify the observations that are required. In others, we lack observations that we know would be helpful. In yet others, we have observations but are not integrating these in modeling and analysis frameworks that could enable forecasting of potential changes. But probably the most important difficulty is that we do not have a national system for ecosystem valuation that takes into account both goods that ecosystems produce that are priced and are traded in markets, and those services that are not priced, but are nevertheless valuable to society. Even services that can in principle result in economic gains, such as wetlands or mangroves protection of shorelines from storm surge and flooding, have not been estimated on large regional or national bases.

Again, in principle it is possible to evaluate both damages and benefits from climate change for any region and/or ecosystem, but such studies will need to be very carefully designed and implemented in order to yield defensible quantification. Until then, we will need to continue to rely on a combination of existing observations made for other purposes and on model output to construct such estimates.

6.4 INTEGRATION OF ECOSYSTEM OBSERVATIONS, MODELING, EXPERIMENTS AND ANALYSIS

The rapid changes in ecosystems that have already been documented pose special challenges to monitoring systems. If their locations cannot be adequately forecast, it is possible for rapid changes to be missed in monitoring data until they become so large that they are obvious. This is especially problematic if the intent of the monitoring program is to provide early warning capabilities. There is currently no analysis in the literature that addresses this problem. A second particular challenge for monitoring ecosystem change due to climate change is the inescapable fact that ecosystems respond to many different factors, of which climate variability and change is only one. Monitoring systems that are established in ways that presuppose one particular driver of change could lead to problematic estimates of change due to other agents.

Ultimately, a national capacity for documenting and evaluating the extent and magnitude of ecosystem changes due to changes in climate will require new system designs that draw on experimentation, modeling and monitoring resources. Expectations derived from modeling time-series can be periodically challenged with observational and experimental data, and the results then fed back to ecosystem models in order to improve their forecasting quantitatively. Such procedures would be analytically similar to data assimilation techniques in wide use in weather and climate modeling, but obviously on very different time scales. It will be necessary for such a system to have systematic sampling of ecosystems with respect to climate variability, and have models that are then capable of ingesting both process observations and observations of ecosystem state and extent.

6.5 OVERARCHING CONCLUSIONS

Climate changes – temperature increases, increasing CO_2 levels, and altered patterns of precipitation – are already affecting U.S. water resources, agriculture, land resources, and biodiversity, and will continue to do so ***(very likely).*** The results of the literature review undertaken for this assessment document case after case of changes in these resources that are the direct result of variability and changes in the climate system, even after accounting for other factors. The number and frequency of forest fires and insect outbreaks are increasing in the interior West, the Southwest, and Alaska. Precipitation, streamflow, and stream temperatures are increasing in most of the continental U.S. The western U.S. is experiencing reduced snowpack and earlier spring runoff peaks. The

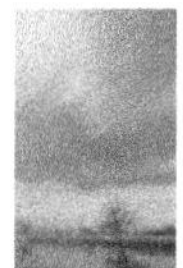

growth of many crops and weeds is being stimulated. Migration of plant and animal species is changing the composition and structure of arid, polar, aquatic, coastal and other ecosystems.

Climate change will continue to have significant effects on these resources over the next few decades and beyond *(very likely).* Warming is very likely to continue in the United States during the next 25-50 years, regardless of the efficacy of greenhouse gas emissions reduction efforts, due to greenhouse gas emissions that have already occurred. U.S. ecosystems and natural resources are already being affected by climate system changes and variability. It is very likely that the magnitude and frequency of ecosystem changes will continue to increase during this period, and it is possible that they will accelerate. As temperature rises, crops will increasingly begin to experience higher temperatures beyond the optimum for their reproductive development. Management of Western reservoir systems is very likely to become more challenging as runoff patterns continue to change. Arid areas are very likely to experience increases erosion and fire risk. In arid region ecosystems that have not co-evolved with a fire cycle, the probability of loss of iconic, charismatic mega flora such as saguaro cacti and Joshua trees will greatly increase.

Many analyses of the impacts of climate change on individual species and ecosystems have been published in the scientific literature, but there is not yet adequate integrated analysis of how climate change could affect ecosystem services.

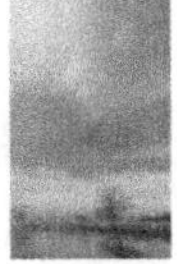

Many other stresses and disturbances are also affecting these resources *(very likely).* For many of the changes documented in this assessment, there are multiple environmental drivers – land use change, nitrogen cycle change, point and non-point source pollution, wildfires, invasive species, and others – that are also changing. Atmospheric deposition of biologically available nitrogen compounds continues to be an important issue, along with persistent, chronic levels of ozone pollution in many parts of the country. It is very likely that these additional atmospheric effects cause biological and ecological changes that interact with changes in the physical climate system. In addition, land cover and land use patterns are changing, e.g., the increasing fragmentation of U.S. forests as exurban development spreads to previously undeveloped areas, further raising fire risk and compounding the effects of summer drought, pests, and warmer winters. There are several dramatic examples of extensive spread of invasive species throughout rangeland and semi-arid ecosystems in western states, and indeed throughout the United States. It is likely that the spread of these invasive species, which often change ecosystem processes, will exacerbate the risks from climate change alone. For example, in some cases invasive species increase fire risk and decrease forage quality.

Climate change impacts on ecosystems will affect the services that ecosystems provide, such as cleaning water and removing carbon from the atmosphere *(very likely)*, but we do not yet possess sufficient understanding to project the timing, magnitude, and consequences of many of these effects. One of the main reasons for needing to understand changes in ecosystems is the need to understand the consequences of those changes for the delivery of services that our society values. Many analyses of the impacts of climate change on individual species and ecosystems have been published in the scientific literature, but there is not yet adequate integrated analysis of how climate change could affect ecosystem services. A comprehensive understanding of the way such services might be affected by climate change will only be possible through quantification of anticipated alteration in ecosystem function and productivity. As described by the Millennium Ecosystem Assessment, some products of ecosystems, such as food and fiber, are priced and traded in markets. Others, such as carbon sequestration capacity, are only beginning to be understood and traded in markets. Still others, such as the regulation of water quality and quantity, and the maintenance of soil fertility, are not priced and traded, but are valuable nonetheless. Yet although these points are recognized and accepted in the scientific literature and increasingly among decision makers, there is no analysis specifically devoted to understanding changes in ecosystem services in the United States from climate change and associated stresses. It is possible to make some generalizations from the existing literature on the physical changes in ecosystems, but interpreting what this means for services provided by ecosystems is very challenging and can only be done for a limited number of cases. This is a significant gap in our knowledge base.

Existing monitoring systems, while useful for many purposes, are not optimized for detecting the impacts of climate change on ecosystems. There are many operational and research monitoring systems that have been deployed in the United States that are useful for studying the consequences of climate change on ecosystems and natural resources. These range from the resource- and species-specific monitoring systems, which land-management agencies depend on, to research networks, such as the Long-Term Ecological Research (LTER) sites, which the scientific community uses to understand ecosystem processes. All of the existing monitoring systems, however, have been put in place for other reasons, and none have been optimized specifically for detecting the effects and consequences of climate change. As a result, it is likely that only the largest and most visible consequences of climate change are being detected. In some cases, marginal changes and improvements to existing observing efforts, such as USDA snow and soil moisture measurement programs, could provide valuable new data detection of climate impacts. But more refined analysis and/or monitoring systems designed specifically for detecting climate change effects would provide more detailed and complete information and probably capture a range of more subtle impacts. This in turn would hold promise of developing early warning systems and more accurate forecasts of potential future changes. But it must be emphasized that improved observations, while needed, are not sufficient for improving understanding of ecological impacts of climate change. Ongoing, integrated and systematic analysis of existing and new observations could enable forecasting of ecological change, thus garnering greater value from observational activities, and contribute to more effective evaluation of measurement needs.

APPENDIX A: ACRONYMS AND GLOSSARY

ACRONYMS

AET	Apparent equivalent temperature
ANPP	Aboveground net primary productivity
AOGCM	Atmosphere-ocean general circulation models
BT	Body temperature
CCSM	Community Climate System Model
CCSP	U.S. Climate Change Science Program
CGC	Canadian Global Coupled Model
DOY	Day of year
ET	Evapotranspiration
ENSO	El Niño-Southern Oscillation
FACE	Free-Air CO_2 Enrichment
GCM	General Circulation Model
GFDL	Geophysical Fluid Dynamics Laboratory
HadCM2	Hadley Centre for Climate Prediction and Research's Climate Model 2
HCN	Historical Climatology Network
HI	Harvest index
HLI	Heat load index
IBP	International Biome Project
IPCC	Intergovernmental Panel on Climate Change
IPCC AR4	Intergovernmental Panel on Climate Change 4th Assessment Report
IPCC TAR	Intergovernmental Panel on Climate Change 3rd Assessment Report
IPM	Integrated pest management
LAI	Leaf area index
LTER	Long Term Ecological Research
LWSI	Livestock weather safety index
NCAR	National Center for Atmospheric Research
NEON	National Ecological Observatory Network
NPP	Net primary productivity
NRCS	Natural Resources Conservation Service
NRCS SCAN	Natural Resources Conservation Service Soil Climate and Analysis Network
NRC	National Research Council
NWS COOP	National Weather Service Cooperative Observer Program
PCMDI	(Lawrence Livermore National Laboratory's) Program for Climate Model Diagnosis and Intercomparison
PDO	Pacific Decadal Oscillation
PE	Potential evaporation
ppb	Parts per billion
ppm	Parts per million
RH	Relative humidity
RMSE	Root mean square error
RR	Respiration rate
SOM	Soil organic matter
SRAD	Solar radiation
SRES	Special Report on Emissions Scenarios
SWE	Snow water equivalent
TBCA	Total carbon allocation belowground
THI	Temperature-humidity index
USDA	United States Department of Agriculture
USGS	United States Geological Survey
USGS HCDN	United States Geological Survey Hydro-Climatic Data Network
VFI	Voluntary feed intake
VIC	Variable Infiltration Capacity
VOC	Volatile organic compound
VPD	Vapor pressure deficit
WS	Wind speed
WUE	Water use efficiency

GLOSSARY

Anthesis
The period during which a flower is fully open and functional.

Boll
The seed-bearing capsule of certain plants, especially cotton and flax.

C_3 species
Almost all plant life on Earth can be divided into two categories based on the way they assimilate carbon dioxide into their systems. During the first steps in CO_2 assimilation, C_3 plants form a pair of three carbon-atom molecules. C_3 species continue to increase photosynthesis with rising CO_2. C_3 plants include more than 95 percent of the plant species on Earth.

C_4 species
C_4 plants initially form four carbon-atom molecules. C_4 plants include such crop plants as sugar cane and corn. They are the second-most prevalent photosynthetic type, and do not assimilate CO_2 as well as C_3 plants.

Carbon sink
A carbon reservoir. Carbon sinks include the oceans, and plants and other organisms that remove carbon from the atmosphere via photosynthetic processes.

Carbon source
The term describing processes that add carbon dioxide to the atmosphere.

Carbon sequestration
The term describing processes that remove carbon dioxide from the atmosphere.

Climate
Climate in a narrow sense is usually defined as the "average weather" or more rigorously as the statistical description in terms of the mean and variability of relevant quantities over a period of time ranging from months to thousands or millions of years. The classical period is 30 years, as defined by the World Meteorological Organization (WMO). These relevant quantities are most often surface variables such as temperature, precipitation, and wind. Climate in a wider sense is the state, including a statistical description, of the *climate system*.

Climate Change
Climate change refers to a statistically significant variation in either the mean state of the *climate* or in its variability, persisting for an extended period (typically decades or longer). Climate change may be due to natural internal processes or *external forcings*, or to persistent *anthropogenic* changes in the composition of the *atmosphere* or in *land use*. Note that the *United Nations Framework Convention on Climate Change* (UNFCCC), in its Article 1, defines "climate change" as: "a change of climate which is attributed directly or indirectly to human activity that alters the composition of the global atmosphere and which is in addition to natural climate variability observed over comparable time periods." The UNFCCC thus makes a distinction between "climate change" attributable to human activities altering the atmospheric composition, and "climate variability" attributable to natural causes. See also *climate variability*.

Climate Variability
Climate variability refers to variations in the mean state and other statistics (such as standard deviations, the occurrence of extremes, etc.) of the *climate* on all *temporal and spatial scales* beyond that of individual weather events. Variability may be due to natural internal processes within the *climate system* (internal variability), or to variations in natural or *anthropogenic external forcing* (external variability). See also *climate change*.

CO_2 enrichment
Addition of CO_2 to the atmosphere.

Coefficient of variation of annual runoff
A measure of the variability of runoff

Complementary hypothesis
This hypothesis states that trends in actual evaporation and pan evaporation should be in opposite directions.

Cucurbits
Any of various mostly climbing or trailing plants of the family Cucurbitaceae, which includes the squash, pumpkin, cucumber, gourd, watermelon, and cantaloupe.

Endophyte
A plant living within another plant, usually as a parasite.

Evaporation paradox
Temperature, precipitation, stream flow and cloud cover records indicate that warmer, rainier weather is now more common in many regions of the world. However, pan evaporation readings, taken at weather stations, indicate that less moisture has been rising back into the air from these pans.

Evapotranspiration
The sum of evaporation and plant transpiration. Evaporation accounts for the movement of water to the air from sources such as the soil, canopy interception, and water bodies. Transpiration accounts for the movement of water within a plant and the subsequent loss of water as vapor through stomata in its leaves.

Free-Air CO_2 Enrichment (FACE)

FACE is a method and infrastructure used to experimentally enrich the atmosphere enveloping portions of a terrestrial ecosystem with controlled amounts of carbon dioxide (and in some cases, other gases), without using chambers or walls.

Forb
A broad-leaved herb (not a grass), especially one growing in a field, prairie, or meadow.

Global dimming
The gradual reduction in the amount of global direct irradiance at the Earth's surface that was observed for several decades after the start of systematic measurements in 1950s.

Herbivores
Animals that feed chiefly on plants.

Homeostasis
The scientific study of periodic biological phenomena, such as flowering, breeding, and migration, in relation to climatic conditions.

Instream flow
The term used to identify a specific stream flow (typically measured in cubic feet per second, or cfs) at a specific location for a defined time, and typically following seasonal variations. Instream flows are usually defined as the stream flows needed to protect and preserve instream resources and values, such as fish, wildlife and recreation. Instream flows are most often described and established in a formal legal document, typically an adopted state rule.

Irrigation Modes

Drip irrigation allows water to drip slowly to the roots of plants through a network of valves, pipes, tubing, and emitters.

Flood irrigation pumps water onto the fields. The water then flows freely along the ground among the crops.

Spray irrigation relies on machinery to spray water in all directions.

Latent heat
The heat required to change the phase of a substance, for example a solid to vapor (sublimation), liquid to vapor (vaporization) or solid to liquid (melting); the temperature does not change during these processes. Heat is released for the reverse processes, for example vapor to solid (frost), liquid to solid (freezing), or vapor to liquid (condensation).

Leaf area index (LAI)
The ratio of total upper leaf surface of a crop divided by the surface area of the land on which the crop grows.

Lignin
An organic substance that, with cellulose, forms the chief part of woody tissue.

Lysimeter
A device for collecting water from the pore spaces of soils, and for determining the soluble constituents removed in the drainage.

Mutualistic relationship
A positive, reciprocal relationship between two species. Through this relationship, both species enhance their survival, growth or fitness.

Net primary productivity (NPP)
The ratio of all biomass accumulation and biomass losses in units of carbon, weight or energy, per land surface unit, over a set time interval (usually a year).

Pan evaporation
Pans used to determine the quantity of evaporation at a given location. These are generally located in agricultural areas, and have been used as an index to potential evaporation.

Panicle
The complete assembly of spikelets on a rice plant.

Phenology
The study of periodic biological phenomena (flowering of plants, breeding, and species migration) in relation to climatic conditions.

Potential Evapotranspiration
A representation of the environmental demand for evapotranspiration and represents the evapotranspiration rate of a short green crop, completely shading the ground, of uniform height and with adequate water status in the soil profile. It is a reflection of the energy available to evaporate water, and of the wind available to transport the water vapor from the ground up into the lower atmosphere.

Runoff ration
The total amount of runoff divided by the total moisture that falls during a precipitation event.

Ruminant
Even-toed, cud-chewing, hoofed mammals of the suborder *Ruminantia*, such as domestic cattle.

Sensible heat
Heat that can be measured by a thermometer.

Spikelet
The individual places on a rice plant where a grain develops.

Stomatal
One of the minute pores in the epidermis of a leaf or stem through which gases and water vapor pass.

Tiller
New shoots that develop at the base of the plant.

APPENDIX B: REFERENCES

CHAPTER 2 REFERENCES

Adams, C.D., S. Spitzer, and R.M. Cowan, 1996: Biodegradation of nonionic surfactants and effects of oxidative pretreatment. *Journal of Environmental Engineering*, **122**, 477-483.

Adams, S.R., K.E. Cockshull, and C.R.J. Cave, 2001: Effect of temperature on the growth and development of tomato fruits. *Annals of Botany*, **88,** 869-877.

Adams, R.M., B.A. McCarl, K. Segerson, C. Rosenzweig, K.J. Bryant, B.L. Dixon, R. Connor, R.E. Evenson, and D. Ojima, 1999: The economic effects of climate change on U.S. agriculture. In: *The Economics of Climate Change* [Mendelsohn, R. and J. Neumann (eds.)]. Cambridge University Press, Cambridge, United Kingdom and New York, NY, USA, pp. 19-54.

Aerts, M., P. Cockrell, G. Nuessly, R. Raid, T. Schueneman, D. Seal, 1999: Crop profile for corn (sweet) in Florida. http://www.impcenters.org/CropProfiles/docs/FLcorn-sweet.html

Afinowicz, J.D., C.L. Munster, B.P. Wilcox, and R.E. Lacey, 2005: A process for assessing wooded plant cover by remote sensing. *Rangeland Ecology & Management*, **58,** 184-190.

Ainsworth, E. A. and S.P. Long, 2005: What have we learned from 15 years of free-air CO_2 enrichment (FACE)? A meta-analytic review of the responses of photosynthesis, canopy properties and plant production to rising CO_2. *New Phytologist*, **165**, 351-372.

Ainsworth, E.A. and A. Rogers, 2007: The response of photosynthesis and stomatal conductance to rising [CO_2]: mechanisms and environmental interactions. *Plant, Cell & Environment*, **30**, 258-270.

Ainsworth, E.A., P.A. Davey, C.J. Bernacchi, O.C. Dermody, E.A. Heaton, D.J. Moore, P.B. Morgan, S.A. Naidu, Hyung-Shim Yoo Ra, Xin-Guand Zhu, P.S. Curtis and S.P. Long, 2002: A meta-analysis of elevated [CO_2] effects on soybean (*Glycine max*) physiology, growth and yield. *Global Change Biology*, **8**, 695-709.

Alagarswamy, G., K.J. Boote, L.H. Allen, Jr., and J.W. Jones, 2006: Evaluating the CROPGRO-Soybean model ability to simulate photosynthesis response to carbon dioxide levels. *Agronomy Journal*, **98**, 34-42.

Alagarswamy, G., and J. T. Ritchie, 1991. Phasic development in CERES-sorghum model, chapter 13, pp 143-152 *In* Hodges, T (ed.) *Predicting Crop Phenology*. CRC Press, Boca Raton.

Allard V., P.C.D. Newton, M. Lieffering J.F. Soussana, P. Grieu, and C. Matthews, 2004: Elevated CO_2 effects on decomposition processes in a grazed grassland. *Global Change Biology*, **10,** 1553-1564.

Allen, L.H. Jr., and K.J. Boote, 2000: Crop ecosystem responses to climatic change: Soybean. Chapter 7. pp. 133-160. *In* K. R. Reddy and H. F. Hodges, *Climate Change and Global Crop Productivity*. CAB International., New York, NY.

Allen, L.H. Jr., D. Pan, K.J. Boote, N.B. Pickering, and J.W. Jones, 2003: Carbon dioxide and temperature effects on evapotranspiration and water-use efficiency of soybean. *Agronomy Journal*, **95,**1071-1081.

Allen, R.G., F.N. Gichuki, and C. Rosenzweig, 1991: CO_2-induced climatic changes and irrigation-water requirements. *Journal of Water Resources Planning and Management*, **117**, 157-178.

Allen R.G., I.A. Walter, R.L. Elliot, T.A. Howell, D. Itenfisu , M.E. Jensen, R.L. Snyder, 2005: *The ASCE Standardized Reference Evapotranspiration Equation*, American Society of Civil Engineers, Reston, VA.

Alocilja, E.C., and J.T. Ritchie, 1991: A model for the phenology of rice. Chapter 16, pp 181-189. *In* Hodges, T (ed.) *Predicting Crop Phenology*. CRC Press, Boca Raton.

Amthor, J.S., 1999: Increasing atmospheric CO_2 concentration, water use, and water stress: scaling up from the plant to the landscape. p. 33-59. In Y. Luo and H.A. Mooney (ed.) *Carbon Dioxide and Environmental Stress*, Academic Press, San Diego.

Amthor, J.S., 2001: Effects of atmospheric CO_2 concentration on wheat yield: review of results from experiments using various approaches to control CO_2 concentration. *Field Crops Research*, **73,** 1-34.

Amundson, J. L., T. L. Mader, R. J. Rasby, and Q. S. Hu, 2005: Temperature and temperature-humidity index effects on pregnancy rate in beef cattle. *Proceedings 17th Intl. Congress on Biometeorology*, September 2005, Dettscher Wetterdienst, Offenbach, Germany.

Amundson, J. L., T. L. Mader, R. J. Rasby, and Q. S. Hu., 2006: Environmental effects on pregnancy rate in beef cattle. *Journal of Animal Science*, **84**, 3415-3420.

Andre, M., and H. du Cloux, 1993: Interaction of CO_2 enrichment and water limitations on photosynthesis and water use efficiency in wheat *Plant Physiology and Biochemistry*, **31**, 103-112.

Archer S., D.S. Schimel, and E.A. Holland, 1995: Mechanisms of shrubland expansion: land use, climate or CO_2? *Climatic Change*, **29**, 91-99.

Ashmore. M.R., 2002: Effects of oxidants at the whole plant and community level. In: JNB Bell, M Treshow, eds, *Air Pollution and Plant Life*, John Wiley, Chichester, pp 89-118.

Ashmore, M.R., 2005: Assessing the future global impacts of ozone on vegetation. *Plant Cell & Environment*, **28**, 949-964.

Augustine, D.J., and S.J. McNaughton, 2004: Temporal asynchrony in soil nutrient dynamics and plant production in a semiarid ecosystem. *Ecosystems* 7:829-840.

Augustine, D.J., and S.J. McNaughton, 2006: Interactive effects of ungulate herbivores, soil fertility, and variable rainfall on ecosystem processes in a semi-arid savanna. *Ecosystems* 9:1-16.

Austin A.T., and L. Vivanco, 2006: Plant litter decomposition in a semi-arid ecosystem controlled by photodegradation. *Nature*, **442**, 555-558.

Ayers, E., D.H. Wall, B.L. Simmons, C.B. Field, J. Roy, D. Milchunas and J.A. Morgan. Belowground grassland herbivores are surprisingly resistant to elevated atmospheric CO_2 concentrations. *Soil Biology and Biochemistry*. (in press, accepted November 29, 2007)

Badeck F.W., A. Bondeau, K. Bottcher, D. Doktor, W. Lucht, J. Schaber, and S. Sitch, 2004: Responses of spring phenology to climate change. *New Phytologist*, **162**, 295-309.

Badu-Apraku, B., R.B. Hunter, and M. Tollenaar, 1983: Effect of temperature during grain filling on whole plant and grain yield in maize (*Zea mays* L.). *Canadian Journal of Plant Science*, **63**, 357-363.

Baker B.B., J.D., Hanson, R.M. Bourdon and J.B. Eckert, 1993: The potential effects of climate change on ecosystem processes and cattle production on U.S. rangelands. *Climatic Change*, **25**, 97-117.

Baker, J.T., and L.H. Allen, Jr., 1993a: Contrasting crop species responses to CO_2 and temperature: rice, soybean, and citrus. *Vegetatio*, **104/105**, 239-260.

Baker, J.T., and L.H. Allen, Jr. 1993b: Effects of CO_2 and temperature on rice: A summary of five growing seasons. *Journal of Agricultural Meteorology.*, **48**, 575-582.

Baker, J.T., L.H. Allen, Jr., and K.J. Boote, 1989: Response of soybean to air temperature and carbon dioxide concentration, *Crop Science*, **29**, 98-105.

Baker, J.T., K.J. Boote, and L.H. Allen, Jr., 1995: Potential climate change effects on rice: Carbon dioxide and temperature. pp. 31-47. *In* C. Rosenzweig, J. W. Jones, and L. H. Allen, Jr. (eds.). *Climate Change and Agriculture: Analysis of Potential International Impacts*, ASA Spec. Pub. No. 59, ASA-CSSA-SSSA, Madison, Wisconsin, USA.

Balaguer, L., J.D. Barnes, A.Panicucci, and A.M. Borland,1995: Production and utilization of assimilates in wheat (*Triticum aestivum* L.) leaves exposed to elevated O_3 and/or CO_2. *New Phytologist.*,**129**, 557-568.

Barnes, J.D., J.H. Ollerenshaw, and C.P. Whitfield, 1995: Effects of elevated CO_2 and/or O_3 on growth, development and physiology of wheat (*Triticum aestivum* L.). *Global Change Biology*, **1**, 101-114.

Bartlett, D.T., L.A. Torrell, N.R. Rimbey, L.W.Van Tassell, and D.W. McCollum, 2002: Valuing grazing use on public land. *Journal of Range Management*, **55**, 426-438.

Batts, G.R., J.I.L. Morison, R.H. Ellis, P. Hadley, and T.R. Wheeler, 1997: Effects of CO_2 and temperature on growth and yield of crops of winter wheat over several seasons. *European Journal of Agronomy*, **7**, 43-52.

Baylis, M., and A.K. Githeko, 2006: T7.3 The effects of climate change on infectious diseases of animals. Foresight: http://www.foresight.gov.uk/previous_projects/detection_and_identification_of_infectious_diseases/Reports_and_Publications/Final_Reports/Index.html

Bender, J., U. Hertstein, and C. Black, 1999: Growth and yield responses of spring wheat to increasing carbon dioxide, ozone and physiological stresses: a statistical analysis of 'ESPACE-wheat' results. *European Journal of Agronomy*, **10**, 185-195.

Bernacchi, C.J., B.A. Kimball, D.R. Quarles, S.P. Long, and D.R. Ort, 2007: Decreases in stomatal conductance of soybean under open-air elevation of CO_2 are closely coupled with decreases in ecosystem evapotranspiration. *Plant Physiology*, **143**, 134-144.

Bernacchi, C.J., A.D.B. Leakey, L.E. Heady, P.B. Morgan, F.G. Dohleman, J.M. McGrath, K.M. Gillespie, V.E. Wittig, A. Rogers, S.P. Long, and D.R. Ort, 2006: Hourly and seasonal variation in photosynthesis and stomatal conductance of soybean grown at future CO_2 and ozone concentrations for 3 years under fully open-air field conditions. *Plant, Cell & Environment*, **29**, 2077-2090.

Bestelmeyer, B.T., J.E. Herrick, J.R. Brown, D.A. Trujillo, and K.M. Havstad. 2004. Land management in the American Southwest: a state-and-transition approach to ecosystem complexity. *Environmental Management* 34:38-51.

Billings S.A., S.M. Schaeffer, and R.D. Evans, 2004: Soil microbial activity and N availability with elevated CO_2 in Mojave Desert soils. *Global Biogeochemical Cycles*, 18, GA1011, doi:10.1029/2003GB002137.

Black, V.J., C.R. Black, J.A. Roberts, and C.A. Stewart, 2000: Impact of ozone on the reproductive development of plants. *New Phytologist*, **147**, 421-447.

Bolhuis, C.G., and W. deGroot, 1959: Observations on the effect of varying temperature on the flowering and fruit set in three varieties of groundnut. *Netherlands Journal of Agricultural Science*, **7**, 317-326.

Bond, W.J., and G.F. Midgley, 2000. A proposed CO_2-controlled mechanism of woody plant invasion in grasslands and savannas. *Global Change Biology*, **6**, 865-869

Booker, F.L., K.O. Burkey, W.A. Pursley, and A.S. Heagle. 2007: Elevated carbon dioxide and ozone effects on peanut: I. Gas-exchange, biomass, and leaf chemistry. *Crop Science*, **47**, 1475-1487.

Boote, K.J., J.W. Jones, and N.B. Pickering, 1996: Potential uses and limitations of crop models. *Agronomy Journal*, **88**, 704-716

Boote, K.J., J.W. Jones, and G. Hoogenboom, 1998: Simulation of crop growth: CROPGRO Model. Chapter 18. pp. 651-692. *In* R.M. Peart and R.B. Curry (eds.). *Agricultural Systems Modeling and Simulation*. Marcel Dekker, Inc, New York.

Boote, K.J., J.W. Jones, W.D. Batchelor, E.D. Nafziger, and O. Myers, 2003: Genetic coefficients in the CROPGRO-soybean model: Links to field performance and genomics. *Agronomy Journal*, **95**, 32-51.

Boote, K.J., L.H. Allen, P.V.V. Prasad, J.T. Baker, R.W. Gesch, A.M. Snyder, D. Pan, and J.M.G. Thomas, 2005: Elevated temperature and CO_2 impacts on pollination, reproductive growth, and yield of several globally important crops. *Journal of Agricultural Meteorology*, **60,** 469-474.

Boote, K.J., N.B. Pickering, and L.H. Allen, Jr., 1997: Plant modeling: Advances and gaps in our capability to project future crop growth and yield in response to global climate change. pp 179-228. *In*: L.H. Allen, Jr., M.B. Kirkham, D.M. Olszyk, and C.E. Whitman (eds.) *Advances in Carbon Dioxide Effects Research. ASA Special Publication No. 61*, ASA-CSSA-SSSA, Madison, WI.

Booth D.T., and S.E. Cox, 2006: Very-large scale aerial photography for rangeland monitoring. *Geocarto International*, **21,** 27-34.

Bowler J.M., and M.C. Press, 1996: Effects of elevated CO_2, nitrogen form and concentration on growth and photosynthesis of a fast- and slow-growing grass. *New Phytologist*, **132,** 391-401.

Briggs, J.M., A.K. Knapp, J.M. Blair, J.L. Heisler, G.A. Hoch, M.S. Lett, and J.K. McCarron, 2005: An ecosystem in transition: Causes and consequences of the conversion of mesic grassland to shrubland. BioScience **55**, 243-254.

Briske D.D., S.D. Fuhlendorf, and F.E. Smeins, 2005: State-and-transition models, thresholds, and rangeland health: A synthesis of ecological concepts and perspectives. *Rangeland Ecology & Management*, **58**, 1-10.

Brown, P.W., 1987. *User's Guide to the Arizona Meteorological Network.* City of Phoenix, Water Conservation and Resource Div. and Arizona Cooperative Extension, Phoenix, AZ.

Brown-Brandl, T.M., R. A. Eigenberg, and J. A. Neinaber. 2006. Heat stress risk factors of feedlot heifers. *Livestock Science* **105**, 57-68.

Brown-Brandl, T.M., J.A. Nienaber, R.A. Eigenberg, G.L. Hahn and H. Freetly, 2003: Thermoregulatory responses of feeder cattle. *J. Therm. Biol.*, **28**, 149-157.

Bunce, J.A., 2000: Acclimation of photosynthesis to temperature in eight cool and warm climate herbaceous C3 species: Temperature dependence of parameters of a biochemical photosynthesis model. *Photosynthesis Research*, **63,** 59-67.

Burkey, K.O., F.L. Booker, W.A. Pursley, and A.S. Heagle, 2007: Elevated carbon dioxide and ozone effects on peanut: II. Seed yield and quality. *Crop Science*, **47**, 1488-1497.

Butterfield, H.S., and C.M. Malmstron, 2006: Experimental use of remote sensing by private range managers and its influence on management decisions. *Rangeland Ecology & Management*, **59**, 541-548.

Caley, C.Y., C.M. Duffus, and B. Jeffcoat, 1990: Effects of elevated temperature and reduced water uptake on enzymes of starch synthesis in developing wheat grains. *Australian Journal of Plant Physiology* **17**, 431-439.

Campbell, B.D., D.M.S. Smith, and G.M. Mckeon, 1997. Elevated CO_2 and water supply interactions in grasslands: A pastures and rangelands management perspective. *Global Change Biology*, **3**, 177-187.

Chapin, F.S. III, G.R. Shaver, A.E. Giblin, K.J. Nadelhoffer, and J.A. Laundre, 1995: Responses of arctic tundra to experimental and observed changes in climate. *Ecology*, **76**, 694-711.

Chowdhury, S.I.C., and I.F. Wardlaw, 1978: The effect of temperature on kernel development in cereals. *Australian Journal Agricultural Research*, **29**, 205-233.

Cleland, E.E., N.R. Chiariello, S.P. Loarie, H.A. Mooney, and C.B. Field, 2006: Diverse responses of phenology to global changes in a grassland ecosystem. *Proceedings of the National Academy of Sciences*, **103**, 13740-13744.

Coakley, S.M., H. Scherm, and S. Chakraborty, 1999: Climate change and plant disease management. *Annual Review of Phytopathology*, **37**, 399-426.

Commuri, P.D., and R.D. Jones, 2001: High temperatures during endosperm cell division in maize: a genotypic comparison under *in vitro* and field conditions. *Crop Science*, **41**, 1122-1130.

Cotrufo, M.F., P. Ineson, and A. Scott. 1998: Elevated CO_2 reduces the nitrogen concentration of plant tissues. *Global Change Biology*, **4**, 43-54.

Coviella, C., and J. Trumble, 1999: Effects of elevated atmospheric carbon dioxide on insect-plant interactions. *Conservation Biology*, **13**, 700-712.

Cox, F.R., 1979: Effect of temperature treatment on peanut vegetative and fruit growth. *Peanut Sci*, **6**, 14-17.

Crafts-Brandner, S.J., and M.E. Salvucci, 2002: Sensitivity of photosynthesis in a C-4 plant, maize, to heat stress. *Plant Physiology*, **129**, 1773-1780.

Craufurd, P.Q., P.V.V. Prasad, and V.G. Kakani, 2003: Heat tolerance in groundnut. *Field Crops Research*, **80,** 63-77.

Curtis, P.S. and X. Wang, 1998: A meta-analysis of elevated CO_2 effects on woody plant mass, form, and physiology. *Oecologia*, **113,** 299-313.

Dahl, B.E., and R.E. Sosebee, 1991: Impacts of weeds on herbage production. In: James, L.F., J.O. Evans, M.H. Ralphs, and R.D. Child (eds.) *Noxious Range Weeds*. Westview, Boulder, pp. 153-164.

Davis, M.S., T.L. Mader, S.M. Holt, and A.M. Parkhurst, 2003: Strategies to reduce feedlot cattle heat stress: effects on tympanic temperature. *Journal of Animal Science*, **81**, 649-661.

Dentener F., D. Stevenson, J. Cofala, R. Mechler, M. Amann, P. Bergamaschi, F. Raes, and R. Derwent, 2005: The impact of air pollutant and methane emission controls on tropospheric ozone and radiative forcing: CTM calculations for the period 1990-2030. *Atmospheric Chemistry and Physics*, **5**, 1731-1755.

Dermody, O., S.P. Long, and E.H. DeLucia, 2006: How does elevated CO_2 or ozone affect the leaf-area index of soybean when applied independently? *New Phytologist*, **169,** 145-155.

Dijkstra, F.A., S.E. Hobbie, and P. Reich, 2006: Soil processes affected by sixteen grassland species grown under different environmental conditions. *Soil Science of America Journal*, **70,** 770-777.

Donnelly, A., M.B. Jones, J.I. Burke, and B. Schnieders, 2000: Elevated CO_2 provides protection from O_3 induced photosynthetic damage and chlorophyll loss in flag leaves of spring wheat (*Triticum aestivum* L., cv. 'Minaret'). *Agriculture, Eosystems & Environment,* **80**, 159-168.

Downs, R.W. 1972: Effect of temperature on the phenology and grain yield of *Sorghum bicolor. Australian Journal Agricultural Research,* **23**, 585-594.

De Koning, A.N.M. 1996: Quantifying the responses to temperature of different plant processes involved in growth and development of glasshouse tomato. *Acta Horticulturae*, **406**, 99-104.

Drake, B.G., M.A. Gonzàlez-Meler, and S.P. Long, 1997: More efficient plants: a consequence of rising atmospheric CO_2? *Annual Review of Plant Physiology and Plant Molecular Biology*, **48**, 609-639.

Duchowski, P., and A. Brazaityte, 2001: Tomato photosynthesis monitoring in investigations on tolerance to low temperatures. *Acta Hort,* **562**, 335-339.

Duff, G.C., and M.L. Galyean, 2007: Board-invited review: Recent advances in management of highly stressed, newly received feedlot cattle. *Journal of Animal Science*, **85**, 823-840.

Dukes, J.S., N.R. Chiariello, E.E. Cleland, L.A. Moore, M.R. Shaw, S. Thayer S, T. Tobeck, H.A. Mooney, and C.B. Field, 2005: Responses of grassland production to single and multiple global environmental changes. *PLoS Biology*, **3**, e319.

Dupuis, L. and C. Dumas, 1990: Influence of temperature stress on *in vitro* fertilization and heat shock protein synthesis in maize (*Zea mays* L.) reproductive systems. *Plant Physiology,* **94**, 665-670.

Edwards, G.E., and N.R. Baker, 1993: Can CO_2 assimilation in maize be predicted accurately from chlorophyll fluorescence analysis. *Photosynthesis Research,* **37**, 89-102.

Eigenberg, R.A., T.M. Brown-Brandl, J.A. Nienaber and G.L. Hahn, 2005: Dynamic response indicators of heat stress in shaded and non-shaded feedlot cattle. Part 2: Predictive relationships. *Biosystems Engineering,* **91**, 111-118.

Egli, D.B., and I.F. Wardlaw, 1980: Temperature response of seed growth characteristics of soybean. *Agronomy Journal*, **72**, 560-564.

Ehleringer, J.R., S.L. Phillips, W.S.F. Schuster, and D.R. Sandquist, 1991: Differential utilization of summer rains by desert plants. *Oecologia*, **88**, 430-434.

Elagoz, V., and W.J. Manning, 2005: Responses of sensitive and tolerant bush beans (Phaseolus vulgaris L.) to ozone in open-top chambers are influenced by phenotypic differences, morphological characteristics, and the chamber environment. *Environmental Pollution,* **136**, 371-383

Epstein, H.E., I.C. Burke, and W.K. Lauenroth, 2002: Regional patterns of decomposition and primary production rates in the U.S. Great Plains. *Ecology*, **83**, 320-327.

Epstein, H.E., R.A. Gill, J.M. Paruelo, W.K. Lauenroth, G.J. Jia, and I.C. Burke, 2002: The relative abundance of three plant functional types in temperature grasslands and shrublands of North and South America: effects of projected climate change. *Journal of Biogeography*, **29**, 875-888.

Everitt, J.H., C. Yang, R.S. Fletcher, and D.L. Drawe, 2006: Evaluation of high-resolution satellite imagery for assessing rangeland resources in south Texas. *Rangeland Ecology & Management,* **59**, 30-37.

Farquhar, G.D, and S. von Cammerer, 1982: Modelling of photosynthetic response to environmental conditions. P. 549-587. In O.L. Lange et al. (eds.) *Encyclopedia of Plant Physiology.* NS. Vol. 12B. Physiological Plant Ecology II. Springer-Verlag, Berlin.

Fay, P.A., J.D. Carlisle, A.K. Knapp, J.M. Blair, and S.L. Collins, 2003: Productivity responses to altered rainfall patterns in a C4-dominated grassland. *Oecologia*, **137,** 245-251.

Field, C.B., C.P. Lund, N.R. Chiariello, and B.E. Mortimer, 1997: CO_2 effects on the water budget of grassland microcosm communities. *Global Change Biology,* **3**, 197-206.

Finnan, J.M., A. Donnelly, J.L. Burke, and M.B. Jones, 2002: The effects of elevated concentrations of carbon dioxide and ozone on potato (*Solanum tuberosum* L.) yield. *Agriculture, Eosystems & Environment*, **88,** 11-22.

Fonseca, A.E., and M.E. Westgate, 2005: Relationship between desiccation and viability of maize pollen. *Field Crops Research,* **94**, 114-125.

Food and Agriculture Organization, 2000: Pastoralism in the new millennium, *Food and Agriculture Organization of the United Nations, Animal Production and Health Paper* **150**, 93pp, Rome, Italy.

Frank, K.L. 2001: Potential effects of climate change on warm season voluntary feed intake and associated production of confined livestock in the United States. M.S. thesis. Kansas State University, Manhattan.

Frank, K.L., T.L. Mader, J.A. Harrington, G.L. Hahn, and M.S. Davis, 2001: Climate change effects on livestock production in the Great Plains. *Proceedings 6th International Livestock Environment Symposium*, American Society of Agricultural Engineers, St. Joseph, MI: 351-358.

Gaughan, J.B., T.L. Mader, S.M. Holt, M.J. Jose, and K.J. Rowan,1999: Heat tolerance of Boran and Tuli crossbred steers. *Journal of Animal Science,* **77**, 2398-2405.

Gaughan, J.B., S.M. Holt, G.L. Hahn, T.L. Mader, and R. Eigenberg, 2000: Respiration rate – is it a good measure of heat stress in cattle? *Asian-Australian Journal of Animal Science.* 13:329-332 (ARD No. 12903).

Gaughan, J.B., W.M. Kreikemeier, and T.L. Mader, 2005: Hormonal growth-promotant effects on grain-fed cattle maintained under different environments. *International Journal of Biometeorology*, **49**, 396-402 (ARD No. 14392).

Gaughan, J.B., J. Goopy and J. Spark, 2002: Excessive heat load index for feedlot cattle. Meat and Livestock-Australia Project Rept, FLOT.316. MLA, Ltd., Locked Bag 991, N. Sydney NSW, 2059 Australia.

Gielen, B., H.J. De Boeck, C.M.H.M. Lemmens, R. Valcke, I. Nijs, and R. Ceulemans, 2005: Grassland species will not necessarily benefit from future elevated air temperatures: a chlorophyll fluorescence approach to study autumn physiology. *Physiologia Plantarum,* **125**, 52-63.

Gill, R.A., L.J. Anderson, H.W. Polley, H.B. Johnson, and R.B. Jackson, 2006: Potential nitrogen constraints on soil carbon sequestration under low and elevated atmospheric CO_2. *Ecology*, **87**, 41-52.

Goho, A. 2004: Gardeners anticipate climate change. *American Gardener*, **83**, 36-41.

Goudriaan, J., and M.H. Unsworth, 1990: Implications of increasing carbon dioxide and climate change for agricultural productivity and water resources. P. 111-130. *In* B. A. Kimball et al. (eds). *Impact of Carbon Dioxide, Trace Gases, and Climate Change on Global Agriculture*. ASA Spec. Publ. **53**. ASA, Madison, WI.

Greer, D.H., W.A. Laing, and B.D. Campbell, 1995: Photosynthetic Responses of Thirteen Pasture Species to Elevated CO_2 and Temperature. *Australian Journal of Plant Physiology* **22**, 713-22.

Grimm, S.S., J.W. Jones, K.J. Boote, and D.C. Herzog, 1994: Modeling the occurrence of reproductive stages after flowering for four soybean cultivars. *Agronomy Journal*, **86**, 31-38.

Grimm, S.S., J.W. Jones, K.J. Boote, and J.D. Hesketh, 1993, Parameter estimation for predicting flowering date of soybean cultivars. *Crop Science*, **33**, 137-144.

Gross, Y., and J. Kigel, 1994: Differential sensitivity to high temperature of stages in the reproduction development of common beans (*Phaseolus vulgaris* L.). *Field Crops Research*, **36**, 201-212.

Hahn, G.L., 1981: Housing and management to reduce climatic impacts on livestock. *Journal of Animal Science*, **52**, 175-186.

Hahn, G.L., 1995: Environmental management for improved livestock performance, health and well-being. *Japanese Journal of Livestock Management*, **30**, 113-127.

Hahn, G.L., 1999: Dynamic responses of cattle to thermal heat loads. *Journal of Animal Science*, **77**, 10-20.

Hahn, G.L., Y.R. Chen, J.A. Nienaber, R.A. Eigenberg and A.M. Parkhurst, 1992: Characterizing animal stress through fractal analysis of thermoregulatory responses. *Journal of Thermal Biology*, **17**, 115-120.

Hahn, G.L. and T.L. Mader, 1997: Heat waves in relation to thermoregulation, feeding behavior and mortality of feedlot cattle. *Proceedings 5th International Livestock Environment Symposium*, American Society of Agricultural Engineers, St. Joseph, MI: 563-571.

Hahn, G.L., T.L. Mader, J.B. Gaughan, Q. Hu and J.A. Nienaber, 1999: Heat waves and their impacts on feedlot cattle. *Proceedings 15th International Congress of Biometeorology and the International Congress on Urban Climatology*, Sydney, Australia.

Hahn, L., T. Mader, D. Spiers, J. Gaughan, J. Nienaber, R. Eigenberg, T. Brown-Brandl, Q. Hu, D. Griffin, L. Hungerford, A. Parkhurst, M. Leonard, W. Adams, and L. Adams, 2001: Heat wave impacts on feedlot cattle: Considerations for improved environmental management. *Proceedings 6th International Livestock Environment Symposium*, American Society of Agricultural Engineers, St. Joseph, MI: 129-130.

Hall, A.E., 1992: Breeding for heat tolerance. P. 129-168. In: *Plant breeding reviewers*. Vol. **10**. John Wiley & Sons, New York.

Hamilton, J.G., O. Dermody, M. Aldea, A.R. Zangerl, A. Rogers, M.R. Berenbaum, and E.H. DeLucia, 2005: Anthropogenic changes in tropospheric composition increase susceptibility of soybean to insect herbivory. *Environmental Entomology*, **34**, 479-455.

Hatfield, J.L., and J.H. Prueger, 2004: Impact of Changing Precipitation Patterns on Water Quality. *Journal Soil and Water Conservation*, **59**, 51-58.

Heagle A.S., V.M. Lesser, J.O. Rawlings, W.W. Heck, and R.B. Philbeck, 1986: Response of soybeans to chronic doses of ozone applied as constant or proportional additions to ambient air. *Phytopathology* **76,** 51-56.

Heagle, A.S., 1989: Ozone and crop yield. *Annual Review Phytopathology*, **27**, 397-423.

Heitschmidt, R.K., and M.R. Haferkamp, 2003: Ecological consequences of drought and grazing on grasslands of the northern Great Plains. In: Weltzin JF, McPherson GR (eds) *Changing Precipitation Regimes and Terrestrial Ecosystems*, University of Arizona Press, Tucson, pp. 107-126.

Henry, H.A.L., J.D. Juarez, C.B. Field, and P.M. Vitousek, 2005: Interactive effects of elevated CO_2, N deposition and climate change on extracellular enzyme activity and soil density fractionation in a California annual grassland. *Global Change Biology*, **11**, 1808-1815.

Herrero, M.P., and R.R. Johnson, 1980: High temperature stress and pollen viability in maize. *Crop Science*, **20**, 796-800.

Hesketh, J.D., D.L. Myhre, and C.R. Willey, 1973: Temperature control of time intervals between vegetative and reproductive events in soybeans. *Crop Science*, **13**, 250-254.

High Plains Regional Climate Center, 2000. http://www.hprcc.unl.edu/

Hileman, D.R., G. Huluka, P.K. Kenjige, N. Sinha, N.C. Bhattacharya, P.K. Biswas, K.F. Lewin, J. Nagy, and G.R. Hendrey, 1994: Canopy photosynthesis and transpiration of field-grown cotton exposed to free-air CO enrichment (FACE) and differential irrigation. *Agricultural and Forest Meteorology*, **70**, 189-207.

Hodges, T., and J.T. Ritchie, 1991: The CERES-Wheat phenology model, chapter 12, pp 115-131. *In* Hodges, T (ed.) *Predicting Crop Phenology*. CRC Press, Boca Raton.

Horie, T., J. T. Baker, H. Nakagawa, T. Matsui, and H. Y. Kim, 2000. Crop ecosystem responses to climatic change: Rice. Chapter 5. pp. 81-106. *In* K. R. Reddy and H. F. Hodges, *Climate Change and Global Crop Productivity*. CAB International., New York, NY.

Hubbard, K.G., D.E. Stooksbury and G.L. Hahn, 1999: A climatological perspective on feedlot cattle performance and mortality related to the Temperature-Humidity Index. *Journal of Production Agriculture*, ***12***, 650-653.

Hunsaker, D.J., B.A. Kimball, P.J. Pinter, Jr., G.W. Wall, and R.L. LaMorte, 1997: Soil water balance and wheat evapotranspiration as affected by elevated CO_2 and variable soil nitrogen. In: *Annual Research Report 1997*. U.S. Water Conservation Laboratory, *ASDA*, ARS, Phoenix, AZ, pp. 67-70.

Hungate, B.A., F.S. Chapin III, H. Zhong, E.A. Holland, and C.B. Field, 1997: Stimulation of grassland nitrogen cycling under carbon dioxide enrichment. *Oecologia*, **109**, 149-153.

Hungate, B.A., C.H. Jaeger III, G. Gamara, F.S. Chapin III, and C.B. Field, 2000: Soil microbiota in two annual grasslands: responses to elevated atmospheric CO_2. *Oecologia*, **124**, 589-598.

Huxman, T.E., and S.D. Smith, 2001. Photosynthesis in an invasive grass and native forb at elevated CO_2 during an El Niño year in the Mojave Desert. *Oecologia*, **128**, 193-201.

Idso, S.B., B.A. Kimball, M.G. Anderson, and J.R. Mauney, 1987: Effects of atmospheric CO_2 enrichment on plant growth: The interactive role of air temperature. *Agriculture, Eosystems & Environment*, **20**, 1-10.

IPCC, 2001: *Climate Change 2001: The Scientific Basis, Contribution from Working Group I to the Third Assessment Report, Intergovernmental Panel for Climate Change*. Cambridge University Press, Cambridge, UK.

IPCC, 2007. *Climate Change 2007: The Physical Science Basis, Contribution from Working Group I to the Fourth Assessment Report, Policy Maker Summary*. Intergovernmental Panel on Climate Change. Cambridge University Press, Cambridge, UK.

Izaurralde, R.C., N.J. Rosenberg, R.A. Brown, and A.M. Thomson, 2003: Integrated assessment of Hadley Centre climate change projections on water resources and agricultural productivity in the conterminous United States. II. Regional agricultural productivity in 2030 and 2095. *Agricultural and Forest Meteorology*, **117**, 97-122.

Jifon, J., and D.W. Wolfe, 2005: High temperature-induced sink limitation alters growth and photosynthetic acclimation response to elevated CO_2 in beans. *Journal of the American Society for Horticultural Science*, **130**, 515-520

Jones, P., J.W. Jones, and L.H. Allen, Jr., 1985: Seasonal carbon and water balances of soybeans grown under stress treatments in sunlit chambers. Transactions *ASAE*, **28**, 2021-2028.

Jones, R.J., S. Ouattar, and R.K. Crookston, 1984: Thermal environment during endosperm cell division and grain filling in maize: Effects on kernel growth and development *in vitro*. *Crop Science*, **24**, 133-137.

Kakani, V.G., K.R. Reddy, S.Koti, T.P. Wallace, P.V.V. Prasad, V.R. Reddy, and D. Zhao, 2005: Differences in *in vitro* pollen germination and pollen tube growth of cotton cultivars in response to high temperature. *Annals of Botany*, **96**, 59-67.

Kandeler, E., A.R. Mosier, J.A. Morgan, D.G. Milchunas, J.Y. King, S. Rudolph, and D. Tscherko, 2006: Response of soil microbial biomass and enzyme activities to the transient elevation of carbon dioxide in a semi-arid grassland. *Soil Biology & Biochemistry*, **38**, 2448-2460.

Kimball, B.A., 1983: Carbon dioxide and agricultural yield. An assemblage of 430 prior observations. *Agronomy Journal*, **75**, 779-788.

Kimball, B.A., 2007: Global change and water resources. In Lascano, R. J., Sojka R. E. (eds.) *Irrigation of Agricultural Crops*. Agronomy Monograph 30, 2nd Edition. ASA-CSSA-SSSA. Madison, WI. pp. 627-653.

Kimball, B.A., and C.J. Bernacchi, 2006: Evapotranspiration, canopy temperature, and plant water relations. In: *Managed Ecosystems and CO_2: Case Studies, Processes, and Pespectives* pp. 311-324. Springer-Verlag, Berlin.

Kimball, B.A., and S.B. Idso, 1983: Increasing atmospheric CO_2: Effects on crop yield, water use, and climate. *Agricultural Water Management*, **7**, 55-72.

Kimball, B.A., and J.R. Mauney, 1993: Response of cotton to varying CO_2, irrigation, and nitrogen: yield and growth. *Agronomy Journal*, **85**, 706-712.

Kimball B.A., K. Kobayashi, and M. Bindi, 2002: Responses of agricultural crops to free-air CO_2 enrichment. *Advances in Agronomy*, **77**, 293-368.

Kimball, B.A., R.L. LaMorte, P.J. Pinter Jr., G.W. Wall, D.J. Hunsaker, F.J. Adamsen, S.W. Leavitt, T.L. Thompson, A.D. Matthias, and T.J. Brooks, 1999: Free-air CO_2 enrichment and soil nitrogen effects on energy balance and evapotranspiration of wheat. *Water Resources Research*, **35**, 1179-1190.

Kim, H.Y., T. Horie, H. Nakagawa, and K. Wada, 1996: Effects of elevated CO_2 concentration and high temperature on growth and yield of rice. II. The effect of yield and its component of Akihikari rice. *Japanese Journal of Crop Science*, **65**, 644-651.

King, K.M. and D.H. Greer, 1986: Effects of carbon dioxide enrichment and soil water on maize. *Agronomy Journal*, **78**, 515-521.

King, J.Y., A.R. Mosier, J.A. Morgan, D.R. LeCain, D.G. Milchunas, and W.J. Parton, 2004. Plant nitrogen dynamics in shortgrass steppe under elevated atmospheric carbon dioxide. *Ecosystems* 7:147-160.

Kiniry, J. R., and R. Bonhomme, 1991: Predicting maize phenology, chapter 11, pp 115-131. *In:* Hodges, T. (ed.) *Predicting Crop Phenology*. CRC Press, Boca Raton.

Knapp, P.A., P.T. Soulè, and H.D. Grissino-Mayer, 2001: Detecting potential regional effects of increased atmospheric CO_2 on growth rates of western juniper. *Global Change Biology*, **7**, 903-917.

Knapp, A.K., J.M. Briggs, and J.K. Koelliker, 2001. Frequency and extent of water limitation to primary production in a mesic temperate grassland. *Ecosystems*, **4**, 19-28.

Knapp, A.K., and M.D. Smith, 2001: Variation among biomes in temporal dynamics of aboveground primary production. *Science*, **291**, 481-484.

Kobza, J. and G.E. Edwards, 1987: Influences of leaf temperature on photosynthetic carbon metabolism in wheat. *Plant Physiology*, **83**, 69-74.

Krug, H., 1997: Environmental influences on development, growth and yield. In: Wien, H.C. (ed.) *The Physiology of Vegetable Crops*. CAB International. Wallingford, UK.

Laing, D.R., P.G. Jones, and J.H. Davis, 1984: Common bean (*Phaseolus vulgaris* L.). pp. 305-351. *In* P.R. Goldsworthy and N.M. Fisher (eds.). *The Physiology of Tropical Field Crops*. John Wiley and Sons, New York.

Lawlor, D.W., and R.A.C. Mitchell, 2000: Crop ecosystem responses to climatic change: Wheat. Chapter 4. pp. 57-80. *In* K. R. Reddy and H. F. Hodges, *Climate Change and Global Crop Productivity*. CAB International, New York, NY.

Lawson, T., J. Craigon, C.R. Black, J.J. Colls, G. Landon, and J.D.B. Weyers, 2002: Impact of elevated CO_2 and O_3 on gas exchange parameters and epidermal characteristics in potato (*Solanum tuberosum* L.). *Journal of Experimental Botany*, **53**, 737-746.

Leakey, A.D.B., M. Uribelarrea, E. A. Ainsworth, S.L. Naidu, A. Rogers, D.R. Ort, and S.P. Long, 2006: Photosynthesis, productivity, and yield of maize are not affected by open-air elevation of CO_2 concentration in the absence of drought. *Plant Physiology*, **140**, 779-790.

Liebig, M.A., J.A. Morgan, J.D. Reeder, B.H. Ellert, H.T. Gollany, and G.S. Schuman, 2005: Greenhouse gas contributions and mitigation potential of agricultural practices in northwestern USA and western Canada. *Soil & Tillage Research*, **83**, 25-52.

Lobell, D.B., and G.P. Asner, 2003: Climate and management contributions to recent trends in U.S. agricultural yields. *Science*, **299**, 1032.

Lobell, D.B., and C.B. Field, 2007: Global scale climate-crop yield relationships and the impact of recent warming. *Environmnetal Research Letters*, **2**, 1-7.

Long, S.P. 1991: Modification of the response of photosynthetic productivity to rising temperature by atmospheric CO_2 concentrations: has its importance been underestimated? *Plant, Cell & Environment*, **14**, 729-739.

Long, S.P., E.A. Ainsworth, A.D.B. Leakey, J. Nosberger, and D.R. Ort, 2006: Food for thought: lower-than-expected crop yield stimulation with rising CO_2 concentrations. *Science*, **213**, 1918-1921.

Luo, Y., D. Hui, and D. Zhang, 2006: Elevated CO_2 stimulate net accumulations of carbon and nitrogen in land ecosystems: a meta-analysis. *Ecology*, **87**, 53-63.

Luo, Y., B. Su, W.S. Currie, J.S. Dukes, A. Finzi, U. Hartwig, B. Hungate, R.E. McMurtrie, R. Oren, W.J. Parton, D.E. Pataki, M.R. Shaw, D.R. Zak, and C.B. Field, 2004: Progressive nitrogen limitation of ecosystem responses to rising atmospheric carbon dioxide. *BioScience*, **54**, 731-739.

Mader, T.L. 2003: Environmental stress in confined beef cattle. *Journal of Animal Science,* **81** (electronic suppl. 2), 110-119.

Mader, T.L., J.M. Dahlquist, and J.B. Gaughan. 1997a: Wind Protection effects and airflow patterns in outside feedlots. *Journal of Animal Science.*, **75**, 26-36.

Mader, T.L., J.M. Dahlquist, G.L. Hahn, and J.B. Gaughan, 1999a: Shade and wind barrier effects on summer-time feedlot cattle performance. *Journal of Animal Science,* **77**, 2065-2072.

Mader, T.L. and M.S. Davis, 2004: Effect of management strategies on reducing heat stress of feedlot cattle: feed and water intake. *Journal of Animal Science,* **82**, 3077-3087.

Mader, T.L., M.S. Davis, and T. Brown-Brandl, 2006: Environmental factors influencing heat stress in feedlot cattle. *Journal of Animal Science*, **84**, 712-719.

Mader, T.L., L.R. Fell, and M.J. McPhee, 1997b: Behavior response of non-Brahman cattle to shade in commercial feedlots. *Proceeding 6th International Livestock Environment Symposium*, American Society of Agricultural Engineers, St. Joseph, MI: 795-802.

Mader, T.L., J.M. Gaughan, and B. A. Young, 1999b: Feedlot diet roughage level of Hereford cattle exposed to excessive heat load. *Professional Animal Scientist,* **15**, 53-62.

Mader, T.L., S.M. Holt, G.L. Hahn, M.S. Davis and D.E. Spiers, 2002: Feeding strategies for managing heat load in feedlot cattle. *Journal of Animal Science*, **80**, 2373-2382.

Mader, T.L. and W.M. Kreikemeier, 2006: Effects of growth-promoting agents and season on blood metabolites and body temperature in heifers. *Journal of Animal Science,* **84**, 1030-1037.

Magliulo, V., M. Bindi, and G. Rana, 2003: Water use of irrigated potato (*Solanum tuberosum* L.) grown under free air carbon dioxide enrichment in central Italy. *Agriculture, Ecosystems and Environment*, **97**, 65-80.

Maroco, J.P., G.E. Edwards and M.S.B. Ku, 1999: Photosynthetic acclimation of maize to growth under elevated levels of carbon dioxide. *Planta*, **210**, 115-125.

Maiti, R.K., 1996: Sorghum science. *Science Publishers, Inc.*, Lebanon, New Hampshire, USA.

Matsui, T., O.S. Namuco, L.H. Ziska and T. Horie, 1997: Effects of high temperature and CO_2 concentration on spikelet sterility in *indica* rice. *Field Crops Research,* **51**, 213-219.

Matsushima, S., T. Tanaka, and T. Hoshino, 1964: Analysis of yield determining process and its application to yield-prediction and culture improvement of lowland rice. LXX. Combined effect of air temperature and water temperature at different stages of growth on the grain yield and its components of lowland rice. *Proceedings of the Crop Science Society of Japan*, **33**, 53-58.

Mauney, J.R., B.A. Kimball, P.J. Pinter, Jr., R.L. LaMorte, K.F. Lewin, J. Nagy, and G.R. Hendrey, 1994: Growth and yield of cotton in response to free-air carbon dioxide enrichment (FACE) environment. *Agricultural and Forest Meterology,* **70**, 49-67.

Medlyn, B.E., C.V.M. Barton, M.S.J. Broadmeadow, R. Ceulemans, P. De Angelis, M. Forstreuter, M. Freeman, S.B. Jackson, S. Kellomaki, E. Laitat, A. Rey, P. Roberntz, B.D. Sigurdsson, J. Strassemeyer, K. Wang, P.S. Curtis, and P.G. Jarvis, 2001: Stomatal conductance of forest species after long-term exposure to elevated CO_2 concentration: a synthesis. *New Phytologist,* **149**, 247-264.

Meeting, F.B., J.L. Smith, J.S. Amthor, and R.C. Izaurralde, 2001: Science needs and new technology for increasing soil carbon sequestration. *Climatic Change,* **51**, 11-34.

Milchunas, D.G., A.R. Mosier, J.A. Morgan, D.R. LeCain, J.Y. King, and J.A. Nelson, 2005: Elevated CO_2 and defoliation effects on a shortgrass steppe: forage quality versus quantity for ruminants. *Agriculture, Ecosystems and Environment*, **111**, 166-184.

Miller, J.E., A.S. Heagle, and W.A. Pursley, 1998: Influence of ozone stress on soybean response to carbon dioxide enrichment: II. Biomass and development. *Crop Science*, **38**, 122-128.

Mills, G., G. Ball, F. Hayes, J. Fuhrer, L. Skarby, B. Gimeno, L. De Temmerman, and A. Heagle, 2000: Development of a multi-factor model for predicting the effects of ambient ozone on the biomass of white clover. *Environmental Pollution*, **109**, 533-542.

Mitchell, M.A., P.J. Kettlewell, R.R. Hunter and A.J. Carlisle, 2001: Physiological stress response modeling – applications to the broiler transport thermal environment. *Proceedings 6th International Livestock Environment Symposium*, American Society of Agricultural Engineers, St Joseph, MI: 550-555.

Mitchell, R.A.C., V.J. Mitchell, S.P. Driscoll, J. Franklin, and D.W. Lawlor, 1993: Effects of increased CO_2 concentration and temperature on growth and yield of winter wheat at two levels of nitrogen application. *Plant Cell & Environment*, **16**, 521-529.

Montaigne, F., 2004: The heat is on: eco-signs. *National Geographic*, **206**, 34-55.

Moore, J.L., S.M. Howden, G.M. McKeon, J.O. Carter, and J.C. Scanlan, 2001: The dynamics of grazed woodlands in southwest Queensland, Australia, and their effect on greenhouse gas emissions. *Environmental International*, **27**, 147-153.

Morgan, J.A. 2005. Rising atmospheric CO_2 and global climate change: Management implications for grazing lands. pp. 245-272 in: S.G. Reynolds and J. Frame (eds) *Grasslands: Developments Opportunities Perspectives.* FAO and Science Pub. Inc.

Morgan, J.A., D.R. LeCain, A.R. Mosier, and D.G. Milchunas, 2001: Elevated CO_2 enhances water relations and productivity and affects gas exchange in C3 and C4 grasses of the Colorado shortgrass steppe. *Global Change Biology*, **7**, 451-466.

Morgan, J.A., D.G. Milchunas, D.R. LeCain, M.S. West and A. Mosier, 2007. Carbon dioxide enrichment alters plant community structure and accelerates shrub growth in the shortgrass steppe. *Proceedings of the National Academy of Sciences* **104**, 14724-14729.

Morgan, J.A., A.R. Mosier, D.G. Milchunas, D.R. LeCain, J.A. Nelson, and W.J. Parton, 2004a: CO_2 enhances productivity, alters species composition, and reduces digestibility of shortgrass steppe vegetation. *Ecological Application*, **14**, 208-219.

Morgan, J.A., D.E. Pataki, C. Körner, H. Clark, S.J. Del Grosso, J.M. Grünzweig, A.J., Knapp, A.R. Mosier, P.C.D. Newton, P.A. Niklaus, J.B. Nippert, R.S. Nowak, W.J. Parton, H.W. Polley, and M.R. Shaw, 2004b: Water relations in grassland and desert ecosystems exposed to elevated atmospheric CO_2. *Oecologia*, **140**, 11-25.

Morgan, P.B., E.A. Ainsworth, and S.P. Long, 2003: How does elevated ozone impact soybean? A meta-analysis of photosynthesis, growth and yield. *Plant, Cell & Environment*, **26**, 1317-1328.

Morgan, P.B., C.J. Bernacchi, D.R. Ort, and S.P. Long, 2004: An in vivo analysis of the effect of season-long open-air elevation of ozone to anticipated 2050 levels on photosynthesis in soybean. *Plant Physiology*, **135**, 2348-2357.

Morgan, P.B., T.A. Mies, G.A. Bollero, R.L. Nelson, and S.P. Long, 2006: Season-long elevation of ozone concentration to projected 2050 levels under fully open-air conditions substantially decreases the growth and production of soybean. *New Phytologist*, **170**, 333-343.

Morison, J.I.L., 1987: Intercellular CO_2 concentration and stomatal response to CO_2. p. 229-251. *In* E. Zeiger, G. D. Farquhar, and I. R. Cowan (eds.) *Stomatal Function.* Stanford Univ. Press, Stanford, CA.

Moura, D.J., I.A. Naas, K.B. Sevegnani and M.E. Corria, 1997: The use of enthalpy as a thermal comfort index. *Proceedings 5th International Livestock Environment Symposium*, American Society of Agricultural Engineers, St. Joseph, MI: 577-583.

Murphy, K.L., I.C. Burke, M.A. Vinton, W.K. Lauenroth, M.R. Aguiar, D.A. Wedin, R.A. Virginia, and P.N. Lowe, 2002: Regional analysis of litter quality in the central grassland region of North America. *Journal of Vegetation Science*, **13**, 395-402.

Muchow, R. C., T. R. Sinclair, and J. M. Bennett, 1990: Temperature and solar-radiation effects on potential maize yield across locations. *Agronomy Journal*, **82**, 338-343.

Nakagawa, H., T. Horie, and H.Y. Kim, 1994: Environmental factors affecting rice responses to elevated carbon dioxide concentrations. *International Rice Research Notes*, **19**, 45-46.

Nelson, J.A., J.A. Morgan, D.R. LeCain, A.R. Mosier, D.G. Milchunas and W.J. Parton, 2004: Elevated CO_2 increases soil moisture and enhances plant water relations in a long-term field study in the semi-arid shortgrass steppe of Northern Colorado. *Plant and Soil*, **259**, 169-179.

Neilson, R.P., 1986. High-resolution climatic analysis and southwest biogeography. *Science*, **232**, 27-34Newman, J.A., M.L. Abner, R.G. Dado, D.J. Gibson, A. Brookings, and A.J. Parsons, 2003: Effects of elevated CO_2, nitrogen and fungal endophyte-infection on tall fescue: growth, photosynthesis, chemical composition and digestibility. *Global Change Biology*, **9**, 425-437.

Newman, Y.C., L.E. Sollenberger, K.J. Boote, L.H. Allen, Jr., J.M. Thomas, and R.C. Littell, 2006: Nitrogen fertilization affects bahiagrass response to elevated atmospheric carbon dioxide. *Agronomy Journal*, **98**, 382-387.

Newman, Y.C., L.E. Sollenberger, K.J. Boote, L.H. Allen, Jr., and R.C. Littell, 2001: Carbon dioxide and temperature effects on forage dry matter production. *Crop Science*, **41**, 399-406.

Newton, P.C.D., H. Clark, C.C. Bell, and E.M. Glasgow, 1996: Interaction of soil moisture and CO_2 on the above-ground growth rate, root length density, and gas exchange of turves from temperature pastures. *Journal of Experimental Botany*, **47**, 771-779.

Niklaus, P.A., J. Alphei, D. Ebersberger, C. Kampichlers, E. Kandeler, and D. Tscherko, 2003: Six years of in situ CO_2 enrichment evoke changes in soil structure and soil biota of nutrient-poor grassland. *Global Change Biology*, **9**, 585-600.

Noormets, A., A. Sôber, E.J. Pell, R.E. Dickson, G.K. Podila, J. Sôber, J.G. Isebrands, and D.F. Karnosky, 2001: Stomatal and non-stomatal limitation to photosynthesis in two trembling aspen (*Populus tremuloides* Michx.) clones exposed to elevated CO_2 and/or O_3, *Plant, Cell, and Environment*, **24**, 327-336.

Norby, R.J., M.F. Cortufo, P. Ineson, E.G. O Neill, and J.G. Canadell, 2001: Elevated CO_2, litter chemistry, and decomposition: a synthesis. *Oecologia*, **127**, 153-165.

NRC (National Research Council), 1981: Effect of environment on nutrient requirements of domestic animals. National Academy Press, Washington, D.C.

NRC, (National Research Council), 1987: Predicting Feed Intake of Food Producing Animals. National Academy Press, Washington, D.C.

NRCS [Natural Resources Conservation Service], 2003: *National Range and Pasture Handbook.* USDA-NRCS, Grazing Lands Technology Institute. Washington, DC.

Oberhuber, W., and G.E. Edwards, 1993: Temperature dependence of the linkage of quantum yield of photosystem II to CO_2 fixation in C4 and C3 plants. *Plant Physiology,* **101**, 507-512.

Ong, C.K. 1986: Agroclimatological factors affecting phenology of groundnut. Pages 115-125 *In:* Agrometeorology of Groundnut: *Proceedings of an International Symposium,* 21-26 Aug. 1985, ICRISAT Sahelian Center, Niamey, Niger. ICRISAT, Patancheru, A.P. 502 324, India.

Ottman, M.J., B.A. Kimball, P.J. Pinter, G.W. Wall, R.L. Vanderlip, S.W. Leavitt, R.L. LaMorte, A.D. Matthias, and T.J. Brooks, 2001: Elevated CO_2 increases sorghum biomass under drought conditions. *New Phytologist,* **15**, 261-273.

Owensby, C.E., P.I. Coyne, and L.M. Auen, 1993: Nitrogen and phosphorus dynamics of a tallgrass prairie ecosystem exposed to elevated carbon dioxide. *Plant, Cell & Environment,* **16**, 843-850.

Owensby, C.E., R.C. Cochran, and L.M. Auen, 1996: Effects of elevated carbon dioxide on forage quality for ruminants. In: Körner, Ch. and F.A. Bazzaz (eds.) *Carbon Dioxide, Populations and Communities.* Academic Press, San Diego, pp. 363-371.

Owensby, C.E., J.M. Ham, A.K. Knapp, and L.M. Auen, 1999: Biomass production and species composition change in a tallgrass prairie ecosystem after long-term exposure to elevated atmospheric CO_2. *Global Change Biology,* **5**, 497-506.

Pan, D., 1996: Soybean responses to elevated temperature and doubled CO_2. Ph.D. dissertation. University of Florida, Gainesville, Florida, USA. 227 p.

Pareulo, J.M., and W.K. Lauenroth, 1996: Relative abundance of plant functional types in grasslands and shrublands of North America. *Ecological Applications,* **6**, 1212-1224.

Parton, W.J., J.A. Morgan, G.Wang, and S. DelGrosso. 2007a: Projected ecoystem impact of the prairie heating and CO_2 enrichment experiment. *New Phytologist* 174, 823-834.

Parton, W., W.L. Silver, I.C. Burke, L. Grassens, M.E. Harmon, W.S. Currie, J.Y. King, E.C. Adair, L.A. Brandt, S.C. Hart, and B. Fasth, 2007b: Global-scale similarities in nitrogen release patterns during long-term decomposition. *Science,* **315**, 361-364.

Parton, W.J., D.S. Schimel, C.V. Cole, and D.S. Ojima, 1987: Analysis of factors controlling soil organic matter levels in Great Plains grasslands. *Soil Science Society of America Journal,* **51**, 1173-1179.

Patterson, D.T., J.K. Westbrook, R.J.C. Joyce, P.D. Lingren, and J. Rogasik, 1999: Weeds, insects and diseases. *Climatic Change,* **43**, 711-727.

Paulsen, G.M., 1994: High temperature responses of crop plants. *In:* K.J. Boote, J.M. Bennett, T.R. Sinclair, and G.M. Paulsen (eds.) *Physiology and Determination of Crop Yield.* ASA-CSSA-SSSA, Madison, WI. Pp. 365-389.

Peat, M.M., S. Sato, and R.G. Gardner, 1998: Comparing heat stress effects on male-fertile and male-sterile tomatoes. *Plant, Cell & Environment,* **21**, 225-231.

Peet, M.M., and D.W. Wolfe, 2000: Crop ecosystem responses to climate change- vegetable crops. In: Reddy K.R., Hodges H.F. (eds) *Climate Change and Global Crop Productivity.* CABI Publishing. New York.

Peng, S., J. Huang, J.E. Sheehy, R.C. Lanza, R.M. Visperas, X. Zhong, G.S. Centeno, G.S. Khush, and KG. Cassman, 2004: Rice yields decline with higher night temperatures from global warming. *Proceedings of the National Academy of Sciences of the United States of America,* http://www.pnas.org/cgi/content/full/101/27/9971, 10 pp.

Peñuelas, J., and M. Estiarte, 1997: Trends in plant carbon concentration and plant demand for N throughout the century. *Oecologia,* **109**, 69-73.

Pepper, D.A., S. Del Grosso, R.E. McMurtrie, and W.J. Parton, 2005: Simulated carbon sink response of shortgrass steppe, tallgrass prairie and forest ecosystems to rising [CO_2], temperature and nitrogen input. *Global Biogeochemical Cycles,* **19**, GB 1004. pp. 20.

Peters, D.P.C., B.T. Bestelmeyer, J.E. Herrick, E.L. Fredrickson, H.C. Monger, and K.M. Havstad, 2006. Disentangling complex landscapes: New insights into arid and semiarid system dynamics. *BioScience* **56**, 491-501.

Pettigrew, W.T., 2008. The effect of higher temperature on cotton lint yield production and fiber quality. *Crop Science,* **48, 278-285.**

Phillips, R.L., O. Beeri, and M. Liebig, 2006. Landscape estimation of canopy C:N ratios under variable drought stress in Northern Great Plains rangelands. *Journal of Geophysical Research.* 111: doi:10.1029/2005JG000135.

Pickering, N.B., J.W. Jones, and K.J. Boote, 1995: Adapting SOYGRO V5.42 for prediction under climate change conditions. *In*: C. Rosenzweig, J.W. Jones, and L.H. Allen, Jr. (eds.). *Climate Change and Agriculture: Analysis of Potential International Impacts,* ASA Spec. Pub. No. 59, ASA-CSSA-SSSA, Madison, WI. pp. 77-98.

Piper, E.L., K.J. Boote, and J.W. Jones, 1998: Evaluation and improvement of crop models using regional cultivar trial data. *Applied Engineering in Agriculture,* **14**, 435-446.

Polley, H.W., 1997: Implications of rising atmospheric carbon dioxide for rangelands. *Journal of Range Management,* **50**, 561-577.

Polley, H.W., W.A. Dugas, P.C. Mielnick, and H.B., Johnson, 2007: C3-C4 composition and prior carbon dioxide treatment regulate the response of grassland carbon and water fluxes to carbon dioxide. *Functional Ecology,* **21**, 11-18.

Polley, H.W., H.B. Johnson, and J.D. Derner, 2003: Increasing CO_2 from subambient to superambient concentrations alters species composition and decreases above-ground biomass in a C3/C4 grassland. *New Phytologist,* **160**, 319-327.

Polley, H.W., H.B. Johnson, and C.R. Tischler, 2002. Woody invasion of grasslands: evidence that CO_2 enrichment indirectly promotes establishment of Prosopis glandulosa. *Plant Ecology,* **164**, 85-94.

Polley, H.W., J.A. Morgan, B.D. Campbell, M. Stafford Smith, 2000: Crop ecosystem responses to climatic change: rangelands. *In*: Reddy, K.R., and H.F. Hodges (eds.) *Climate change and global crop productivity.* CABI, Wallingford, Oxon, UK, pp. 293-314.

Prasad, P.V.V., K.J. Boote, and L.H. Allen, Jr., 2006a: Adverse high temperature effects on pollen viability, seed-set, seed yield and harvest index of grain-sorghum [*Sorghum bicolor* (L.) Moench] are more severe at elevated carbon dioxide due to high tissue temperature. *Agricultural and Forest Meteorology,* **139**, 237-251.

Prasad, P.V.V., K.J. Boote, L.H. Allen, Jr., J.E. Sheehy, and J.M.G. Thomas, 2006b: Species, ecotype and cultivar differences in spikelet fertility and harvest index of rice in response to high temperature stress. *Field Crops Research,* **95**, 398-411.

Prasad, P.V.V., K.J. Boote, L.H. Allen, Jr., and J.M.G. Thomas, 2002: Effects of elevated temperature and carbon dioxide on seed-set and yield of kidney bean (*Phaseolus vulgaris* L.). *Global Change Biol,* **8**, 710-721.

Prasad, P. V. V., K. J. Boote, L. H. Allen, Jr., and J. M. G. Thomas, 2003: Supra-optimal temperatures are detrimental to peanut (*Arachis hypogaea* L) reproductive processes and yield at ambient and elevated carbon dioxide. *Global Change Biology,* **9**, 1775-1787.

Prasad, P.V.V., P.Q. Craufurd, V.G. Kakani, T.R. Wheeler, and K.J. Boote, 2001: Influence of high temperature during pre- and post-anthesis stages of floral development on fruit-set and pollen germination in peanut. *Australian Journal of Plant Physiology,* **28**, 233-240.

Rae, A.M., R. Ferris, M.J. Tallis, and G. Taylor, 2006: Elucidating genomic regions determining enhanced leaf growth and delayed senescence in elevated CO_2. *Plant, Cell & Environment,* **29**, 1730-1741.

Read, J.J., J.A. Morgan, N.J. Chatterton, and P.A Harrison, 1997: Gas exchange and carbohydrate and nitrogen concentrations in leaves of *Pascopyrum smithii* (C3) and *Bouteloua gracilis* (C4) at different carbon dioxide concentrations and temperatures. *Ann. Bot,* **79**, 197-206

Reddy, K.R., G.H. Davidonis, A.S. Johnson, and B.T. Vinyard, 1999: Temperature regime and carbon dioxide enrichment alter cotton boll development and fiber properties. *Agronomy Journal* **91**, 851-858.

Reddy, K.R., H.F. Hodges, and B.A. Kimball, 2000: Crop ecosystem responses to climatic change: Cotton. Chapter 8. pp. 161-187. *In:* K. R. Reddy and H. F. Hodges, *Climate Change and Global Crop Productivity*. CAB International, New York, NY.

Reddy, K.R., H.F. Hodges, and J.M. McKinion, 1995: Carbon dioxide and temperature effects on Pima cotton growth. *Agriculture, Ecosystems & Environment,* **54**, 17-29.

Reddy, K.R., H.F. Hodges, and J.M. McKinion, 1997: A comparison of scenarios for the effect of global climate change on cotton growth and yield. *Australian Journal of Plant Physiology,* **24**, 707-713.

Reddy, K.R., H.F. Hodges, J.M. McKinion, and G.W. Wall, 1992a: Temperature effects on Pima cotton growth and development. *Agronomy Journal,* **84**, 237-243.

Reddy, K.R., H.F. Hodges, and V.R. Reddy, 1992b: Temperature effects on cotton fruit retention. *Agronomy Journal,* **84**, 26-30.

Reddy, K.R., P.V.V Prasad, and V.G. Kakani, 2005: Crop responses to elevated carbon dioxide and interactions with temperature: *Cotton. J. of Crop Improvement,* **13**, 157-191.

Reddy, V.R., K.R. Reddy, and H.F. Hodges, 1995: Carbon dioxide enrichment and temperature effects on cotton canopy photosynthesis, transpiration, and water use efficiency. *Field Crops Research,* **41**, 13-23.

Reich, P.B., S.E. Hobbie, T. Lee, D.S. Ellsworth, J.B. West, D. Tilman, J.M.H. Knops, S. Naeem, and J. Trost, 2006a: Nitrogen limitation constrains sustainability of ecosystem response to CO_2. *Nature,* **440**, 922-924.

Reich, P.B., B.A. Hungate, and Y. Luo, 2006b: Carbon-nitrogen interactions in terrestrial ecosystems in response to rising atmospheric carbon dioxide. *Annual Review of Ecological System,* **37**, 611-636.

Ritchie, J.T. 1972: Model for predicting evaporation from a row crop with incomplete cover. *Water Resources Research,* **8**, 1204-1213.

Rubatzky, V.E., M. Yamaguchi. 1997. World Vegetables.2nd Edition. Chapman and Hall. New York. Chapter 6, pp. 59-65.

Rudorff, B.F.T., C.L. Mulchi, C.S.T. Daughtry, and E.H. Lee, 1996: Growth, radiation use efficiency, and canopy reflectance of wheat and corn grown under elevated ozone and carbon dioxide atmospheres. *Remote Sensing of the Environment,* **55**, 163-173.

Runge, E. C. A. 1968: Effect of rainfall and temperature interactions during the growing season on corn yield. *Agronomy Journal,* **60**, 503-507.

Russelle, M.P., M.H. Entz, and A.J. Franzluebbers, 2007: Reconsidering integrated crop-livestock systems in North America. *Agronomy Journal,* **99**, 325-334.

Rustad, L.E., J.L. Campbell, G.M. Marion, R.J. Norby, M.J. Mitchell, A.E. Hartley, J.H.C. Cornelissen, and J. Gurevitch, 2001: A meta-analysis of the response of soil respiration, net mitrogen mineralization, and aboveground plant growth to experimental ecosystem warming. *Oecologia,* **126**, 543-562.

Salem, M.A., V.G. Kakani, S. Koti, and K.R. Reddy, 2007: Pollen-based screening of soybean genotypes for high temperature, *Crop Science,* **47**, 219-231.

Samson, F. and F. Knopf, 1994. Prairie conservation in North America. *BioScience,* **44,** 418-421.

Sasek, T.W., and B.R. Strain, 1990: Implications of atmospheric CO_2 enrichment and climatic change for the geographical distribution of two introduced vines in the USA. *Climatic Change,* **16**, 31-51.

Satake, T, and S. Yoshida, 1978: High temperature-induced sterility in *indica* rice at flowering. Japanese Journal of Crop Science, **47**, 6-17.

Sato, S., M.M. Peet, and J.F. Thomas, 2000: Physiological factors limit fruit set of tomato (*Lycopersicon esculentum* Mill.) under chronic high temperature stress. *Plant, Cell & Environment,* **23**, 719-726.

Sau, F., K.J. Boote, W.M. Bostick, J.W. Jones, and M.I. Minguez, 2004: Testing and improving evapotranspiration and soil water balance of the DSSAT crop models. *Agronomy Journal*, **96**, 1243-1257.

Schlesinger, W.H. 2006: Carbon trading. *Science*, **314**, 1217.

Schoper, J.B., R.J. Lambert, B.L. Vasilas, and M.E. Westgate, 1987: Plant factors controlling seed set in maize. *Plant Physiology*, **83**, 121-125.

Schuman, G.E., J.E. Herrick, and H.H. Janzen, 2001: The dynamics of soil carbon in rangelands. pp. 267-290, *In:* R.F. Follett, J.M. Kimble and R. Lal (eds). *The Potential of U.S. Grazing Lands to Sequester Carbon and Mitigate the Greenhouse Effect*. Boca Raton, FL: Lewis Publishers.

Semmartin, M., M.R. Aguiar, R.A. Distel, A.S. Moretto, and C.M. Ghersa, 2004: Litter quality and nutrient cycling affected by grazing-induced species replacements along a precipitation gradient. *Oikos: A Journal of Ecology*, **107**, 148-160.

Sexton, P.J., J.W. White, and K.J. Boote, 1994: Yield-determining processes in relation to cultivar seed size of common bean. *Crop Science*, **34**, 84-91.

Schaeffer, S.M., S.A. Billings, and R.D. Evans, 2007: Laboratory incubations reveal potential responses of soil nitrogen cycling to changes in soil C and N availability in Mojave Desert soils exposed to elevated atmospheric CO_2. *Global Change Biology*, **13**, 854-865.

Shaw, M.R., E.S. Zavaleta, N.R. Chiariello, E.E. Cleland, H.A. Mooney, and C.B. Field, 2002: Grassland responses to global environmental changes suppressed by elevated CO_2. *Science*, **298**, 1987-1990.

Sherry, R.A., X. Zhou, S. Gu, J.A. Arnone III, D.S. Schimel, P.S. Verburg, L.L. Wallace, and Y. Luo, 2007: Divergence of reproductive phenology under climate warming. *Proceedings of the National Academy Of Sciences*, **104**, 198-202.

Six, J., R.T. Conant, E.A. Paul, and K. Paustian, 2002: Stabilization mechanisms of soil organic matter: Implications for C-saturation of soils. *Plant Soil*, **241**, 155-176.

Smith, S.D., T.E. Huxman, S.F. Zitzer, T.N. Charlet, D.C. Housman, J.S. Coleman, L.K. Fenstermaker, J.R. Seemann, and R.S. Nowak, 2000: Elevated CO_2 increases productivity and invasive species success in an arid ecosystem. *Nature*, **408**, 79-82.

Snyder, A.M. 2000: The effects of elevated carbon dioxide and temperature on two cultivars of rice. Master's Thesis, University of Florida, Gainesville, Florida, USA. 167 pp.

Sofield, I., L.T. Evans, M.G. Cook, and I.F. Wardlaw, 1977: Factors influencing the rate and duration of grain filling in wheat. *Australian Journal of Plant Physiology*, **4**, 785-797.

Sofield, I., L.T. Evans, and I.F. Wardlaw, 1974: The effects of temperature and light on grain filling in wheat. P. 909-915. *In* R. L. Bieleski et al. (eds.) *Mechanisms of Regulation of Plant Growth*. Bull. 12. R. Soc. N.Z., Wellington, N.Z.

Sonnemann, I., V. Wolters. 2005. The microfood web of grassland soils respond to a moderate increase in atmospheric CO_2. *Global Change Biology* 11: 1148-1155.

Sprott, L.R., G.E. Selk, and D.C. Adams, 2001: Review: Factors affecting decisions on when to calve beef females. *Professional Animal Scientist*, **17**, 238-246.

Stockle, C.O., P.T. Dyke, J.R. Williams, C.A. Jones, and N.J. Rosenberg, 1992a: A method for estimating the direct and climatic effects of rising atmospheric carbon dioxide on growth and yield of crops: Part II – Sensitivity analysis at three sites in the Midwestern USA. *Agricultural Systems*, **38**, 239-256.

Stockle, C.O., J.R. Williams, N.J. Rosenberg, and C.A. Jones, 1992b: A method for estimating the direct and climatic effects of rising atmospheric carbon dioxide on growth and yield of crops: Part 1 – Modification of the EPIC model for climate change analysis. *Agricultural Systems*, **38**, 225-238.

Stephenson, N.L. 1990: Climatic control of vegetation distribution: the role of the water balance. *American Naturalist*, **135**, 649-670.

Stivers, L., 1999: Crop Profiles for Corn (Sweet) in New York. http://pestdata.ncsu.edu/cropprofiles/docs/nycorn-sweet.html

Sustainable Rangeland Roundtable Members (2006) Progress Report http://sustainablerangelands.warnercnr.colostate.edu/Images/ProgressReport.pdf

Suter, D., J. Nösberger, and A. Lüscher, 2001: Response of perennial ryegrass to Free-Air CO_2 Enrichment (FACE) is related to the dynamics of sward structure during regrowth. *Crop Science*, **41**, 810-817.

Svejcar, T.J., J. Bates, R.F. Angell, and R. Miller, 2003: The influence of precipitation timing on the sagebrush steppe ecosystem. *In*: Weltzin JF, McPherson GR (eds) *Changing Precipitation Regimes and Terrestrial Ecosystems*, University of Arizona Press, Tucson, pp. 90-106.

Tashiro, T., and I.F. Wardlaw, 1990: The response to high temperature shock and humidity changes prior to and during the early stages of grain development in wheat. *Australian Journal of Plant Physiology*, **17**, 551-561.

Temple, P.J. 1990: Growth form and yield responses of 4 cotton cultivars to ozone. *Agronomy Journal*, **82**, 1045-1050.

Terri, J.A., and L.G. Stowe. 1976. Climatic patterns and the distribution of C4 grasses in North America. *Oecologia* 23:1-12.

Thomas, J.M.G., 2001: Impact of elevated temperature and carbon dioxide on development and composition of soybean seed. Ph.D. Dissertation. University of Florida. Gainesville, Florida, USA. 185 pp.

Thomson A.M., R.A. Brown, N.J. Rosenberg, R.C. Izaurralde, and V.W. Benson, 2005: Climate change impacts for the conterminous USA: An integrated assessment Part 3. Dryland production of grain and forage crops. *Climatic Change*, **69**, 43-65.

Thornley, J.H.M., and M.G.R. Cannell, 1997: Temperate grassland responses to climate change: an analysis using the Hurley Pasture Model. *Annals of Botany*, **80**, 205-221.

Thornley, J.H.M., and M.G.R. Cannell, 2000: Dynamics of mineral N availability in grassland ecosystems under increased [CO_2]: hypotheses evaluated using the Hurley Pasture model. *Plant Soil*, **224**, 153-170.

Tingey, D.T., K.D. Rodecap, E.H. Lee, W.E. Hogsett, and J.W. Gregg, 2002: Pod development increases the ozone sensitivity of Phaseolus vulgaris. *Water Air and Soil Pollution,* **139**, 325-341.

Tommasi, P.D., V. Magliulo, R. Dell'Aquila, F. Miglietta, A. Zaldei, and G. Gaylor, 2002: Water consumption of a CO_2 enriched poplar stand. *Atti del Convegno CNR-ISAFOM*, Ercolano, Italy.

Triggs, J.M., B.A. Kimball, P.J. Pinter Jr, G.W. Wall, M.M. Conley, T.J. Brooks, R.L. LaMorte, N.R. Adam, M.J. Ottman, A.D. Matthias, S.W. Leavitt, and R.S. Cerveny, 2004: Free-air carbon dioxide enrichment (FACE) effects on energy balance and evapotranspiration of sorghum. *Agricultural and Forest Meteorology,* **124**, 63-79.

Tubiello, F.N., J.S. Amthor, K.J. Boote, M. Donatelli, W. Easterling, G. Fischer, R.M. Gifford, M. Howden, J. Reilly, and C. Rosenzweig, 2007: Crop response to elevated CO_2 and world food supply: A comment on "Food for Thought..." by Long et al., Science 312:1918-1921, 2006. *European J. Agronomy,* 26, 215-223.

Van Groenigen, K.J., J. Six, B.A. Hungate, M.A, Graaff, N. van Breemen, and C. van Kessel, 2006: Element interactions limit soil carbon storage. *Proceedings of the National Academy Of Sciences,* **103**, 6571-6574.

Van Kooten, G.C. 2006: Economic of forest and agricultural carbon sinks. Chapter 19 *In*: Bhatti, J.S., R. Lal, M.J. Apps, and M.A. Price (eds), *Climate Change and Managed Ecosystems,* 375-395, Taylor & Francis Group, New York.

Villalobos, F.J. and E. Fereres, 1990: Evaporation measurements beneath corn, cotton, and sunflower canopies. *Agronomy Journal*, **82**, 1153-1159.

Vu, J.C.V., J.T. Baker, A.H. Pennanen, L.H. Allen, Jr., G. Bowes, and K.J. Boote, 1998: Elevated CO_2 and water deficit effects on photosynthesis, ribulose bisphosphate carboxylase-oxygenase, and carbohydrate metabolism in rice. *Physiologia Plantarum,* **103**, 327-339.

Wall, G.W., T.J. Brooks, R. Adam, A.B. Cousins, B.A. Kimball, P.J. Pinter, R.L. LaMorte, L. Trigs, M.J. Ottman, S.W. Leavitt, A.D. Matthias, D.G. Williams, and A.N. Webber, 2001: Elevated atmospheric CO_2 improved sorghum plant water status by ameliorating the adverse effects of drought. *New Phytologist,* **152**, 231-248.

Wall, G.W., R.L. Garcia, B.A. Kimball, D.J. Hunsaker, P.J. Pinter, Jr., S.P. Long, C.P. Osborne, D.L. Hendrix, F. Wechsung, G. Wechsung, S.W. Leavitt, R.L. LaMorte, and S.B. Idso, 2006: Interactive effects of elevated carbon dioxide and drought on wheat. *Agronomy Journal*, **98**, 354-381.

Walther G.R., E. Post, P. Convey, A. Menzel, C. Parmesan, T. Beaber, J.M. Fromenline, O. Hoegh-Goldberg, F. Baukin. 2002. Ecological responses to recent climate change. *Nature,* 416:389-395.

Wan, S., D. Hui, L. Wallace, and Y. Luo, 2005: Direct and indirect effects of experimental warming on ecosystem carbon processes in a tallgrass prairie. *Global Biogeochemical Cycles*, **19**, 2014, doi:10.1029/2004GB002315.

Wand, S.J.E., G.F. Midgley, M.H. Jones, and P.S. Curtis., 1999: Responses of wild C4 and C3 grasses (Poaceae) species to elevated atmospheric CO_2 concentration: a meta-analytic test of current theories and perceptions. *Global Change Biology*, **5**, 723-741.

Wardle, D.A., R.D. Bardgett, J.N. Klironomos, H. Setälä, W.H. van der Putten, and D.H. Wall,. 2004: Ecological linkages between aboveground and belowground biota. *Science*, **304**, 1629-1633.

Weatherly H.E., S.F. Zitzer, J.S. Coleman, and J.A. Arnone III, 2003: In situ litter decomposition and litter quality in a Mojave Desert ecosystem: effects of elevated atmospheric CO_2 and interannual climate variability. *Global Change Biology,* **9**, 1223-1233.

Weber K.T., 2006: Challenges of integrating geospatial technologies into rangeland research and management. *Rangeland Ecology & Management*, **59**, 38-43.

Weltzin J.F., and G.R. McPherson, 1997: Spatial and temporal soil moisture resource partitioning by trees and grasses in a temperate savanna, Arizona, USA. *Oecologia*, **112**, 156-164.

Weltzin, J.F., and G.R. McPherson, 2003: Response of southwestern oak savannas to potential future precipitation regimes. *In*: Weltzin JF, McPherson GR (eds) *Changing Precipitation Regimes and Terrestrial Ecosystems*, University of Arizona Press, Tucson, pp. 127-146.

Westwood, M.N., 1993: *Temperate Zone Pomology.* Timber Press. Portland, OR.

Whitney, S., J. Whalen, M. VanGessel, B. Mulrooney, 2000: Crop profiles for corn (sweet) in Delaware. http://www.impcenters.org/CropProfiles/docs/DEcorn-sweet.html

Williams, J.H., J.H.H. Wilson, and G.C. Bate, 1975: The growth of groundnuts (*Arachis hypogaea* L. cv. Makulu Red) at three altitudes in Rhodesia. *Rhodesian Journal of Agricultural Resources*, **13**, 33-43.

Williams, J.R., 1995: The EPIC model, 1995. In: Singh, V.P. (Ed.), Computer Models of Watershed Hydrology. Water Resources Publications. Highlands Ranch, CO, pp. 909-1000.

Wilsey, B.J., 1996: Urea additions and defoliation affect plant responses to elevated CO_2 in a C3 grass from Yellowstone National Park. *Oecologia*, **108**, 321-327.

Wilsey, B.J., 2001: Effects of elevated CO_2 on the response of Phleum pratense and Poa pratensis to aboveground defoliation and root-feeding nematodes. *International Journal of Plant Science*,. **162**, 1275-1282.

Wolfe, D.W., 1994: Physiological and growth responses to atmospheric CO_2 concentration. *In*: Pessarakli M (ed) *Handbook of Plant and Crop Physiology.* Marcel Dekker. New York.

Wolfe, D.W., M.D. Schwartz, A.N. Lakso, Y. Otsuki, R.M. Pool, and N.J. Shaulis, 2005: Climate change and shifts in spring phenology of three horticultural woody perennials in northeastern USA. *International Journal of Biometeorology,* **49**, 303-309.

Wullschleger, S.D., and R.J. Norby, 2001: Sap velocity and canopy transpiration in a sweetgum stand exposed to free-air CO_2 enrichment (FACE). *New Phytologist*, **150**, 489-498.

Yoshimoto, M., H. Oue, and K. Kobayashi, 2005: Responses of energy balance, evapotranspiration, and water use efficiency of canopies to free-air CO_2 enrichment. *Agricultural and Forest Meteorology*, **133**, 226-246.

Young, J.A., 1991: Cheatgrass. *In*: James, L.F., J.O. Evans, M.H. Ralphs, and R.D. Child, (eds.) *Noxious Range Weeds*. Westview Press, Boulder, pp. 408-418.

Zavaleta, E.S., M.R. Shaw, N.R. Chiariello, B.D. Thomas, E.E. Cleland, C.B. Field, and H.A. Mooney, 2003a: Grassland responses to three years of elevated temperature, CO_2, precipitation, and N deposition. *Ecological Monographs*, **73**, 585-604.

Zavaleta, E.S., B.D.Thomas, N.R. Chiariello, G.P. Asner, M.R. Shaw, and C.B. Field, 2003b: Plants reverse warming effect on ecosystem water balance. *Proceedings National Academy of Sciences*, USA, **100**, 9892-9893.

Ziska, L.H. 2003: Evaluation of the growth response of six invasive species to past, present and future carbon dioxide concentrations. *Journal of Experimental Botany*, **54**, 395-404.

Ziska, L.H. and J.A. Bunce, 1997: Influence of increasing carbon dioxide concentration on the photosynthetic and growth stimulation of selected C4 crops and weeds. *Photosynthesis Research*, **54**, 199-208.

Ziska, L.H., and K. George, 2004: Rising carbon dioxide and invasive, noxious plants: potential threats and consequences. *World Resource Rev*, **16**, 427-447.

Ziska, L.H., J.B. Reeves, and B. Blank, 2005: The impact of recent increases in atmospheric CO_2 on biomass production and vegetative retention of Cheatgrass (*Bromus tectorum*): implications for fire disturbance. *Global Change Biology*, **11**, 1325-1332.

Ziska, L.H., G.B. Runion. 2006. Future weed, pest and disease problems for plants. In: Newton P., A. Carman, G. Edwards, P. Niklaus (eds.) Agroecosystems in a Changing Climate. CRC. New York. Chapter 11, pp. 262-287.

Ziska, L.H., J.R. Teasdale, and J.A. Bunce, 1999: Future atmospheric carbon dioxide may increase tolerance to glyphosate. *Weed Sci*, **47**, 608-615.

Ziska, L.H., W. Weerakoon, O.S. Namuco, and R. Pamplona, 1996: The influence of nitrogen on the elevated CO_2 response in field-grown rice. *Australian Journal of Plant Physiology*, **23**, 45-52.

CHAPTER 3 REFERENCES

Aber, J., W. McDowell, K. Nadelhoffer, A. Magill, G. Berntson, M. Kamakea, S. McNulty, W. Currie, L. Rustad, and I. Fernandez, 1998. Nitrogen saturation in temperate forest ecosystems – Hypotheses revisited. *BioScience*, **48**, 921-934.

Abrahams, A.D., A.J. Parsons, and S.H. Luk, 1988. Hydrologic and sediment responses to simulated rainfall on desert hill slopes in southern Arizona. *Catena*, **15**,103-117.

Adams, A.B., R.B. Harrison, R.S. Sletten, B.D. Strahm, E.C. Turnblom, and C.M. Jensen, 2005. Nitrogen-fertilization impacts on carbon sequestration and flux in managed coastal Douglas-fir stands of the Pacific Northwest. *Forest Ecology and Management*, **220**, 313-325.

Albaugh, T.J., H.L. Allen, P.M. Dougherty, L.W. Kress, and J.S. King, 1998. Leaf area and above- and belowground growth responses of loblolly pine to nutrient and water additions. *Forest Science*, **44**,317-328.

Amiro, B.D., J.B. Todd, B.M. Wotton, K.A. Logan, M.D. Flannigan, B.J. Stocks, J.A. Mason, D.L. Martell, and K.G. Hirsch, 2001. Direct carbon emissions from Canadian forest fires, 1959-1999. Canadian *Journal of Forest Research* **31**, 512-525.

Amthor, J.S., 2000. The McCree-de Wit-Penning de Vries-Thornley respiration paradigms: 30 years later. *Annals of Botany*, **86**, 1-20.

Anderson, J., and R.S. Inouye, 2001. Landscape-scale changes in plant species abundance and biodiversity of a sagebrush steppe over 45 years. *Ecological Monographs*, **71**, 531-556.

Archer, S., 1994. Woody plant encroachment into southwestern grasslands and savannas: rates, patterns and proximate causes. Pages 13-68 *In*: M. Vavra, W. Laycock, and R. Pieper, (eds), *Ecological implications of livestock herbivory in the West*. Society for Range Management, Denver, CO.

Archer, S. ,1996. Assessing and interpreting grass-woody plant dynamics. Pages 101-134 *In*: J. Hodgson and A. Illius, (eds), *The ecology and management of grazing systems*. CAB International, Wallingford, Oxon, United Kingdom.

Archer, S., and C. J. Stokes, 2000. Stress, disturbance and change in rangeland ecosystems. Pages 17-38 *In:* O. Arnalds and S. Archer (eds). *Rangeland desertification*. Kluwer Academic Publishers, Dordrecht, The Netherlands.

Archer, S., D.S. Schimel, and E.A. Holland, 1995. Mechanisms of shrubland expansion: land use, climate or CO_2? *Climatic Change*, **29**, 91-99.

Archer, S., T.W. Boutton, and K.A. Hibbard. 2001, Trees in grasslands: biogeochemical consequences of woody plant expansion. Pages 115-138 *In*: E.-D. Schulze, M. Heimann, S. Harrison, E. Holland, J. Lloyd, I. Prentice, and D. Schimel, (eds), *Global biogeochemical cycles in the climate system*. Academic Press, San Diego.

Arriaga, L., A.E. Castellanos, E. Moreno, and J. Alaron, 2004. Potential ecological distribution of alien invasive species and risk assessment: a case study of buffelgrass in arid regions of Mexico. *Conservation Biology*, **18**, 1504-1514.

Ashmore, M.R., 2002. Effects of oxidants at the whole plant and community level. Pages 89-118 *In*: J.N.B. Bell and M. Treshow, (eds), *Air pollution and plant life*. John Wiley, Chichester, UK.

Ashmore, M.R., 2005. Assessing the future global impacts of ozone on vegetation. *Plant Cell and Environment*, **28**, 949-964.

Asner, G., and S. Archer, In Press: Global Biogeochemical cycles, livestock and carbon. *In*: H.A. Mooney, H. Steinfeld, F. Schneider, L.E. Neville, (eds*), Livestock in a Changing Landscape: Drivers, Consequences and Responses*. United Nations Island Press, Washington, D.C.

Asner, G.P., C.E. Borghi, and R.A. Ojeda, 2003. Desertification in central Argentina: changes in ecosystem carbon and nitrogen from imaging spectroscopy. *Ecological Applications*, **13**, 629-648.

Asner, G.P., S. Archer, R.F. Hughes, J.Ansley, and C.A. Wessman, 2003b. Net changes in regional woody vegetation cover and carbon storage in Texas drylands. *Global Change Biology,* **9**, 1937-1999.

Asner, G., and S. Archer, 2008. Global Biogeochemical cycles, livestock and carbon. *In*: H.A. Mooney, H. Steinfeld, F. Schneider, L.E. Neville, (eds*), Livestock in a Changing Landscape: Drivers, Consequences and Responses.* United Nations Island Press, Washington, D.C.

Atkin, O.K., and M.G. Tjoelker, 2003. Thermal acclimation and the dynamic response of plant respiration to temperature. *Trends in Plant Science,* **8**, 343-351.

Atkin, O.K., E.J. Edwards, and B.R. Loveys, 2000. Response of root respiration to changes in temperature and its relevance to global warming. *New Phytologist,* **147**, 141-154.

Auble, G.T., J.M. Friedman, and M.L. Scott, 1994. Relating riparian vegetation to present and future streamflows. *Ecological Applications*, **4**, 544-554.

Austin AT, L. Yahdjian, J.M. Stark, J. Belnap, A. Porporato U. Norton, D.A. Ravetta, S.M. Schaeffer, 2004. Water pulses and biogeochemical cycles in arid and semiarid ecosystems. *Oecologia,* **141**, 221-235.

Ayres, M.P., and M.J. Lombardero, 2000. Assessing the consequences of global change for forest disturbance from herbivores and pathogens. *Science of the Total Environment,* **262**, 263-286.

Bachelet, D., R.P. Neilson, J.M. Lenihan, and R.J. Drapek, 2001. Climate change effects on vegetation distribution and carbon budget in the United States. *Ecosystems,* **4**, 164-185.

Baldocchi, D., E. Falge, L.H. Gu, R. Olson, D. Hollinger, S. Running, P. Anthoni, C. Bernhofer, K. Davis, R. Evans, J. Fuentes, A. Goldstein, G. Katul, B. Law, X.H. Lee, Y. Malhi, T. Meyers, W. Munger, W. Oechel, K.T.P. U.K. Pilegaard, H.P. Schmid, R. Valentini, S. Verma, T. Vesala, K. Wilson, and S. Wofsy. 2001. FLUXNET: A new tool to study the temporal and spatial variability of ecosystem-scale carbon dioxide, water vapor, and energy flux densities. *Bulletin of the American Meteorological Society,* **82**, 2415-2434.

Barnett,T.P., D.W. Pierce, H.G. Hidalgo, C. Bonfils, B.D. Santer, T. Das, G. Bala, A.W. Wood, T, Nozawa, A.A. Mirin, D.R. Cayan, and M.D. Dettinger, 2008. Human-induced changes in the hydrology of the western United States. *ScienceExpress.* 10.1126/science.1152538.

Baldwin, C.K., F.H. Wagner, U. Lall, 2003. Water resources. Pages 79-112 *In*: F.H. Wagner, (ed). *Rocky Mountain/Great Basin Regional Climate-Change Assessment.* Report of the US Global Change Research Program. Utah State University, Logan, UT, 240pp.

Bale, J.S., G.J. Masters, I.D. Hodkinson, C. Awmack, T.M. Bezemer, V.K. Brown, J. Butterfield, A. Buse, J.C. Coulson, J. Farrar, J.E.G. Good, R. Harrington, S. Hartley, T.H. Jones, R.L. Lindroth, M.C. Press, I. Symrnioudis, A.D. Watt, and J.B. Whittaker, 2002. Herbivory in global climate change research: direct effects of rising temperature on insect herbivores. *Global Change Biology,* **8**, 1-16.

Beatley, J., 1967. Survival of winter annuals in northern Mojave Desert. *Ecology,* **48**, 745-759.

Bebi, P., D. Kulakowski, and T.T. Veblen, 2003. Interactions between fire and spruce beetles in a subalpine rocky mountain forest landscape. *Ecology,* **84**, 362-371.

Bechtold, W.A., and P.L. Patterson, (eds), 2005. *Forest inventory and analysis national sample design and estimation procedures, General Technical Report SRS-80.* USDA Forest Service, Asheville, NC, USA.

Benavides-Solorio, J., and L.H. MacDonald. 2001. Post-fire runoff and erosion from simulated rainfall on small plots, Colorado Front Range. *Hydrological Processes*, **15**, 2931-2952.

Bennett, I., 1959. *Glaze- its meterology and climatology, geographic distribution, and economic effects, Technical Report EP-105.* U.S. Army Quartermaster Research and Engineering Command, Natick, MA.

Berg, E.E., J.D. Henry, C.L. Fastie, A.D. De Volder, and S.M. Matsuoka, 2006. Spruce beetle outbreaks on the Kenai Peninsula, Alaska, and Kluane National Park and Reserve, Yukon Territory: Relationship to summer temperatures and regional differences in disturbance regimes. *Forest Ecology and Management,* **227**, 219-232.

Bestelmeyer, B.T., J.P. Ward, and K.M. Havstad, 2006. Soil-geomorphic heterogeneity governs patchy vegetation dynamics at an arid ecotone. *Ecology,* **87**, 063-973.

Bethlahmy, N., 1974. More streamflow after a bark beetle epidemic. *Journal of Hydrology*, **23**,185-189.

Bigler, C., D. Kulakowski, and T.T. Veblen, 2005. Multiple disturbance interactions and drought influence fire serverity in Rocky Mountain subalpine forests. *Ecology*, **86**, 3018-3029.

Birdsey, R., K. Pregitzer, and A. Lucier, 2006. Forest carbon management in the United States: 1600-2100. *Journal of Environmental Quality,* **35**, 1461-1469.

Birdsey, R. A., and G.M. Lewis, 2002. *Carbon in U.S. Forests and Wood Products, 1987-1997: State-by-State Estimates, GTR-NE-310.* United States Department of Agriculture, Forest Service, Northeasten Research Station, Newtown Square, PA.

Bisal, F., 1960. The effect of raindrop size and impact velocity on sand splash. *Canadian Journal of Soil Science,* **49**, 242-245.

Black, T.A., W.J. Chen, A.G. Barr, M.A. Arain, Z. Chen, Z. Nesic, E.H. Hogg, H.H. Neumann, and P.C. Yang, 2000. Increased carbon sequestration by a boreal deciduous forest in years with a warm spring. *Geophysical Research Letters*, **27**, 1271-1274.

Boisvenue, C., and S.W. Running, 2006. Impacts of climate change on natural forest productivity - evidence since the middle of the 20th century. *Global Change Biology,* **12**, 862-882.

Bond, W.J., and G.F. Midgley, 2000. A proposed CO_2-controlled mechanism of woody plant invasion in grasslands and savannas. *Global Change Biology,* **6**, 865-869.

Boutton, T.W., S.R. Archer, and A.J. Midwood, 1999. Stable isotopes in ecosystem science: structure, function and dynamics of a subtropical savanna. *Rapid Communications in Mass Spectrometry*, **13**, 1263-1277.

Bowers, J.E., 2005. Effects of drought on shrub survival and longevity in the northern Sonoran Desert. *Journal of the Torrey Botanical Society* **132**, 421-431.

Bradley, B.A., R.A. Houghton, J.F. Mustard, and S.P. Hamburg, 2006. Invasive grass reduces aboveground carbon stocks in shrublands of the Western U.S.. *Global Change Biology,* **12**, 1815.

Bragg, D.C., M.G. Shelton, and B. Zeide, 2003. Impacts and management implications of ice storms on forests in the southern United States. *Forest Ecology and Management,* **186**, 99-123.

Brauman, K.A., G.C. Daily, T.K. Duarte, and H.A. Mooney, 2007. The Nature and Value of Ecosystem Services: An Overview Highlighting Hydrologic Services. *Annual Review of Environment and Resources,* **32**, 67-98

Breshears, D.D., J.J. Whicker, M.P. Johansen, and J.E. Pinder, 2003. Wind and water erosion and transport in semi-arid shrubland, grassland and forest ecosystems: quantifying dominance of horizontal wind-driven transport. *Earth Surface Processes and Landforms,* **28**, 1189-1209.

Breshears, D.D., N. S. Cobb, P.M. Rich, K.P. Price, C.D. Allen, R.G. Balice, W.H. Romme, J.H. Kastens, M.L. Floyd, J. Belnap, J.J. Anderson, O.B. Myers, and C.W. Meyer, 2005. Regional vegetation die-off in response to global-change-type drought. *Proceedings of the National Academy of Sciences of the United States of America,* **102**, 15144-15148.

Brock, J.H, 1994. Tamarix spp. (salt cedar), an invasive exotic woody plant in arid and semi-arid riparian habitats of western USA. Pages 27-44 *In*: L. C. de Wall et al., (eds.), *Ecology and Management of Invasive Riverside Plants.* John Wiley, Hoboken, New Jersey.

Brooks, M.L., 2003. Effects of increased soil nitrogen on the dominance of alien annual plants in the Mojave Desert. *Journal of Applied Ecology,* **40**, 344-353.

Brooks, M.L., and K.H. Berry, 2006. Dominance and environmental correlates of alien annual plants in the Mojave Desert, USA. *Journal of Arid Environments*, **67**, 100-124.

Brooks, M.L., C.M. D'Antonio, D.M. Richardson, J.B. Grace, J.E. Keeley, J.M. DiTomaso, R.J. Hobbs, M. Pellant, and D. Pyke, 2004. Effects of invasive alien plants on fire regimes. *BioScience*, **54**, 677-688.

Brooks, R.T., 2004. Early regeneration following the presalvage cutting of hemlock from hemlock-dominated stands. *Northern Journal of Applied Forestry*, **21**, 12-18.

Brown, D. E., editor. 1994. *Biotic communities of the American Southwest United States and Mexico.* University of Utah Press, Salt Lake City.

Brown, T.J., B.L. Hall, and A.L. Westerling, 2004. The impact of twenty-first century climate change on wildland fire danger in the western United States: An applications perspective. *Climatic Change,* **62**, 365-388.

Browning, D., S.R. Archer, G.P. Asner, M.P. McClaran, and C.A. Wessman, 2008. Woody plants in grasslands: post-encroachment stand dynamics. *Ecological Applications,* In Press.

Bruhn, D., J.W. Leverenz, and H. Saxe, 2000. Effects of tree size and temperature on relative growth rate and its components of Fagus sylvatica seedlings exposed to two partial pressures of atmospheric [CO_2]. *New Phytologist,* **146**, 415-425.

Bunn, S.E., M.C. Thoms, S.K. Hamilton, and S.J. Capon, 2006. Flow variability in dryland rivers: boom, bust, and the bits in between. *River Research and Applications,* **22**, 179-186.

Butin, E., A.H. Porter, and J. Elkinton, 2005. Adaptation during biological invasions and the case of Adelges tsugae. *Evolutionary Ecology Research,* **7**, 887-900.

Byrne, T., C. Stonestreet, and B. Peter, 2006. Characteristics and utilization of post-mountain pine beetle wood in solid wood products. Pages 233-253 *In*: L. Safranyik and B. Wilson, (eds.), *The Mountain Pine Beetle: A Synthesis of Biology, Management, and Impacts on Lodgepole Pine.* Pacific Forestry Centre, Canadian Forest Service, Natural Resources Canada, Victoria, BC, Canada.

Calkin, D.E., K.M. Gebert, J.G. Jones, and R.P. Neilson, 2005. Forest service large fire area burned and suppression expression trends, 1970-2002. *Journal of Forestry,* **103**, 179-183.

Canadell, J., R.B. Jackson, J.R. Ehleringer, H.A. Mooney, O.E. Sala, and E.D. Schulze, 1996. Maximum rooting depth of vegetation types at the global scale. *Oecologia,* **108**, 583-595.

Cannell, M.G. R., J.H.M. Thornley, D.C. Mobbs, and A.D. Friend, 1998. UK conifer forests may be growing faster in response to increased N deposition, atmospheric CO_2 and temperature. *Forestry*, **71**, 277-296.

Carroll, A.L., S.W. Taylor, J. Regniere, and L. Safranyik, 2004. Effects of climate change on range expansion by the mountain pine beetle in British Columbia. Pages 223-232 *In*: *Mountain Pine Beetle Symposium: Challenges and Solutions.* Natural Resources Canada, Canadian Forest Service, Pacific Forestry Centre, Kelowna, BC.

Canadell, J.G., D.E. Pataki, R. Gifford, R.A. Houghton, Y. Luo, M.R. Raupach, P. Smith, and W. Steffen, 2007. Saturation of the terrestrial carbon sink. Pages 59-78 *In:* J.G. Canadell, D.E. Pataki, and L.F. Pitelka, (eds), *Terrestial ecosystems in a changing world.* Springer-Verlag, Berlin.

Cayan, D.R., S.A. Kammerdiener, M.D. Dettinger, J.M. Caprio, and D.H. Peterson, 2001. Changes in the onset of spring in the western United States. *Bulletin American Meteorological Society* , **82**, 399-415.

CCSP 4.2. 2008. Thresholds of change in ecosystems. U.S. Climate Change Science Program Synthesis and Assessment Product 4.2.

Chadwick, O A., L.A. Derry, P.M. Vitousek, B.J. Huebert, and L.O. Hedin, 1999. Changing sources of nutrients during four million years of ecosystem development. *Nature,* **397**, 491-497.

Chambers, J.Q., J.I. Fisher, H. Zeng, E.L. Chapman, D.B. Baker, and G.C. Hurtt, 2007. Hurricane Katrina's carbon footprint on U.S. Gulf Coast forests. *Science,* **318**, 1107.

Chavez, P.S., Jr., D.J. Mackinnon, R.L. Reynolds, and M.G. Velasco, 2002. Use of satellite and ground-based images to monitor dust storms and map landscape vulnerability to wind erosion. Page 98 *In*: *Proceedings of ICAR5/GCTE-SEN Joint Conference, International Center for Arid and Semiarid Lands Studies*, Texas Tech University, Lubbock, Texas, USA.

Chomette, O., M. Legrand, and B. Marticorena, 1999. Determination of the wind speed threshold for the emission of desert dust using satellite remote sensing in the thermal infrared. *Journal of Geophysical Research,* **104**, 31207-31215.

Christensen, N.S., A.W. Wood, N. Voisin, D.P. Lettenmaier, and R.N. Palmer, 2004. The effects of climate change on the hydrology and water resources of the Colorado River basin. *Climatic Change*, **62**, 337-363.

Chuine, I., and E.G. Beaubien, 2001. Phenology is a major determinant of tree species range. *Ecology Letters*, **4**, 500-510.

Clarke, P.J., P.K. Latz, and D.E. Albrecht, 2005. Long-term changes in semi-arid vegetation: Invasion of an exotic perennial grass has larger effects than rainfall variability. *Journal of Vegetation Science*, **16**, 237-248.

Cleverly, J.R., C.N. Dahm, J.R. Thibault, D.E. McDonnell, and J.E.A. Coonrod, 2006. Riparian ecohydrology: regulation of water flux from the ground to the atmosphere in the Middle Rio Grande, New Mexico *Hydrological Processes*, **20**, 3207-3225.

Cohen, S., K. Miller, K. Duncan, E. Gregorich, P. Groffman, P. Kovacs, V. Magaña, D. McKnight, E. Mills, and D. Schimel. 2001. North America. *In*: J.J. MCCarthy, O.F. Canziani, N. A. Leary, D. J. Dokken, and K. S. White, (eds), *Climate Change 2001: Impacts, Adaptation and Vulnerability*. Intergovernmental Panel on Climate Change, Washington, D.C.

Cobb, R.C., D.A. Orwig, and S. Currie, 2006. Decomposition of green foliage in eastern hemlock forests of southern New England impacted by hemlock woolly adelgid infestations. *Canadian Journal of Forest Research-Revue Canadienne De Recherche Forestiere*, **36**, 1331-1341.

Cole, K., 1985. Past rates of change, species richness and a model of vegetation inertia in the Grand Canyon, Arizona. *American Naturalist*, **125**, 289-303.

Colorado State Forest Service, 2007. 2006 Report on the Health of Colorado's Forests. Colorado Department of Natural Resources, Division of Forestry.

Cooper, C.F., 1983. Carbon storage in managed forests. *Canadian Journal of Forest Research* **13**, 155-66.

Conant, R.T., J.M. Klopatek, R.C. Malin, and C.C. Klopatek, 1998. Carbon pools and fluxes along an environmental gradient in northern Arizona. *Biogeochemistry*, **43**, 43-61.

Conil, S., and A. Hall, 2006. Local regimes of atmospheric variability: A case study of southern California. *Journal of Climate*, **19**, 4308-4325.

Constantz, J., A.E. Stewart, R. Niswonger, and L. Sarma, 2002. Analysis of temperature profiles for investigating stream losses beneath ephemeral channels. *Water Resources Research*, **38**, 52.51 - 52.13.

Constantz, J., and C.L. Thomas, 1997. Streambed temperature profiles as indicators of percolation characteristics beneath arroyos in the Middle Rio Grande basin, USA. *Hydrological Processes*, **11**, 1621-1634.

Cornelis, W.M., D. Gabriels, and R. Hartmann, 2004. A parameterisation for the threshold shear velocity to initiate deflation of dry and wet sediment. *Geomorphology*, **59**, 43-51.

Costanza, R., R. dArge, R. deGroot, S. Farber, M. Grasso, B. Hannon, K. Limburg, S. Naeem, R.V. Oneill, J. Paruelo, R.G. Raskin, P. Sutton, and M. vandenBelt, 1997. The value of the world's ecosystem services and natural capital. *Nature*, **387**, 253-260.

Cowley, D.E., 2006. Strategies for ecological restoration of the Middle Rio Grande in New Mexico and recovery of the endangered Rio Grande silvery minnow. *Reviews in Fisheries Science*, **14**, 169-186.

Curtis, P.S., and X. Wang, 1998. A meta-analysis of elevated CO_2 effects on woody plant mass, form and physiology. *Oecologia*, **113**, 299-313.

da Silva, R.R., G. Bohrer, D. Werth, M.J. Otte, and R. Avissar, 2006. Sensitivity of ice storms in the southeastern United States to Atlantic SST - Insights from a case study of the December 2002 storm. *Monthly Weather Review*, **134**, 1454-1464.

Dahm, C.N., J.R. Cleverly, J.E. A. Coonrod, J.R. Thibault, D.E. McDonnell, and D.J. Gilroy, 2002. Evapotranspiration at the land/water interface in a semi-arid drainage basin. *Freshwater Biology*, **47**, 831-843.

Daily, G.C., T. Soderqvist, S. Aniyar, K. Arrow, P. Dasgupta, P.R. Ehrlich, C. Folke, A. Jansson, B.O. Jansson, N. Kautsky, S. Levin, J. Lubchenco, K.G. Maler, D. Simpson, D. Starrett, D. Tilman, and B. Walker, 2000. Ecology - The value of nature and the nature of value. *Science*, **289**, 395-396.

Dale, V.H., 1997. The relationship between land-use change and climate change. *Ecological Applications*, **7**, 753-769.

Dale, V.H., L.A. Joyce, S. McNulty, R.P. Neilson, M.P. Ayres, M.D. Flannigan, P.J. Hanson, L.C. Irland, A.E. Lugo, C.J. Peterson, D. Simberloff, F.J. Swanson, B.J. Stocks, and B.M. Wotton, 2001. Climate change and forest disturbances. *BioScience*, **51**, 723-734.

Danby, R.K., and D.S. Hik, 2007. Responses of white spruce (Picea glauca) to experimental warming at a subarctic alpine treeline. *Global Change Biology*, **13**, 437-451.

Daniels, T., 1999. *When city and country collide*. Island Press, Washington, DC.

D'Antonio, C.M., and P.M. Vitousek, 1992. Biological invasions by exotic grasses, the grass fire cycle, and global change. *Annual Review of Ecology and Systematics*, **23**, 63-87.

Davidson, E.A., and I.A. Janssens, 2006. Temperature sensitivity of soil carbon decomposition and feedbacks to climate change. *Nature*, **440**, 165-173.

de Graaff, M.A., K.J. van Groenigen, J. Six, B. Hungate, and C. van Kessel, 2006. Interactions between plant growth and soil nutrient cycling under elevated CO_2: a meta-analysis. *Global Change Biology*, **12**, 2077-2091.

Denning, A. S., editor. 2005. *Science Implementation Strategy for the North American Carbon Program*. Report of the NACP Implementation Strategy Group of the U.S. Carbon Cycle Interagency Working Group. U.S. Carbon Cycle Science Program, Washington, DC.

D'Odorico, P., F. Laio, and L. Ridolfi, 2006. A probabilistic analysis of fire-induced tree-grass coexistence in savannas. *The American Naturalist*, **167**, E79-E87.

Dole, K.P., M.E. Loik, and L.C. Sloan, 2003. The relative importance of climate change and the physiological effects of CO_2 on freezing tolerance for the future distribution of Yucca brevifolia. *Global and Planetary Change*, **36**, 137-146.

Donner, B. and S. Running, 1986. Water stress response after thinning Pinus contorta stands in Montana. *Forest Science,* **32**, 614-625.

Drezner, T.D., 2006. Saguaro (Carnegiea gigantea) densities and reproduction over the northern Sonoran Desert. *Physical Geography*, **27**, 505-518.

Duce, R.A., and N.W. Tindale, 1991. Atmospheric transport of iron and its deposition in the ocean. *Limnology and Oceanography,* **36**, 1715-1726.

Dugas, W.A., R.A. Hicks, and R.P. Gibbens, 1996. Structure and function of C-3 and C-4 Chihuahuan Desert plant communities. Energy balance components. *Journal of Arid Environments,* **34**, 63-79.

Duffy, P.A., J.E. Walsh, J.M. Graham, D.H. Mann, and T.S. Rupp, 2005. Impacts of large-scale atmospheric-ocean variability on Alaskan fire season severity. *Ecological Applications*, **15**, 1317-1330.

Easterling, D.R., 2002. Recent changes in frost days and the frost-free season in the United States. *Bulletin of the American Meteorological Society,* **83**, doi: 10.1175/1520-0477.

Ehleringer, J.R., T.E. Cerling, and B.R. Helliker, 1997. C-4 photosynthesis, atmospheric CO_2 and climate. *Oecologia*, **112**, 285-299.

Emmerich, W.E., 2007. Ecosystem water use efficiency in a semiarid shrubland and grassland community. *Rangeland Ecology & Management,* **60**, 464-470.

Ellison, W.D., 1944. Studies of raindrop erosion. *Agricultural Engineering*, **25**, 131-136, 181-182.

Eschtruth, A.K., N.L. Cleavitt, J.J. Battles, R.A. Evans, and T.J. Fahey, 2006. Vegetation dynamics in declining eastern hemlock stands: 9 years of forest response to hemlock woolly adelgid infestation. *Canadian Journal of Forest Research-Revue Canadienne De Recherche Forestiere*, **36**, 1435-1450.

Fagre, D.B., D.L. Peterson, and A.E. Hessl, 2003. Taking the pulse of mountains: Ecosystem responses to climatic variability. *Climatic Change*, **59**, 263-282.

Fang, C.M., P. Smith, J.B. Moncrieff, and J.U. Smith, 2005. Similar response of labile and resistant soil organic matter pools to changes in temperature. *Nature*, **433**, 57-59.

Farid, A., D.C. Goodrich, and S. Sorooshian, 2006. Using airborne lidar to discern age classes of cottonwood trees in a riparian area. *Western Journal of Applied Forestry,* **21**, 149-158.

Fearnside, P.M., 2002. Time preference in global warming calculations: a proposal for a unified index. *Ecological Economics,* **41**, 21-31.

Feng, S., and Q. Hu, 2004. Changes in agro-meteorological indicators in the contiguous United States: 1951-2000. *Theoretical and Applied Climatology*, **78**, 247-264.

Fenn, M.E., J.S. Baron, E.B. Allen, H.M. Reuth, K.R. Nydick, L. Geiser, W.D. Bowman, J.O. Sickman, T. Meixner, D.W. Johnson, and P. Neitlich, 2003. Ecological effects of nitrogen deposition in the western United States. *BioScience*, **53**, 404-420.

Ferguson, A., 2004. Challenges and solutions - An industry perspective. Pages 223-232 *In*: *Mountain Pine Beetle Symposium: Challenges and Solutions.* Natural Resources Canada, Canadian Forest Service, Pacific Forestry Centre, Kelowna, BC.

Field, C.B., D.B. Lobell, H.A. Peters, and N.R. Chiariello, 2007. Feedbacks of terrestrial ecosystems to climate change. *Annual Review of Environment and Resources,* **32**, 1-29.

Field, C.B., L.D. Mortsch, M. Brklacich, D.L. Forbes, P. Kovacs, J.A. Patz, S.W. Running, and M.J. Scott, 2007: *Climate change 2007: Impacts, adaptation and vulnerability*. Cambridge University Press, Cambridge, UK.

Finzi, A.C., D.J.P. Moore, E.H. DeLucia, J. Lichter, K.S. Hofmockel, R.B. Jackson, H.S. Kim, R. Matamala, H.R. McCarthy, R. Oren, J.S. Pippen, and W.H. Schlesinger, 2006. Progressive nitrogen limitation of ecosystem processes under elevated CO_2 in a warm-temperate forest. *Ecology*, **87**, 15-25.

Finzi, A.C., E.H. DeLucia, J.G. Hamilton, D.D. Richter, and W.H. Schlesinger, 2002. The nitrogen budget of a pine forest under free air CO_2 enrichment. *Oecologia*, **132**, 567-578.

Fisher, J.I., A.D. Richardson, and J.F. Mustard, 2007. Phenology model from surface meteorology does not capture satellite-based greenup estimations. *Global Change Biology*, **13**, 707-721.

Flanner, M.G., C.S. Zender, J.T. Randerson, and P.J. Rasch, 2007. Present day climate forcing and response from black carbon in snow. *Journal of Geophysical Research-Atmospheres*: (In Press)

Flannigan, M.D., B.J. Stocks, and B.M. Wotton, 2000. Climate change and forest fires. *Science of the Total Environment*, **262**, 221-229.

Flannigan, M.D., K.A. Logan, B.D. Amiro, W.R. Skinner, and B.J. Stocks, 2005. Future area burned in Canada. *Climatic Change*, **72**, 1-16.

Fleischner, T.L., 1994. Ecological costs of livestock grazing in western North America. *Conservation Biology*, **8**, 629-644.

Fleming, R.A., 2000. Climate change and insect disturbance regimes in Canada's boreal forests. *World Resources Review*, **12**, 520-555.

Flowers, R.W., S.M. Salom, and L.T. Kok, 2006. Competitive interactions among two specialist predators and a generalist predator of hemlock woolly adelgid, Adelges tsugae (Hemiptera : Adelgidae) in south-western Virginia. *Agricultural and Forest Entomology,* **8**, 253-262.

Franklin, K.A., K. Lyons, P.L. Nagler, D. Lampkin, E.P. Glenn, F. Molina-Freaner, T. Markow, and A.R. Huete, 2006. Buffelgrass (Pennisetum ciliare) land conversion and productivity in the plains of Sonora, Mexico. *Biological Conservation*, **127**, 62-71.

Fredrickson, E., K.M. Havstad, and R. Estell, 1998. Perspectives on desertification: south-western United States. *Journal of Arid Environments*, **39**, 191-207.

Fries, A., D. Lindgren, C.C. Ying, S. Ruotsalainen, K. Lindgren, B. Elfving, and U. Karlmats, 2000. The effect of temperature on site index in western Canada and Scandinavia estimated from IUFRO Pinus contorta provenance experiments. *Canadian Journal of Forest Research*, **30**, 921-929.

Galloway, J.N., F.J. Dentener, D.G. Capone, E.W. Boyer, R.W. Howarth, S.P. Seitzinger, G.P. Asner, C.C. Cleveland, P.A. Green, E.A. Holland, D.M. Karl, A.F. Michaels, J.H. Porter, A.R. **Townsend**, and C.J. Vorosmarty, 2004. Nitrogen cycles: past, present, and future. *Biogeochemistry*, **70**, 153-226.

Geron, C., A. Guenther, J. Greenberg, T. Karl, and R. Rasmussen, 2006. Biogenic volatile organic compound emissions from desert vegetation of the southwestern U.S. *Atmospheric Environment*, **40**, 165-1660.

Gibson, K.E. 2006. Mountain pine beetle conditions in whitebark pine stands in the Greater Yellowstone Ecosystem, 2006. R1Pub06-03, USDA Forest Service, Northern Region, Missoula. Forest Health Protection Report.

Gill, R.A. and R.B. Jackson, 2000. Global patterns of root turnover for terrestrial ecosystems. *New Phytologist*, **147**:13-31.

Gillett, N.P., A.J. Weaver, F.W. Zwiers, and M.D. Flannigan, 2004. Detecting the effect of climate change on Canadian forest fires. *Geophysical Research Letters*, **31**, 1-4, L18211, doi:10.1029/2004GL020876, 2004.

Gillette, D.A., and A.M. Pitchford, 2004. Sand flux in the northern Chihuahuan Desert, New Mexico, USA, and the influence of mesquite-dominated landscapes. *Journal of Geophysical Research-Earth Surface*, **109**, F04003.

Gillson, L. and M.T. Hoffman, 2007. Rangeland ecology in a changing world. *Science*, **315**, 53-54.

Gitlin, A.R., C.M. Sthultz, M.A. Bowker, S. Stumpf, K.L. Paxton, K. Kennedy, A. Munoz, J.K. Bailey, and T.G. Whitham, 2006. Mortality gradients within and among dominant plant populations as barometers of ecosystem change during extreme drought. *Conservation Biology*, **20**, 1477-1486.

Gonzelez-Meller, M.A., L. Taneva, and R.J. Trueman, 2004. Plant respiration and elevated atmospheric CO_2 concentration: Cellular responses and global significance. *Annals of Botany*, **94**, 647-656.

Goodrich, D.C., R. Scott, J. Qi, et al. 2000. Seasonal estimates of riparian evapotranspiration using remote and in situ measurements. *Agricultural and Forest Meteorology*, **105**, 281-309.

Goslee, S.C., W.A. Niering, D.L. Urban, and N.L. Christensen, 2005. Influence of environment, history and vegetative interactions on stand dynamics in a Connecticut forest. *Journal of the Torrey Botanical Society*, **132**, 471-482.

Gower, S.T., K.A. Vogt, and C.C. Grier, 1992. Carbon dynamics of Rocky Mountain Douglas-fir: influence of water and nutrient availability. *Ecological Monographs*, **62**, 43-65.

Gregoire, T.G., and H.T. Valentine, In Press. Sampling strategies for natural resources and the environment. Chapman&Hall/CRC Press.

Grice, A.C., 2006. The impacts of invasive plant species on the biodiversity of Australian rangelands. The *Rangeland Journal*, **28**, 1-27

Griffin, D.W., V.H. Garrison, J.R. Herman, and E.A. Shinn, 2001. African desert dust in the Caribbean atmosphere: microbiology and public health. *Aerobiologia*, **17**, 203-213.

Groffman, P., J. Baron, T. Blett, A. Gold, I. Goodman, L. Gunderson, B. Levinson, M. Palmer, H. Paerl, G. Peterson, N. Poff, D. Rejeski, J. Reynolds, M. Turner, K. Weathers, and J. Wiens, 2006. Ecological thresholds: The key to successful environmental management or an important concept with no practical application? *Ecosystems*, **9**, 1-13.

Grulke, N.E., and P.R. Miller, 1994. Changes in gas exchange characteristics during the life span of giant sequoia: implications for response to current and future concentrations of atmospheric ozone. *Tree Physiology*, **14**, 659-668.

Grunzweig, J.M., T. Lin, E. Rotenberg, A. Schwartz, and D. Yakir, 2003. Carbon sequestration in arid-land forest. *Global Change Biology*, **9**, 791-799.

Guenther, A., S. Archer, J. Greenberg, P. Harley, D. Helmig, L. Klinger, L. Vierling, M. Wildermuth, P. Zimmerman, and S. Zitzer, 1999. Biogenic hydrocarbon emissions and land cover/ climate change in a subtropical savanna. *Physics and Chemistry of the Earth* (B), **24**, 659-667.

Hall, F.C., 2002. *Photo point monitoring handbook: Part A- Field Procedures*. USDA Forest Service Pacific Northwest Station Gen Tech Rep PNW-GTR-526.

Hamilton, S.K., S.E. Bunn, M.C. Thoms, and J. Marshall, 2005. Persistence of aquatic refugia between flow pulses in a dryland river system (Cooper Creek, Australia). *Limnology and Oceanography*, **50**, 743-754.

Hansen, A.J., and D.G. Brown, 2005. Land-use change in rural America: rates, drivers, and consequences. *Ecological Applications*, **15**, 1849-1850.

Hansen, A.J., R.R. Neilson, V.H. Dale, C.H. Flather, L.R. Iverson, D.J. Currie, S. Shafer, R. Cook, and P.J. Bartlein. 2001a. Global change in forests: Responses of species, communities, and biomes. *BioScience*, **51**, 765-779.

Hansen, E.M., and B. Bentz, 2003. Comparison of reproductive capacity among univoltine, semivoltine, and re-emerged parent spruce beetles (Coleoptera: Scolytidae). *Canadian Entomologist*, **135**, 697-712.

Hansen, M.E., B.J. Bentz, and D.L. Turner, 2001b. Temperature-based model for predicting univoltine brood proportions in spruce beetle (Coleoptera: Scolytidae). *Canadian Entomologist*, **133**, 827-841.

Hanson, P.J., and J.F. Weltzin, 2000. Drought disturbance from climate change: response of United States forests. *Science of the Total Environment*, **262**, 205-220.

Hanson, P.J., S.D. Wullschleger, R.J. Norby, T.J. Tschaplinski, and C.A. Gunderson, 2005. Importance of changing CO_2, temperature, precipitation, and ozone on carbon and water cycles of an upland-oak forest: Incorporating experimental results into model simulations. *Global Change Biology*, **11**, 1402-1423.

Hanson, P.J., D.E. Todd, Jr., and J.S. Amthor, 2001. A six-year study of sapling and large-tree growth and mortality responses to natural and induced variability in precipitation and throughfall. *Tree Physiology*, **21**, 345-358.

Harden, J.W., S.E. Trumbore, B.J. Stocks, A. Hirsch, S.T. Gower, K.P. O'Neill, and E.S. Kasischke, 2000. The role of fire in the boreal carbon budget. *Global Change Biology*, **6**, 174-184.

Hargrove, W.W., F.M. Hoffman, and B.E. Law, 2003. New analysis reveals representativeness of the AmeriFlux network. *EOS Transactions,* **84**, 529-535.

Harley, P.C., R.K. Monson, and M.T. Lerdau, 1999. Ecological and evolutionary aspects of isoprene emission from plants. *Oecologia,* **118**, 109-123.

Hart, R.H., and W.A. Laycock, 1996. Repeat photography on range and forest lands in the western United States. *Journal of Range Management,* **49**, 60-67.

Hastings, A., K. Cuddington, K.F. Davies, C.J. Dugaw, S. Elmendorf, A. Freestone, S. Harrison, M. Holland, J. Lambrinos, U. Malvadkar, B.A. Melbourne, K. Moore, C. Taylor, and D. Thomson, 2005a. The spatial spread of invasions: new developments in theory and evidence. *Ecology Letters,* **8**, 91-101.

Hastings, S.J., W.C. Oechel, and A. Muhlia-Melo, 2005b. Diurnal, seasonal and annual variation in the net ecosystem CO_2 exchange of a desert shrub community (Sarcocaulescent) in Baja California, Mexico. *Global Change Biology*, **11**, 927-939.

Hereford, R., R.H. Webb, and C.I. Longpré, 2006. Precipitation history and ecosystem response to multi-decadal precipitation variability in the Mojave Desert region, 1893-2001. *Journal of Arid Environments*, **67**, 13-34.

Hershey, R.L. and S.A. Mizell, 1995. Water chemistry of spring discharge from the carbonate-rock province of Nevada and California. Volume 1. *Desert Research Institute Publication* No. 41140.

Hershler, R. and D.W. Sada, 2002. Biogeography of Great Basin freshwater snails of the genus *Pyrgulopsis*. Pages 255-276. *In*: R. Hershler, D.B. Madsen, and D.R. Currey (eds.). *Great Basin Aquatic Systems History*. Smithsonian Contributions to Earth Sciences, Number 33.Hicke, J.A., G.P. Asner, J.T. Randerson, C. Tucker, S. Los, R. Birdsey, J.C. Jenkins, and C. Field, 2002a. Trends in North American net primary productivity derived from satellite observations, 1982-1998. *Global Biogeochemical Cycles*, **16**, 1-16, 1018, doi:10.1029/2001GB001550, 2002.

Hicke, J.A., G.P. Asner, J.T. Randerson, C. Tucker, S. Los, R. Birdsey, J.C. Jenkins, C. Field, and E. Holland, 2002b. Satellite-derived increases in net primary productivity across North America, 1982-1998. *Geophysical Research Letters,* **29**, 1-4, 1427, doi:10.1029/2001GL013578, 2002.

Hicke, J.A., J.A. Logan, J. Powell, and D.S. Ojima, 2006. Changing temperatures influence suitability for modeled mountain pine beetle (Dendroctonus ponderosae) outbreaks in the western United States. *Journal of Geophysical Research-Biogeosciences,* **111**, G02019, doi:02010.01029/02005JG000101.

Hinzman, L.D., N.D. Bettez, W.R. Bolton, F.S. Chapin, M.B. Dyurgerov, C.L. Fastie, B. Griffith, R.D. Hollister, A. Hope, H.P. Huntington, A.M. Jensen, G.J. Jia, T. Jorgenson, D.L. Kane, D.R. Klein, G. Kofinas, A. H. Lynch, A.H. Lloyd, A.D. McGuire, F.E. Nelson, W.C. Oechel, T.E. Osterkamp, C.H. Racine, V.E. Romanovsky, R.S. Stone, D.A. Stow, M. Sturm, C.E. Tweedie, G.L. Vourlitis, M.D. Walker, D.A. Walker, P.J. Webber, J.M. Welker, K. Winker, and K. Yoshikawa, 2005. Evidence and implications of recent climate change in northern Alaska and other arctic regions. *Climatic Change*, **72**, 251-298.

Hobbs, R.J., and L.F. Huenneke, 1992. Disturbance, diversity, and invasion - implications for conservation. *Conservation Biology*, **6**, 324-337.

Hobbs, R.J., S. Arico, J. Aronson, J.S. Baron, P. Bridgewater, V.A. Cramer, P.R. Epstein, J.J. Ewel, C.A. Klink, A.E. Lugo, D. Norton, D. Ojima, D.M. Richardson, E.W. Sanderson, F. Valladares, M. Vila, R. Zamora, and M. Zobel, 2006. Novel ecosystems: theoretical and management aspects of the new ecological world order. *Global Ecology and Biogeography*, **15**, 1-7.

Holechek, J.L., R.D. Pieper, and C.H. Herbel, 2003. *Range management: principles and practices.* Fifth edition. Prentice-Hall, London.

Hollinger, D.Y., J. Aber, B. Dail, E.A. Davidson, S.M. Goltz, H. Hughes, M.Y. Leclerc, J.T. Lee, A.D. Richardson, C. Rodrigues, N.A. Scott, D. Achuatavarier, and J. Walsh, 2004. Spatial and temporal variability in forest-atmosphere CO_2 exchange. *Global Change Biology*, **10**, 1689-1706.

Holmgren, M., and M. Scheffer, 2001. El Niño as a window of opportunity for the restoration of degraded arid ecosystems. *Ecosystems*, **4**, 151-159.

Holmgren, M., P. Stapp, C.R. Dickman, C. Gracia, S. Graham, J. Gutierrez, C. Hice, et. al., 2006. Extreme climatic events shape arid and semi-arid ecosystems. *Frontiers in Ecology and the Environment*, **4**, 87-95.

Holsten, E.H., R.A. Werner, and R.L. Develice, 1995. Effects of a spruce beetle (Coleoptera: Scolytidae) outbreak and fire on Lutz spruce in Alaska. *Environmental Entomology*, **24**, 1539-1547.

Holsten, E.H., R.W. Thier, A.S. Munson, and K.E. Gibson, 1999. *The Spruce Beetle. Forest Insect and Disease Leaflet 127*, USDA Forest Service.

Holzapfel, C., and B.E. Mahall, 1999. Bidirectional facilitation and interference between shrubs and annuals in the Mojave Desert. *Ecology*, **80**, 1747-1761.

Hooper, D.U., and L. Johnson, 1999. Nitrogen limitation in dryland ecosystems: response to geographical and temporal variations in precipitation. *Biogeochemistry*, **46**, 247-293.

Horton, J.L., T.E. Kolb, and S.C. Hart, 2001a. Responses of riparian trees to interannual variation in ground water depth in a semi-arid river basin. *Plant Cell and Environment*, **24**, 293-304.

Horton, J. L., T.E. Kolb, and S.C. Hart, 2001b. Physiological response to groundwater depth varies among species and with river flow regulation. *Ecological Applications*, **11**, 1046-1059.

Huenneke, L.F., J.P. Anderson, M. Remmenga, and W.H. Schlesinger, 2002. Desertification alters patterns of aboveground net primary production in Chihuahuan ecosystems. *Global Change Biology*, **8**, 247-264.

Hughes, L. 2000, Biological consequences of global warming: is the signal already apparent? *Trends in Ecology & Evolution,* **15**, 56-61.

Hummel, S., and J.K. Agee 2003. Western spruce budworm defoliation effects on forest structure and potential fire behavior. *Northwest Science*, **77**, 159-169.

Hunter, R., 1991. Bromus invasions on the Nevada Test Site - present status of B. rubens and B. tectorum with notes on their relationship to disturbance and altitude. *Great Basin Naturalist*, **51**, 176-182.

Huxman T.E., J.M. Cable, D.D. Ignace, A.J. Eilts, N. English, J. Weltzin, D.G. Williams, 2004. Response of net ecosystem gas exchange to a simulated precipitation pulse in a semiarid grassland: the role of native versus non-native grasses and soil texture. *Oecologia*, **141**, 295-305.

Huxman, T.E., and S.D. Smith, 2001. Photosynthesis in an invasive grass and native forb at elevated CO_2 during an El Niño year in the Mojave Desert. *Oecologia*, **128**, 193-201.

Huxman, T.E., K.A. Snyder, D.T. Tissue, A.J. Leffler, K. Ogle, W.T. Pockman, D.R. Sandquist, D.L. Potts, and S. Schwinning, 2004. Precipitation pulses and carbon fluxes in semiarid and arid ecosystems. *Oecologia*, **141**, 254-268.

Hyvonen, R., G.I. Agren, S. Linder, T. Persson, M.F. Cotrufo, A. Ekblad, M. Freeman, A. Grelle, I. A. Janssens, P. G. Jarvis, S. Kellomaki, A. Lindroth, D. Loustau, T. Lundmark, R.J. Norby, R. Oren, K. Pilegaard, M.G. Ryan, B. D. Sigurdsson, M. Stromgren, M. van Oijen, and G. Wallin, 2007. The likely impact of elevated [CO_2], nitrogen deposition, increased temperature and management on carbon sequestration in temperate and boreal forest ecosystems: a literature review. *New Phytologist*, **173**, 463-480.

Ibarra, F.A., J.R. Cox, M.H. Martin, T.A. Crowl, and C.A. Call, 1995. Predicting buffelgrass survival across a geographical and environmental gradient. *Journal of Range Management*, **48**, 53-59.

IPCC, 2007. *Climate Change 2007: The Physical Science Basis IPCC WGI Fourth Assessment Report, Policy Maker Summary*, Intergovernmental Panel on Climate Change, Working Group I, Fourth Assessment Report.

Irvine, J., B.E. Law, M.R. Kurpius, P.M. Anthoni, D. Moore, and P.A. Schwarz, 2004. Age-related changes in ecosystem structure and function and effects on water and carbon exchange in ponderosa pine. *Tree Physiology*, **24**, 753-763.

Ivans, S., L. Hipps, A. Leffler, and C.V. Ivans, 2006. Response of water vapor and CO_2 fluxes in semiarid lands to seasonal and intermittent precipitation pulses. *Journal of Hydrometeorology*, **7**, 995-1010.

Jackson, R.B., J.L. Banner, E.G. Jobbagy, W.T. Pockman, and D.H. Wall, 2002. Ecosystem carbon loss with woody plant invasion of grassland. *Nature*, **418**, 623-626.

Jackson, R.B., and W.H. Schlesinger, 2004. Curbing the U.S. carbon deficit. *Proceedings of the National Academy of Science*, **101**, 15827-15829.

Jarvis P., A. Rey, C. Petsikos, L. Wingate, M. Rayment, J. Pereira, J. Banza, J. David, F. Miglietta, M. Borghetti, G. Manca, R. Valentini, 2007. Drying and wetting of Mediterranean soils stimulates decomposition and carbon dioxide emission: The "Birch effect." *Tree Physiology*, **27**, 929-940.

Jastrow, J.D., R.M. Miller, R. Matamala, R.J. Norby, T.W. Boutton, C.W. Rice, and C.E. Owensby, 2005. Elevated atmospheric carbon dioxide increases soil carbon. *Global Change Biology*, **11**, 2057-2064.

Jepsen, R., R. Langford, J. Roberts, and J. Gailani, 2003. Effects of arroyo sediment influxes on the Rio Grande River channel near El Paso, Texas. *Environmental & Engineering Geoscience*, **9**, 305-312.

Jickells, T.D., Z.S. An, K.K. Andersen, et al., 2005. Global iron connections between desert dust, ocean biogeochemistry and climate. *Science*, **308**, 67-71.

Jobbagy, E.G., and R.B. Jackson, 2000. The vertical distribution of soil organic carbon and its relation to climate and vegetation. *Ecological Applications*, **10**, 423-436.

Johnson, D.W, 2006. Progressive N limitation in forests: review and implications for long-term responses to elevated CO_2. *Ecology* **87**, 64-75.

Karlsson, P.E., J. Uddling, S. Braun, M. Broadmeadow, S. Elvira, B.S. Gimeno, D. Le Thiec, E. Oksanen, K. Vandermeiren, M. Wilkinson, and L. Emberson, 2004. New critical levels for ozone effects on young trees based on AOT40 and simulated cumulative leaf uptake of ozone. *Atmospheric Environment*, **38**, 2283-2294.

Kashian, D.M., W.H. Romme, D.B. Tinker, M.G. Turner, and M.G. Ryan, 2006. Carbon storage on landscapes with stand-replacing fires. *BioScience*, **56**, 598-606.

Kasischke, E.S., and M.R. Turetsky, 2006. Recent changes in the fire regime across the North American boreal region - Spatial and temporal patterns of burning across Canada and Alaska. *Geophysical Research Letters*, **33**, 1-5, L09703, doi:10.1029/2006GL025677, 2006.

Katz, G.L., and P.B. Shafroth, 2003. Biology, ecology and management of Elaeagnus angustifolia L. (Russian olive) in western North America. *Wetlands* **23**, 763-777.

Keeley, J.E., and C.J. Fotheringham, 2001. Historic fire regime in Southern California shrublands. *Conservation Biology*, **15**, 1536-1548.

Keeley, J.E., C.J. Fotheringham, and M. Morais, 1999. Reexamining fire suppression impacts on brushland fire regimes. *Science*, **284**, 1829-1832.

Kharin, V.V., F.W. Zwiers, X.B. Zhang, and G.C. Hegerl, 2007. Changes in temperature and precipitation extremes in the IPCC ensemble of global coupled model simulations. *Journal of Climate*, **20**, 1419-1444.

King, J.S., P.J. Hanson, E. Bernhardt, P. DeAngelis, R.J. Norby, and K.S. Pregitzer, 2004. A multiyear synthesis of soil respiration responses to elevated atmospheric CO_2 from four forest FACE experiments. *Global Change Biology*, **10**, 1027-1042.

Kirschbaum, M.U.F., 2004. Soil respiration under prolonged soil warming: are rate reductions caused by acclimation or substrate loss? *Global Change Biology*, **10**, 1870-1877.

Kirschbaum, M.U.F., 2005. A modeling analysis of the interaction between forest age and forest responsiveness to increasing CO_2 concentration. *Tree Physiology*, **25**, 953-963.

Kirschbaum, M.U.F., 2006. Temporary carbon sequestration cannot prevent climate change. *Mitigation and Adaptation Strategies for Global Change*, **11**, 1151-1164.

Kitzberger, T., P.M. Brown, E.K. Heyerdahl, T.W. Swetnam, and T.T. Veblen, 2007. Contingent Pacific-Atlantic Ocean influence on multi-century wildfire synchrony over western North America. *Proceedings National Academy of Science*, **104**, 543-548.

Knapp, A.K., and M.D. Smith, 2001. Variation among biomes in temporal dynamics of aboveground primary production. *Science*, **291**, 481-484.

Knapp, A.K., P.A. Fay, J.M. Blair, S.L. Collins, M.D. Smith, J.D. Carlisle, C.W. Harper, B.T. Danner, M.S. Lett, and J.K. McCarron, 2002. Rainfall variability, carbon cycling, and plant species diversity in a mesic grassland. *Science*, **298**, 2202-2205.

Knapp, P.A., 1995. Intermountain West lightning-caused fires - climatic predictors of area burned. *Journal of Range Management*, **48**, 85-91.

Knapp, P.A., 1996. Cheatgrass (Bromus tectorum L) dominance in the Great Basin Desert. *Global Environmental Change*, **6**, 37-52.

Knapp, P.A, 1998. Spatio-temporal patterns of large grassland fires in the Intermountain West, USA. *Global Ecology and Biogeography Letters*, **7**, 259-272.

Knochenmus, L., J. Wilson, R. Laczniak, D. Sweetkind, J. Thomas, L. Justet and R. Hershey, 2007. Ground-water conditions, Pages 58-61. *In:* A. Welch and D. Bright (eds.). *Water resources of the Basin and Range carbonate aquifer system in White Pine County, Nevada, and adjacent areas in Nevada and Utah*. U.S. Geological Survey Open-File Report 2007-1156.

Koch, F.H., H.M. Cheshire, and H.A. Devine, 2006. Landscape-scale prediction of hemlock woolly adelgid, Adelges tsugae (Homoptera: Adelgidae), infestation in the southern Appalachian Mountains. *Environmental Entomology*, **35**, 1313-1323.

Koricheva, J., S. Larsson, and E. Haukioja, 1998. Insect performance on experimentally stressed woody plants: A meta-analysis. *Annual Review of Entomology*, **43**, 195-216.

Körner, C. 2000, Biosphere responses to CO_2 enrichment. *Ecological Applications*, **10**, 1590-1619.

Körner, C. 2006, Plant CO_2 responses: an issue of definition, time and resource supply. *New Phytologist*, **172**, 393-411.

Körner, C., R. Asshoff, O. Bignucolo, S. Hättenschwiler, S.G. Keel, S. Pelaez-Riedl, S. Pepin, R.T.W. Siegwolf, and G. Zotz, 2005. Carbon flux and growth in mature deciduous forest trees exposed to elevated CO_2. *Science*, **309**, 1360-1362.

Krieger, D.J. 2001. The economic value of forest ecosystem services: a review. The Wilderness Society, Washington, DC.

Kruger, E.L., J.C. Volin, and R.L. Lindroth, 1998. Influences of atmospheric CO_2 enrichment on the responses of sugar maple and trembling aspen to defoliation. *New Phytologist*, **140**, 85-94.

Kulakowski, D., and T.T. Veblen, 2007. Effect of prior disturbances in the extent and severity of a 2002 wildfire in Colorado subalpine forests. *Ecology*, **88**, 759-769.

Kupfer, J.A., and J.D. Miller, 2005. Wildfire effects and post-fire responses of an invasive mesquite population: the interactive importance of grazing and non-native herbaceous species invasion. *Journal of Biogeography*, **32**, 453-466.

Kurc, S.A., and E.E. Small, 2004. Dynamics of evapotranspiration in semiarid grassland and shrubland during the summer monsoon season, central New Mexico. *Water Resources Research*, **40**:W09305, doi: 09310.01029/02004WR003068.

Kurc S.A., E.E. Small, 2007. Soil moisture variations and ecosystem-scale fluxes of water and carbon in semiarid grassland and shrubland. *Water Resources Research*, **43**, W06416.

Lal, R. 2001. Potential of desertification control to sequester carbon and mitigate the greenhouse effect. *Climatic Change*, **51**, 35-72.

Kurz, W.A., and M.J. Apps, 1999. A 70-year retrospective analysis of carbon fluxes in the canadian forest sector. *Ecological Applications*, **9**, 526-547.

Lane, L.J., and M.R. Kidwell, 2003. Hydrology and soil erosion. Pages 92-100 *In*: Santa Rita Experimental Range: 100 years (1903 to 2003) of accomplishments and contributions. Proc. RMRS-P-30, U.S. Department of Agriculture, Forest Service, Rocky Mountain Research Station, Ogden, UT, Tucson, AZ.

Lavee, H., A.C. Imeson, and P. Sarah, 1998. The impact of climate change on geomorphology and desertification along a Mediterranean-arid transect. *Land Degradation and Development*, **9**, 407-422.

Leathers, C.R., 1981. Plant components of desert dust in Arizona and their significance for man. Pages 191-206 *In*: T. L. Péwé, (ed), *Desert Dust: Origin, Characteristics, and Effect on Man.* Geological Society of America, Boulder, Colorado.

Leith, H., 1975. Modelling the primary productivity of the world. Pages 237-263 *In*: H. Leith and R. H. Whittaker, (eds), *Primary productivity of the biosphere*. Springer-Verlag, New York.

Lichter, J., S.H. Barron, C.E. Bevacqua, A.C. Finzli, K.E. Irving, E.A. Stemmler, and W.H. Schlesinger, 2005. Soil carbon sequestration and turnover in a pine forest after six years of atmospheric CO_2 enrichment. *Ecology*, **86**, 1835-1847.

Loehle, C. and D.C. LeBlanc, 1996. Model-Based Assessments of Climate Change Effects on Forests: A Critical Review. *Ecological Modelling*, **90**, 1-31.

Logan, J.A., and J.A. Powell, 2001. Ghost forests, global warming and the mountain pine beetle (Coleoptera: Scolytidae). *American Entomologist*, **47**, 160-173.

Logan, J.A., J. Regniere, and J.A. Powell, 2003b. Assessing the impacts of global warming on forest pest dynamics. *Frontiers in Ecology and the Environment*, **1**, 130-137.

Logan, J., J. Regniere, and J.A. Powell, 2003a. Assessing the impacts of global warming on forest pest dynamics. *Frontiers in Ecology and the Environment*, **1**, 130-137.

Loik, M.E., T.E. Huxman, E.P. Hamerlynck, and S.D. Smith, 2000. Low temperature tolerance and cold acclimation for seedlings of three Mojave Desert Yucca species exposed to elevated CO_2. *Journal of Arid Environments*, **46**, 43-56.

Long, S.P., 1991. Modification of the response of photosynthetic productivity to rising temperature by atmospheric CO_2 concentrations: has its importance been underestimated? *Plant, Cell, and Environment*, **14**, 729-740.

Luk, S.H., A.D. Abrahams, and A.J. Parsons, 1993. Sediment sources and sediment transport by rill flow and interrill flow on a semiarid piedmont slope, southern Arizona. *Catena*, **20**, 93-111.

Luo, Y.Q., D.F. Hui, and D.Q. Zhang, 2006. Elevated CO_2 stimulates net accumulations of carbon and nitrogen in land ecosystems: A meta-analysis. *Ecology*, **87**, 53-63.

Luo, Y., B. Su, W.S. Currie, J.S. Dukes, A. Finzi, U. Hartwig, B. Hungate, R.E. McMurtrie, R. Oren, W.J. Parton, D.E. Pataki, M.R. Shaw, D.R. Zak, and C.B. Field, 2004. Progressive nitrogen limitation of ecosystem responses to rising atmospheric carbon dioxide. *BioScience*, **54**, 731-739.

Lynch, H.J., R.A. Renkin, R.L. Crabtree, and P.R. Moorcroft, 2006. The influence of previous mountain pine beetle (Dendroctonus ponderosae) activity on the 1988 Yellowstone fires. *Ecosystems*, **9**, 1318-1327.

MacMahon, J., and F. Wagner, 1985. The Mojave, Sonoran and Chihuahuan Deserts of North America. *In*: Noy-Meir, I., Evanari, M., Goodall, D.W. (eds.), Hot Deserts and Arid Shrublands. *Ecosystems of the World*, **12A**, Elsevier.

Magill, A.H., J.D. Aber, W.S. Currie, K.J. Nadelhoffer, M.E. Martin, W.H. McDowell, J.M. Melillo, and P. Steudler, 2004. Ecosystem response to 15 years of chronic nitrogen additions at the Harvard Forest LTER, Massachusetts, USA. *Forest Ecology and Management*, **196**, 7-28.

Magnani, F., M. Mencuccini, M. Borghetti, P. Berbigier, F. Berninger, S. Delzon, A. Grelle, P. Hari, P.G. Jarvis, P. Kolari, A.S. Kowalski, H. Lankreijer, B.E. Law, A. Lindroth, D. Loustau, G. Manca, J.B. Moncrieff, M. Rayment, V. Tedeschi, R. Valentini, and J. Grace, 2007. The human footprint in the carbon cycle of temperate and boreal forests. *Nature*, **447**, 848-850.

Maier, C.A., T.J. Albaugh, H.L. Allen, and P.M. Dougherty, 2004. Respiratory carbon use and carbon storage in mid-rotation loblolly pine (Pinus taeda L.) plantations: The effect of site resources on the stand carbon balance. *Global Change Biology*, **10**, 1335-1350.

Malmström, C.M., and K.F. Raffa, 2000. Biotic disturbance agents in the boreal forest: considerations for vegetation change models. *Global Change Biology*, **6**, 35-48.

Matyssek, R., and H. Sandermann, 2003. Impact of ozone on trees: an ecophysiological perspective. Pages 349-404 *In*: K. Esser, U. Lüttge, W. Beyschlag, and F. Hellwig, (eds), *Progress in Botany*, Vol. **64**. Springer-Verlag, Heidelberg, Germany.

Mau-Crimmins, T., H.R. Schussman, H.R., Geiger, E.L., 2006. Can the invaded range of a species be predicted sufficiently using only native-range data?: Lehmann lovegrass (Eragrostis lehmanniana) in the southwestern United States. *Ecological Modelling*, **193**, 736-746.

May, R.M., 1977. Thresholds and breakpoints in ecosystems with a multiplicity of stable states. *Nature*, **269**, 471-477.

McAuliffe, J.R., 2003. The interface between precipitation and vegetation: the importance of soils in arid and semiarid environments. Pages 9-27 *In*: J.F. Weltzin, McPherson, GR., (eds). *Changing Precipitation Regimes and Terrestrial Ecosystems*. University of Arizona Press, Tucson, AZ, USA.

McAuliffe, J.R., L.A. Scuderi, and L.D. McFadden, 2006. Tree-ring record of hill slope erosion and valley floor dynamics: Landscape responses to climate variation during the last 400 years in the Colorado Plateau, northeastern Arizona. *Global and Planetary Change*, **50**, 184-201.

McCarthy, H.R., R. Oren, A.C. Finzi, and K.H. Johnsen, 2006a. Canopy leaf area constrains $[CO_2]$-induced enhancement of productivity and partitioning among aboveground carbon pools. *Proceedings of the National Academy of Sciences of the United States of America*, **103**, 19356-19361.

McCarthy, H.R., R. Oren, H.S. Kim, K.H. Johnsen, C. Maier, S.G. Pritchard, and M.A. Davis, 2006b. Interaction of ice storms and management practices on current carbon sequestration in forests with potential mitigation under future CO_2 atmosphere. *Journal of Geophysical Research-Atmospheres*, **111**, 1-10, D15103, doi:10.1029/2005JD006428, 2006.

McClaran, M.P., 2003. A century of vegetation change on the Santa Rita Experimental Range. Pages 16-33 *In*: Santa Rita Experimental Range: 100 years (1903 to 2003) of accomplishments and contributions. Proc. RMRS-P-30, U.S. Department of Agriculture, Forest Service, Rocky Mountain Research Station, Ogden, UT, Tucson, AZ.

McKeen, S.A., G. Wotawa, D.D. Parrish, J.S. Holloway, M.P. Buhr, G. Hubler, F.C. Fehsenfeld, and J.F. Meagher, 2002. Ozone production from Canadian wildfires during June and July of 1995. *Journal of Geophysical Research-Atmospheres*, **107**.

McMurtrie, R.E., B.E. Medlyn, and R.C. Dewar, 2001. Increased understanding of nutrient immobilization in soil organic matter is critical for predicting the carbon sink strength of forest ecosystems over the next 100 years. *Tree Physiology*, **21**, 831-839.

McNulty, S.G., 2002. Hurricane impacts on U.S. forest carbon sequestration. *Environmental Pollution*, **11**, S17-S24.

Melillo, J.M., P.A. Steudler, J.D. Aber, K. Newkirk, H. Lux, F.P. Bowles, C. Catricala, A. Magill, T. Ahrens, and S. Morrisseau, 2002. Soil warming and carbon-cycle feedbacks to the climate system. *Science*, **298**, 2173-2176.

Menzel, A., and P. Fabian, 1999. Growing season extended in Europe. *Nature*, **397**, 659-659.

Millennium-Ecosystem-Assessment. 2005. Ecosystems *and Human Well-being: Synthesis*. Island Press, Washington, DC.

Miller, N.L., and N.J. Schlegel, 2006. Climate change projected fire weather sensitivity: California Santa Ana wind occurrence. *Geophysical Research Letters*, **33**, 1-5, L15711, doi:10.1029/2006GL025808, 2006.

Miller, R.F., and J.A. Rose, 1999. Fire history and western juniper encroachment in sagebrush steppe. *Journal of Range Management*, **52**, 550-559.

Miller, R.F., J.D. Bates, T.J. Svejcar, F.B. Pierson, and L.E. Eddleman, 2005. *Biology, ecology and management of western juniper*. Technical Bulletin 152, Agricultural Experiment Station, Oregon State University.

Miller, S.D., 2003. A consolidated technique for enhancing desert dust storms with MODIS., **30**, 12.11-12.14.

Milly, P.C. D., K.A. Dunne, and A.V. Vecchia, 2005. Global pattern of trends in streamflow and water availability in a changing climate. *Nature*, **438**, 347-350.

Miriti, M.N., 2007. Twenty years of changes in spatial association and community structure among desert perennials. *Ecology*, **88**, 1177-1190.

Monger, H.C., and J.J. Martinez-Rios, 2000. Inorganic carbon sequestration in grazing lands. Pages 87-118 *In*: R. Follett, J. Kimble, and R. Lal, editors. *The Potential of U.S. Grazing Lands to Sequester Carbon and Mitigate the Greenhouse Effect.* Lewis Publishers, Boca Raton, Florida.

Morris, G.A., S. Hersey, A.M. Thompson, S. Pawson, J.E. Nielsen, P.R. Colarco, W.W. McMillan, A. Stohl, S. Turquety, J. Warner, B.J. Johnson, T.L. Kucsera, D.E. Larko, S.J. Oltmans, and J.C. Witte, 2006. Alaskan and Canadian forest fires exacerbate ozone pollution over Houston, Texas, on 19 and 20 July 2004. *Journal of Geophysical Research-Atmospheres*, **111**, 1-10, D24S03, doi:10.1029/2006JD007090, 2006.

Morgan, J.A., D.G. Milchunas, D.R. LeCain, M. West, and A.R. Mosier, 2007. Carbon dioxide enrichment alters plant community structure and accelerates shrub growth in the shortgrass steppe. *Proceedings of the National Academy of Sciences*, **104**, 14724-14729.

Mote, P.W., A.F. Hamlet, M.P. Clark, and D.P. Lettenmaier, 2005. Declining mountain snowpack in western North America. *Bulletin of the American Meteorological Society*, **86**,39-49.

Murray, B.C., B.A. McCarl, and Heng-Chi Lee, 2004. Estimating Leakage from Forest Carbon Sequestration Programs. *Land Economics*, **80**, 109-124.

Mueller, R.C., C.M. Scudder, M.E. Porter, R.T. Trotter, C.A. Gehring, and T.G. Whitham, 2005. Differential tree mortality in response to severe drought: evidence for long-term vegetation shifts. *Journal of Ecology*, **93**, 1085-1093.

Nagel, J.M., T.E. Huxman, K.L. Griffin, and S.D. Smith, 2004. CO_2 enrichment reduces the energetic cost of biomass construction in an invasive desert grass. *Ecology*, **85**, 100-106.

NASA-Office-of-Earth-Science, 2004. *Earth science applications plan.* NASA, Washington, D.C.

National-Ecological-Observatory-Network, 2006. *Integrated science and education plan for the National Ecological Observatory Network.* Available at: http://www.neoninc.org/ NEON, Inc., Washington, DC.

Neilson, R.P., 1986. High resolution climatic analysis and southwest biogeography. *Science*, **232**, 27-34.

Nelson, A., 1992. Characterizing exurbia. *Journal of Planning Literature*, **6**, 350-368.

Nettleton, W.D., and M.D. Mays, 2007. Estimated Holocene soil carbon-soil degradation in Nevada and western Utah, USA. *Catena*, **69**, 220-229.

Newman, B.D., B.P. Wilcox, S.R. Archer, D.D. Breshears, C.N. Dahm, C.J. Duffy, N.G. McDowell, F.M. Phillips, B.R. Scanlon, and E.R. Vivoni, 2006. Ecohydrology of water-limited environments: a scientific vision. *Water Resources Research*, **42**, W06302, doi:06310.01029/02005WR004141.

Norby, R.J., E.H. DeLucia, B. Gielen, C. Calfapietra, C. P. Giardina, J.S. King, J. Ledford, H.R. McCarthy, D.J. P. Moore, R. Ceulemans, P. De Angelis, A.C. Finzi, D.F. Karnosky, M.E. Kubiske, M. Lukac, K.S. Pregitzer, G.E. Scarascia-Mugnozza, W.H. Schlesinger, and R. Oren, 2005. Forest response to elevated CO_2 is conserved across a broad range of productivity. *Proceedings of the National Academy of Sciences of the United States of America*, **102**, 18052-18056.

Norby, R.J., J. Ledford, C.D. Reilly, N.E. Miller, and E.G. O'Neill, 2004. Fine-root production dominates response of a deciduous forest to atmospheric CO_2 enrichment. *Proceedings of the National Academy of Sciences of the United States of America*, **101**, 9689-9693.

Novak, S.J., and R.N. Mack, 2001. Tracing plant introduction and spread: Genetic evidence from Bromus tectorum (Cheatgrass). *BioScience*, **51**, 114-122.

NWRC. 2007. Northwest Watershed Research Center (NWRC) and the Reynolds Creek Experimental Watershed (RCEW) USDA-ARS NW Watershed Research, 800 Park Blvd. Plaza IV, S 105 Boise, ID 83712. http://www.nwrc.ars.usda.gov/

Okin, G.S., D.A. Gillette, and J.E. Herrick, 2006. Multi-scale controls on and consequences of aeolian processes in landscape change in arid and semi-arid environments. *Journal of Arid Environments*, **65**,253-275.

Okin, G.S., and M.C. Reheis, 2002. An ENSO predictor of dust emission in the southwestern United States. *Geophysical Research Letters*, **29**, 46.41-46.43.

Okin, G.S., J.E. Herrick, and D.A. Gillette, 2006. Multi-scale controls on and consequences of aeolian processes in landscape change in arid and semiarid environments. *Journal of Arid Environments*, **65**, 253-275.

Olson, D.M., E. Dinerstein, E.D. Wikramanayake, N.D. Burgess, G.V.N. Powell, E.C. Underwood, J.A. D'Amico, H.E.S. I. Itoua, J.C. Morrison, C.J. Loucks, T.F. Allnutt, T.H. Ricketts, Y. Kura, J.F. Lamoreux, W.W. Wettengel, P. Hedao, and K.R. Kassem, 2001. Terrestrial ecoregions of the world: a new map of life on Earth. *BioScience*, **51**, 933-938.

Onken, B., and R. Reardon (compilers), 2005. Third Symposium on Hemlock Wooly Adelgid in the Eastern United States. USDA Forest Service Forest Health Technology Enterprise Team, FHTET-2005-01, Morgantown, WV. http://na.fs.fed.us/fhp/hwa/pub/2005_proceedings/index.shtm

Oren, R., D.S. Ellsworth, K.H. Johnsen, N. Phillips, B.E. Ewers, C. Maier, K.V.R. Schafer, H. McCarthy, G. Hendrey, S.G. McNulty, and G.G. Katul, 2001. Soil fertility limits carbon sequestration by forest ecosystems in a CO_2-enriched atmosphere. *Nature*, **411**, 469-472.

Orwig, D.A., D.R. Foster, and D.L. Mausel, 2002. Landscape patterns of hemlock decline in New England due to the introduced hemlock woolly adelgid. *Journal of Biogeography*, **29**, 1475-1487.

Overpeck, T., D. Rind, and R. Goldberg, 1990. Climate-induced changes in forest disturbance and vegetation. *Nature*, **343**, 51-53.

Owens, M.K., and G.W. Moore, 2007. Saltcedar water use: realistic and unrealistic expectations. *Rangeland Ecology and Management*, **60**, 553-557.

Owensby, C.E., P.I. Coyne, J.M. Hamm, L.M. Auen, and A.K. Knapp. 1993. Biomass production in a tallgrass prairie ecosystem exposed to ambient and elevated CO_2. *Ecological Applications*, **3**, 666-681.

Painter, T.H., A.P. Barrett, C. Landry, J. Neff, M.P. Cassidy, C. Lawrence, K.E. McBride, and G.L. Farmer, 2007. Impact of disturbed desert soils on duration of mountain snowcover. *Geophysical Research Letters* (*In Press*).

Palmroth, S., R. Oren, H.R. McCarthy, K.H. Johnsen, A.C. Finzi, J.R. Butnor, M.G. Ryan, and W.H. Schlesinger, 2006. Aboveground sink strength in forests controls the allocation of carbon below ground and its [CO_2] - induced enhancement. *Proceedings of the National Academy of Sciences of the United States of America*, **103**, 19362-19367.

Pan, Y., R. Birdsey, J. Hom, K. McCullough, and K. Clark, 2006. Improved estimates of net primary productivity from MODIS satellite data at regional and local scales. *Ecological Applications*, **16**, 125-132.

Parker, B. L., M. Skinner, S. Gouli, T. Ashikaga, and H.B. Teillon, 1999. Low lethal temperature for hemlock woolly adelgid (Homoptera : Adelgidae). *Environmental Entomology*, **28**, 1085-1091.

Parmesan, C., and G. Yohe, 2003. A globally coherent fingerprint of climate change impacts across natural systems. *Nature*, **421**, 37-42.

Parsons, A.J., A.D. Abrahams, and J. Wainwright, 1994. Rainsplash and erosion rates in an inter-rill area on semiarid grassland, southern Arizona. *Catena*, **22**, 215-226.

Parsons, A.J., A.D. Abrahams, and J. Wainwright, 1996. Responses of interrill runoff and erosion rates to vegetation change in southern Arizona. *Geomorphology*, **14**, 311-317.

Parsons, A.J., A.D. Abrahams, and S.H. Luk, 1991. Size characteristics of sediment in inter-rill overland-flow on a semiarid hill slope, southern Arizona. *Earth Surface Processes and Landforms*, **16**, 143-152.

Patrick L., J. Cable, D. Potts, D. Ignace, G. Barron-Gafford, A. Griffith, H. Alpert, N. Van Gestel, T. Robertson, T.E. Huxman, J. Zak, M.E. Loik, D. Tissue, 2007. Effects of an increase in summer precipitation on leaf, soil and ecosystem fluxes of CO_2 and H_2O in a stool grassland in Big Bend National Park, Texas. *Oecologia*, **151**, 704-718.

Penuelas, J., and I. Filella, 2001. Phenology - Responses to a warming world. *Science*, **294**, 793-795.

Pereira, J.S., J.A. Mateus, L.M. Aires, G. Pita, C. Pio, J.S. David, V. Andrade, J. Banza, T.S. David, T.A. Paco, A. Rodrigues, 2007. Net ecosystem exchange in three contrasting Mediterranean ecosystems – the effect of drought. *Biogeosciences*, **4**, 791-802.

Peters, R., 1992. Conservation of biological diversity in the face of climate change. Pages 3-30 *In:* P. RL and T. Lovejoy, (eds), *Global Warming and Biological Diversity*. Yale University Press, New Haven, CT, USA.

Pfister, G.G., L.K. Emmons, P.G. Hess, R. Honrath, J.F. Lamarque, M.V. Martin, R.C. Owen, M.A. Avery, E.V. Browell, J.S. Holloway, P. Nedelec, R. Purvis, T.B. Ryerson, G.W. Sachse, and H. Schlager, 2006. Ozone production from the 2004 North American boreal fires. *Journal of Geophysical Research-Atmospheres*, **111**, 1-13, D24S07, doi:10.1029/2006JD007695, 2006.

Phillips, N., and R. Oren, 2001. Intra- and inter-annual variation in transpiration of a pine forest. *Ecological Applications*, **11**, 385-396.

Pierce, J.L., G.A. Meyer, and A.J.T. Jull, 2004. Fire-induced erosion and millennial scale climate change in northern ponderosa pine forests. *Nature*, **432**, 87-90.

Piketh, S.J., P.D. Tyson, and W. Steffen, 2000. Aeolian transport from southern Africa and iron fertilization of marine biota in the South Indian Ocean. *South African Journal of Geology*, **96**, 244-246.

Plume, R.W. and S.M. Carlton, 1988. Hydrogeology of the Great Basin region of Nevada, Utah, and adjacent states. *U.S. Geological Survey Hydrologic Investigations Atlas* HA-694-A.

Polhemus, D.A. and J.T. Polhemus, 2001. Basin and ranges. The biogeography of aquatic true bugs (Insecta: Hemiptera) in the Great Basin. Pages 235-254. *In*: R. Hershler, D.B. Madsen, and D.R. Currey (eds.). *Great Basin Aquatic Systems History*. Smithsonian Contributions to Earth Sciences, Number 33.

Polley, H.W., H.B. Johnson, and C.R. Tischler, 2003. Woody invasion of grasslands: evidence that CO_2 enrichment indirectly promotes establishment of Prosopis glandulosa. *Plant Ecology*, **164**, 85-94.

Polley, H.W., H.S. Mayeux, H.B. Johnson, and C.R. Tischler, 1997. Atmospheric CO_2, soil water, and shrub/grass ratios on rangelands. *Journal of Range Management*, **50**, 278-284.

Polley, H., H. Johnson, and J. Derner, 2002. Soil- and plant-water dynamics in a C3/C4 grassland exposed to a subambient to superambient CO_2 gradient. *Global Change Biology*, **8**, 1118-1129.

Poorter, H., and M.L. Navas, 2003. Plant growth and competition at elevated CO_2: on winners, losers and functional groups. *New Phytologist*, **157**, 175-198.

Potts, D.F., 1984. Hydrologic impacts of a large-scale mountain pine beetle (Dendroctonus ponderosae Hopkins) epidemic. *Water Resources Bulletin*, **20**, 373-377.

Pregitzer, K.S., A.J. Burton, D.R. Zak, and A.F. Talhelm, 2008. Simulated chronic nitrogen deposition increases carbon storage in northern temperate forests. *Global Change Biology*, **14**, 142-153.

Prieur-Richard, A.H., and S. Lavorel, 2000. Invasions: perspective of diverse plant communities. *Australian Ecology*, **25**, 1-7.

Raich, J.W., and W.H. Schlesinger, 1992. The global carbon dioxide flux in soil respiration and its relationship to vegetation and climate. *Tellus*, **44B**, 81-89.

Randerson, J.T., H. Liu, M.G. Flanner, S.D. Chambers, Y. Jin, P.G. Hess, G. Pfister, M.C. Mack, K.K. Treseder, L.R. Welp, F.S. Chapin, J.W. Harden, M.L. Goulden, E. Lyons, J.C. Neff, E.A.G. Schuur, and C.S. Zender. 2006. The impact of boreal forest fire on climate warming. *Science*, **314**, 1130-1132.

Raphael, M.N. 2003. The Santa Ana winds of California. *Earth Interactions*, **7**, 1-13.

Raupach, M.R., P.J. Rayner, D.J. Barrett, R.S. DeFries, M. Heimann, D.S. Ojima, S. Quegan, and C.C. Schmullius, 2005. Model-data synthesis in terrestrial carbon observation: methods, data requirements and data uncertainty specifications. *Global Change Biology*, **11**, 378-397.

Ravi, S., P. D'Odorico, T.M. Over, and T.M. Zobeck, 2006. On the effect of air humidity on soil susceptibility to wind erosion: the case of air-dry soils. *Geophysical Research Letters*, **31**, Art. No. L09501.

Regab, R., and C. Prudhomme, 2002. Climate change and water resource management in arid and semi-arid regions: prospective and challenges for the 21st century. *Biosystems Engineering*, **81**, 3-34.

Reheis, M.C. 2006. A 16-year record of eolian dust in Southern Nevada and California, USA: controls on dust generation and accumulation. *Journal of Arid Environments*, **67**, 487-520.

Reich, P.B., S.E. Hobbie, T. Lee, D.S. Ellsworth, J.B. West, D. Tilman, J.M.H. Knops, S. Naeem, and J. Trost, 2006. Nitrogen limitation constrains sustainability of ecosystem response to CO_2. *Nature*, **440**, 922-925.

Reynolds, J.F., D.W. Hilbert, and P.R. Kemp, 1993. Scaling ecophysiology from the plant to the ecosystem: A conceptual framework. Pages 127-140 *In*: J.R. Ehleringer and C.B. Field, (eds). *Scaling physiological processes: Leaf to globe*. San Diego: Academic Press.

Richards, K.R., and C. Stokes, 2004. A review of forest carbon sequestration cost studies: a dozen years of research. *Climatic Change*, **63**,1-48.

Richardson, D.M., P. Pysek, M. Rejmanek, M.G. Barbour, F.D. Panetta, and C.J. West, 2000. Naturalization and invasion of alien plants: concepts and definitions. *Diversity and Distributions*, **6**, 93-107.

Ries, J.B., and I. Marzolff, 2003. Monitoring of gully erosion in the Central Ebro Basin by large-scale aerial photography taken from a remotely controlled blimp. *Catena*, **50**, 309-328.

Roden, J.S., G.G. Lin, and J.R. Ehleringer, 2000. A mechanistic model for interpretation of hydrogen and oxygen isotope ratios in tree-ring cellulose. *Geochimica et Cosmochimica Acta*, **64**, 21-35.

Romme, W.H., J. Clement, J. Hicke, D. Kulakowski, L.H. MacDonald, T.L. Schoennagel, and T.T. Veblen, 2006. Recent Forest Insect Outbreaks and Fire Risk in Colorado Forests: A Brief Synthesis of Relevant Research. Colorado Forest Restoration Institute, Colorado State University.

Roshier, D.A., P.H. Whetton, R.J. Allan, and A. I. Robertson. 2001. Distribution and persistence of temporary wetland habitats in arid Australia in relation to climate. *Austral Ecology*, **26**, 371-384.

Ross, R.M., R.M. Bennett, C.D. Snyder, J.A. Young, D.R. Smith, and D.P. Lemarie, 2003. Influence of eastern hemlock (Tsuga canadensis L.) on fish community structure and function in headwater streams of the Delaware River basin. *Ecology of Freshwater Fish*, **12**, 60-65.

Ross, R.M., L.A. Redell, R.M. Bennett, and J.A. Young, 2004. Mesohabitat use of threatened hemlock forests by breeding birds of the Delaware river basin in northeastern United States. *Natural Areas Journal*, **24**, 307-315.

Rundel, P., and A. Gibson, 1996. *Ecological communities and processes in a Mojave Desert ecosystem: Rock Valley, Nevada.* Cambridge University Press, New York.

Running, S.W., P.E. Thornton, R. Nemani, and J.M. Glassy, 2000. Global terrestrial gross and net primary productivity from the earth observing system. Pages 44-57 *In*: O.Sala, R. Jackson, and H.Mooney, (eds.), *Methods in Ecosystem Science*. Springer-Verlag, New York.

Running, S.W., R.R. Nemani, F.A. Heinsch, M.S. Zhao, M. Reeves, and H. Hashimoto, 2004. A continuous satellite-derived measure of global terrestrial primary production. *BioScience*, **54**,547-560.

Rustad, L.E., J.L. Campbell, G.M. Marion, R.J. Norby, M.J. Mitchell, A.E. Hartley, J.H.C. Cornelissen, and J. Gurevitch, 2001. A meta-analysis of the response of soil respiration, net nitrogen mineralization, and aboveground plant growth to experimental ecosystem warming. *Oecologia*, **126**, 543-562.

Ryan, M.G., D. Binkley, J.H. Fownes, C.P. Giardina, and R.S. Senock, 2004. An experimental test of the causes of forest growth decline with stand age. *Ecological Monographs*, **74**, 393-414.

Ryan, M.G., D. Binkley, and J.H. Fownes, 1997. Age-related decline in forest productivity: pattern and process. *Advances in Ecological Research*, **27**, 213-262.

Ryan, M.G., S. Linder, J.M. Vose, and R.M. Hubbard, 1994. Dark respiration in pines. Pages 50-63 *In*: H.L. Gholz, S. Linder, and R.E. McMurtrie, (eds). *Ecological Bulletins 43, Environmental constraints on the structure and productivity of pine forest ecosystems: a comparative analysis*. Munksgaard, Uppsala.

Sada, D.W. and R. Hershler, 2007. *Desert Research Institute Springs Database*. Desert Research Institute, Reno, NV.

Sada, D. W., and G. L. Vinyard, 2002. Anthropogenic changes in historical biogeography of Great Basin aquatic biota. Pages 277-295 *In:* R. Hershler, D. B. Madsen, and D. Currey, (eds). *Great Basin Aquatic Systems History*. Smithsonian Contributions to the Earth Sciences.

Sada, D. W., J. E. Williams, J. C. Silvey, A. Halford, J. Ramakka, P. Summers, and L. Lewis, 2001. *Riparian area management: A guide to managing, restoring, and conserving springs in the western United States*. Technical Reference 1737-17, Bureau of Land Management, BLM/ST/ST-01/001+1737, Denver, CO, 70 pp.

Sage, R.F., 1996. Atmospheric modification and vegetation responses to environmental stress. *Global Change Biology*, **2**, 79-83.

Sakai, A., and C.J. Weiser, 1973. Freezing resistance of trees in North-America with reference to tree regions. *Ecology*, **54**, 118-126.

Sala, O.E., F.S. Chapin, J.J. Armesto, E. Berlow, J. Bloomfield, R. Dirzo, E. Huber-Sanwald, L.F. Huenneke, R.B. Jackson, A. Kinzig, R. Leemans, D.M. Lodge, H.A. Mooney, M. Oesterheld, N.L. Poff, M.T. Sykes, B.H. Walker, M. Walker, and D.H. Wall, 2000. Global biodiversity scenarios for the year 2100. *Science*. **287**, 1770-1774.

Sala, A., G. Peters, L. McIntyre, and M. Harrington, 2005. Physiological responses of ponderosa pine in western Montana to thinning, prescribed fire and burning season. *Tree Physiology*, **25**, 339-348.

Salo, L.F., 2005. Red brome (Bromus rubens subsp. madritensis) in North America: possible modes for early introductions, subsequent spread. *Biological Invasions*, **7**, 165-180.

Salo, L.F., G.R. McPherson, and D.G. Williams, 2005. Sonoran desert winter annuals affected by density of red brome and soil nitrogen. *American Midland Naturalist*, **153**, 95-109.

Saxe, H., M.G.R. Cannell, Ø. Johnsen, M.G. Ryan, and G. Vourlitis, 2001. Tree and forest functioning in response to global warming. *New Phytologist*, **149**, 369-399.

Scanlon, B.R., D.G. Levitt, R.C. Reedy, K.E. Keese, and M.J. Sully, 2005. Ecological controls on water-cycle response to climate variability in deserts. *Proceedings National Academy of Science*, **102**, 6033-6038.

Schäfer, K.V.R., R. Oren, D.S. Ellsworth, C.T. Lai, J.D. Herrick, A.C. Finzi, D.D. Richter, and G.G. Katul, 2003. Exposure to an enriched CO_2 atmosphere alters carbon assimilation and allocation in a pine forest ecosystem. *Global Change Biology*, **9**, 1378-1400.

Schlesinger, W.H., 1977. Carbon balance in terrestrial detritus. *Annual Review of Ecology and Systematics*, **8**, 51-81.

Schlesinger, W.H., 1982. Carbon storage in the caliche of arid soils: A case study from Arizona. *Soil Science*, **133**, 247-255.

Schlesinger, W.H. 1985. The formation of caliche in soils of the Mojave Desert, California. Geochimica et Cosmochimica Acta 49: 57-66.

Schlesinger, W.H. 1990. Evidence from chronosequence studies for a low carbon-storage potential of soils. Nature 348: 232-234.

Schlesinger, W.H. 2001. Carbon sequestration in soils: Some cautions amidst optimism. Agriculture, Ecosystems and Environment 82: 121-127.

Schlesinger, W.H., 2000. Carbon sequestration in soils: some cautions amidst optimism. *Agriculture, Ecosystems & Environment*, **82**,121-127.

Schlesinger, W.H., and C.S. Jones, 1984. The comparative importance of overland runoff and mean annual rainfall to shrub communities of the Mojave Desert. *Botanical Gazette*, **145**, 116-124.

Schlesinger, W.H., and A.M. Pilmanis, 1998. Plant-soil interactions in deserts. *Biogeochemistry*, **42**,169-187.

Schlesinger, W.H., and J.Lichter, 2001. Limited carbon storage in soil and litter of experimental forest plots under increased atmospheric CO_2. *Nature*, **411**, 466-469.

Schlesinger, W.H., J.A. Raikes, A.E. Hartley, and A.E. Cross, 1996. On the spatial pattern of soil nutrients in desert ecosystems. *Ecology*, **77**, 364-374.

Schlesinger, W.H., J.F. Reynolds, G.L. Cunningham, L.F. Huenneke, W.M. Jarrell, R.A. Virginia, and W.G. Whitford, 1990. Biological feedbacks in global desertification. *Science*, **247**, 1043-1048.

Schlesinger, W.H., S.L. Tartowski, and S.M. Schmidt, 2006. Nutrient cycling within an arid ecosystem. Pages 133-149 *In*: K.M. Havstad, L.F. Huenneke and W.H. Schlesinger, (eds.), *Structure and Function of a Chihuahuan Desert Ecosystem: The Jornada Basin LTER*. Oxford University Press, Oxford.

Schmidtling, R.C., 1994. Use of provenance tests to predict response to climatic-change - Loblolly-pine and Norway spruce. *Tree Physiology*, **14**, 805-817.

Schreuder, H.T., and C.E. Thomas., 1991. Establishing cause-effect relationships using forest survey data. *Forest Science*, **37**, 1497-1512.

Schutzenhofer, M.R., and T.J. Valone, 2006. Positive and negative effects of exotic Erodium cicutarium on an arid ecosystem. *Biological Conservation*, **132**, 376-381.

Schwartz, M.D., R. Ahas, and A. Aasa, 2006. Onset of spring starting earlier across the Northern Hemisphere. *Global Change Biology*, **12**, 343-351.

Scott, R.L., T.E. Huxman, D.G. Williams, and D.C. Goodrich, 2006. Ecohydrological impacts of woody plant encroachment: seasonal patterns of water and carbon dioxide exchange within a semiarid riparian environment. *Global Change Biology*, **12**, 311-324.

Seager, R., M. Ting, I. Held, Y. Kushnir, J. Lu, G. Vecchi, H.P. Huang, N. Harnik, A. Leetmaa, N.C. Lau, C. Li, J. Velez, and N. Naik, 2007. Model projections of an imminent transition to a more arid climate in Southwestern North America. *Science*, **316**, 1181-1184.

Sharkey, T.D., and S.S. Yeh, 2001. Isoprene emission from plants. *Annual Review of Plant Physiology and Plant Molecular Biology*, **52**, 407-436.

Shore, T.L., B.G. Riel, L. Safranyik, and A. Fall, 2006. Decision support systems. Pages 193-230 *In*: L. Safranyik and W.R. Wilson, (eds.), *The mountain pine beetle: a synthesis of biology, management, and impacts on lodgepole pine*. Natural Resources Canada, Canadian Forest Service, Pacific Forestry Centre, Victoria, British Columbia.

Sims, D.A., A.F. Rahman, B.Z. El Masri, D.D. Baldocchi, L.B. Flanagan, A.H. Goldstein, D.Y. Hollinger, L. Mission, R.K. Monson, W.C. Oechel, H.P. Schmid, and L. Xu, 2006. On the use of MODIS EVI to assess gross primary productivity of North American ecosystems. *Journal of Geophysical Research*, **111**,G04015, doi:04010.01029/02006JG000162.

Skinner, M., B. L. Parker, S. Gouli, and T. Ashikaga, 2003. Regional responses of hemlock woolly adelgid (Homoptera : Adelgidae) to low temperatures. *Environmental Entomology*, **32**, 523-528.

Small, M.J., C.J. Small, and G.D. Dreyer, 2005. Changes in a hemlock-dominated forest following woolly adelgid infestation in southern New England. *Journal of the Torrey Botanical Society*, **132**, 458-470.

Smith, E. 1999, Atlantic and east coast hurricanes 1900-98: A frequency and intensity study for the twenty-first century. *Bulletin of the American Meteorological Society*, **80**, 2717-2720.

Smith, P., 2004. How long before a change in soil organic carbon can be detected? *Global Change Biology*, **10**, 1878-1883.

Smith, S.D., T.E. Huxman, S.F. Zitzer, T.M. Charlet, D.C. Housman, J.S. Coleman, L.K. Fenstermaker, J.R. Seemann, and R.S. Nowak, 2000. Elevated CO_2 increases productivity and invasive species success in an arid ecosystem. *Nature*, **408**, 79-82.

Snyder, C.D., J.A. Young, D.P. Lemarie, and D.R. Smith, 2002. Influence of eastern hemlock (Tsuga canadensis) forests on aquatic invertebrate assemblages in headwater streams. *Canadian Journal of Fisheries and Aquatic Sciences*, **59**, 262-275.

Sokolik, I.N., and O.B. Toon, 1996. Direct radiative forcing by anthropogenic airborne mineral aerosols. *Nature*, **381**, 681-683.

Souto, D., T. Luther, B. Chianese, 1996, Past and current status of HWA in eastern and Carolina hemlock stands. *In:* Salom, S.M., Tignor, T.C., Reardon, R.C. (Eds.), *Proceedings of the First Hemlock Woolly Adelgid Review, USDA Forest Service, Morgantown, WV, pp. 9-15.* http://www.na.fs.fed.us/fhp/hwa/maps/hwaprojectedspreadmap.htm

Stadler, B., T. Muller, and D. Orwig, 2006. The ecology of energy and nutrient fluxes in hemlock forests invaded by hemlock woolly adelgid. *Ecology*, **87**, 1792-1804.

Stadler, B., T. Muller, D. Orwig, and R. Cobb, 2005. Hemlock woolly adelgid in new england forests: Canopy impacts transforming ecosystem processes and landscapes. *Ecosystems*, **8**, 233-247.

Stanturf, J.A., S.L. Goodrick, and K.W. Outcalt, 2007. Disturbance and coastal forests: A strategic approach to forest management in hurricane impact zones. *Forest Ecology and Management. (In Press)*

Stednick, J.D., 1996. Monitoring the effects of timber harvest on annual water yield. *Journal of Hydrology*, **176**, 79-95.

Stewart, I.T., D.R. Cayan, and M.D. Dettinger, 2004. Changes in snowmelt timing in western North America under a 'business as usual' climate change scenario. *Climate Change*, **62**, 217-232.

Stohlgren, T.J., D. Binkley, G.W. Chong, M.A. Kalkhan, L.D. Schell, K.A. Bull, Y. Otsuki, G. Newman, M. Bashkin, and Y. Son, 1999. Exotic plant species invade hot spots of native plant diversity. *Ecological Monographs*, **69**, 25-46.

Stohlgren, T.J., K.A. Bull, Y. Otsuki, C.A. Villa, and M. Lee, 1998. Riparian zones as havens for exotic plant species in the central grasslands. *Plant Ecology*, **138**, 113-125.

Stoy, P.C., G. G. Katul, M. B. S. Siqueira, J. Y. Juang, K. A. Novick, J. M. Uebelherr, and R. Oren. 2006. An evaluation of models for partitioning eddy covariance-measured net ecosystem exchange into photosynthesis and respiration. *Agricultural and Forest Meteorology*, **141**, 2-18.

Stromberg, J.C., R. Tiller, and B. Richter, 1996. Effects of groundwater decline on riparian vegetation of semiarid regions: the San Pedro, Arizona. *Ecological Applications*, **6**, 13-131.

Sullivan, K.A., and A.M. Ellison, 2006. The seed bank of hemlock forests: implications for forest regeneration following hemlock decline. *Journal of the Torrey Botanical Society*, **133**, 393-402.

Svejcar, T., J. Bates, R. Angell, and R. Miller, 2003. The influence of precipitation timing on the sagebrush steppe ecosystem. *In*: J. F. Weltzin and G. R. McPherson, (eds), *Changing Precipitation Regimes and Terrestrial Ecosystems: A North American Perspective*. University of Arizona Press, Tucson.

Swap, R., M. Garstang, S. Greco, R. Talbot, and P. Kallberg, 1992. Saharan dust in the Amazon Basin. *Tellus Series B-Chemical and Physical Meteorology*, **44**,133-149.

SWRC, 2007. Southwest Watershed Research Center and Walnut Gulch Experimental Watershed. http://www.ars.usda.gov/SP2UserFiles/Place/53424500/SWRCWGEW_2007.pdf edition. Southwest Watershed Research Center, 2000 E. Allen Road, Tucson, AZ, http://www.ars.usda.gov/SP2UserFiles/Place/53424500/SWRCWGEW_2007.pdf

Taylor, D.W, 1985. Evolution of freshwater drainages and mollusks in western North America. Pages 265-321. *In:* C.J. Smiley and A.J. Leviton (eds.). *Late Cenozoic history of the Pacific Northwest*. American Association for the Advancement of Science, San Francisco.

Taylor, S.W., A.L. Carroll, R.I. Alfaro, and L. Safranyik, 2006. Forest, climate, and mountain pine beetle outbreak dynamics in western Canada. Pages 67-94 *In*: L. Safranyik and W.R. Wilson, (eds),. *The mountain pine beetle: a synthesis of biology, management, and impacts on lodgepole pine.* Natural Resources Canada, Canadian Forest Service, Pacific Forestry Centre, Victoria, British Columbia.

The-Heinz-Center, 2002. *The state of the nation's ecosystems.* Cambridge University Press.

Thomas, C.D., A.M.A. Franco, and J.K. Hill, 2006. Range retractions and extinction in the face of climate warming. *Trends in Ecology & Evolution*, **21**, 415-416.

Thomas, J.M., A.H. Welch, and M.D. Dettinger, 1996. *Geochemistry and isotope hydrology of representative aquifers in the Great Basin region of Nevada, Utah, and adjacent states.* U.S. Geological Survey Professional Paper 1409-C.

Thornton, P.E., H. Hasenauer, and M.A. White, 2000. Simultaneous estimation of daily solar radiation and humidity from observed temperature and precipitation: an application over complex terrain in Austria. *Agricultural and Forest Meteorology*, **104**, 255-271.

Throop, H.L., E.A. Holland, W.J. Parton, D.S. Ojima, and C.A. Keough, 2004. Effects of nitrogen deposition and insect herbivory on patterns of ecosystem-level carbon and nitrogen dynamics: results from the CENTURY model. *Global Change Biology*, **10**, 1092-1105.

Thurow, T. L., 1991. Hydrology and erosion. Pages 141-160 *In*: R. K. Heitschmidt and J. W. Stuth (eds). *Grazing Management: An Ecological Perspective*. Timber Press, Portland, OR.

Tickner, D.P., P.G. Angold, A.M. Gurnell, and J.O. Mountford, 2001. Riparian plant invasions: hydrogeomorphological control and ecological impacts. *Progress in Physical Geography*, **25**, 22-52.

Tingley, M.W., D. A. Orwig, and R. Field, 2002. Avian response to removal of a forest dominant: consequences of hemlock woolly adelgid infestations. *Journal of Biogeography*, **29**, 1505-1516.

Townsend, P.A., K.N. Eshleman, and C. Welcker, 2004. Relationships between stream nitrogen concentrations and intensity of forest disturbance following gypsy moth defoliation in 2000-2001. *Ecological Applications*, **14**, 504-516.

Tran, J.K., T. Ylioja, R. Billings, J. Régnière, and M.P. Ayres. in press. Testing a climatic model to predict populations dynamics of a forest pest, Dendroctonus frontalis (Coleptera: Scolydidae). *Ecological Applications.*

Tucker, C.J., D.A. Slayback, J.E. Pinzon, S.O. Los, R.B. Myneni, and M.G. Taylor, 2001. Higher northern latitude normalized difference vegetation index and growing season trends from 1982 to 1999. *International Journal of Biometeorology*, **45**, 184-190.

Turetsky, M.R., J.W. Harden, H.R. Friedli, M. Flannigan, N. Payne, J. Crock, and L. Radke, 2006. Wildfires threaten mercury stocks in northern soils. *Geophysical Research Letters,* **33**, 1-6, L16403, doi:10.1029/2005GL025595, 2006.

Turner, D.P., S.V. Ollinger, and J.S. Kimball, 2004. Integrating remote sensing and ecosystem process models for landscape-to regional-scale analysis of the carbon cycle. *BioScience*, **54**, 573-584.

Turner, M.G., W.H. Romme, and R.H. Gardner, 1999. Prefire heterogeneity, fire severity, and early postfire plant reestablishment in subalpine forests of Yellowstone National Park, Wyoming. *International Journal of Wildland Fire*, **9**, 21-36.

Turner, R.M., J.E. Bowers, and T.L. Burgess, 1995. *Sonoran Desert Plants: An Ecological Atlas.* University of Arizona Press, Tucson.

Ungerer, M.J., M.P. Ayres, and M.A.J. Lombardero, 1999. Climate and the northern distribution limits of Dendroctonus frontalis Zimmermann (Coleoptera:Scolytidae). *Journal of Biogeography*, **26**, 1133-1145.

United States- Department of Agriculture, 2003. *National report on sustainable forests – 2003. Forest Service Report FS-766.* USDA Forest Service, Washington, DC.

Unland, H.E., P.R. Houser, S.W.J., and Z.L. Yang, 1996. Surface flux measurement and modeling at a semi-arid Sonoran Desert site. *Agricultural and Forest Meteorology*, **82**,119-153.

USDA Forest Service, 2005. *Forest Insect and Disease Conditions in the United States, 2004.* Washington, D.C.

USDA Forest Service and U.S. Geological Survey, 2002. Forest Cover Types: National Atlas of the United States, Reston, VA http://nationalatlas.gov/articles/biology/a_forest.html

Valentin, C., J. Poesen, and Y. Li. 2005, Gully erosion: impacts, factors and control. *Catena*, **63**, 132-153.

Van Auken, O.W., 2000. Shrub invasions of North American semiarid grasslands. *Annual Review of Ecology & Systematics*, **31**, 197-215.

Van de Koppel, J., M. Reitkerk, F.V. Langevelde, L. Kumar, C.A. Klausmier, J.M. Fryxell, J.W. Hearne, J.V. Andel, N.D. Ridder, A. Skidmore, L. Stroosnijder, and H.T. Prins, 2002. Spatial heterogeneity and irreversible vegetation change in semiarid grazing systems. *American Naturalist*, **159**, 209-218.

Venable, D.L., and C.E. Pake, 1999. Population ecology of Sonoran Desert annual plants. Pages 115-142 *In*: R. H. Robichaux, (ed), *The ecology of Sonoran Desert plants and plant communities.* University of Arizona Press, Tucson.

Wagner, F.H., (ed), 2003. *Preparing for a changing climate: the potential consequences of climate variability and change, Rocky Mountains, Great Basin.* Report of the Rocky Mountain/ Great Basin Regional Assessment Team, U.S. Global Change Research Program, Utah State University, Logan.

Wainwright, J.A., A.J. Parsons, W.H. Schlesinger, and A.D. Abrahams, 2002. Hydrology-vegetation interactions in areas of discontinuous flow on a semi-arid bajada, southern New Mexico. *Journal of Arid Environments*, **51**, 219-258.

Wainwright, J.A., A.J. Parsons, W.H. Schlesinger, and A.D. Abrahams, 2002. Hydrology-vegetation interactions in areas of discontinuous flow on a semi-arid bajada, southern New Mexico. *Journal of Arid Environments*, **51**,219-258.

Wainwright, J., A.J. Parsons, and A.D. Abrahams 2000. Plot-scale studies of vegetation, overland flow and erosion interactions: case studies from Arizona and New Mexico. *Hydrological Processes*, **14**,2921-2943.

Walther, G.R., 2007. Tackling ecological complexity in climate impact research. *Science*, **315**, 606-607.

Walther, G.R., E. Post, P. Convey, A. Menzel, C. Parmesan, T.J.C. Beebee, J.M. Fromentin, O. Hoegh-Guldberg, and F. Bairlein, 2002. Ecological responses to recent climate change. *Nature*, **416**, 389-395.

Ward, J.K., D.T. Tissue, R.B. Thomas, and B.R. Strain, 1999. Comparative responses of model C3 and C4 plants to drought in low and elevated CO_2. *Global Change Biology*, **5**, 857-867.

Waring, R.H., 1987. Characteristics of trees predisposed to die. *BioScience*, **37**, 569-574.

Warren, M.S., J.K. Hill, J.A. Thomas, J. Asher, R. Fox, B. Huntley, D.B. Roy, M. G. Telfer, S. Jeffcoate, P. Harding, G. Jeffcoate, S.G. Willis, J.N. Greatorex-Davies, D. Moss, and C.D. Thomas, 2001. Rapid responses of British butterflies to opposing forces of climate and habitat change. *Nature*, **414**, 65-69.

Webb, R.H., 1996. *Grand Canyon, a Century of Change: Rephotography of the 1889-1890 Stanton Expedition.* University of Arizona Press, Tucson.

Webb, R.H., and S.A. Leake, 2006. Ground-water surface-water interactions and long-term change in riverine riparian vegetation in the southwestern United States. *Journal of Hydrology*, **320**, 302-323.

Webb, R.H., S.A. Leake, and R.M. Turner, 2007. *The Ribbon of Green: Change in Riparian Vegetation in the Southwestern United States.* University of Arizona Press, Tucson.

Webb, W.L., W.K. Lauenroth, S.R. Szarek, and R.S. Kinerson, 1983. Primary production and abiotic controls in forests, grasslands, and desert ecosystems in the United States. *Ecology*, **64**, 134-151.

Weiss, J., and J.T. Overpeck, 2005. Is the Sonoran Desert losing its cool? *Global Change Biology*, **11**, 2065-2077.

Wells, O.O., and P.C. Wakeley, 1966. Geographic variation in survival, growth, and fusiform rust infection of planted loblolly pine. *Forest Science Monographs*, **11**, 1-40.

Wells, S.G., L.D. McFadden, J. Poths, and C.T. Olinger, 1995. Cosmogenic ^{3}He surface-exposure dating of stone pavements: implications for landscape evolution in deserts. *Geology*, **23**, 613-616.

Weltzin, J.F., and G.R. McPherson, 2000. Implications of precipitation redistribution for shifts in temperate savanna ecotones. *Ecology,* **81**, 1902-1913.

Wessman, C., S. Archer, L. Johnson, and G. Asner, 2004. Woodland expansion in U.S. grasslands: assessing land-cover change and biogeochemical impacts. Pages 185-208 *In*: G. Gutman, Janetos, A.C., Justice, C.O., Moran, E.F., Mustard, J.F., Rindfuss, R.R., Skole, D., Turner II, B.L., Cochrane, M.A., (eds.), *Land Change Science: Observing, Monitoring and Understanding Trajectories of Change on the Earth's Surface.* Kluwer Academic Publishers, Dordrecht.

West, N., (ed), 1983. *Temperate deserts and semi-deserts.* Ecosystems of the World 5, Elsevier Scientific Publishing Co.

West, N.E., and T.P. Yorks, 2006. Long-term interactions of climate, productivity, species richness, and growth form in relictual sagebrush steppe plant communities. *Western North American Naturalist*, **66**, 502-526.

Westerling, A.L., D.R. Cayan, T.J. Brown, and B.L. Hall, 2004. Climate, Santa Ana winds, and wildfires in Southern California. *EOS*, transactions **85**, 289-300.

Westerling, A.L., H.G. Hidalgo, D.R. Cayan, and T.W. Swetnam, 2006. Warming and earlier spring increase western U.S. forest wildfire activity. *Science*, **313**, 940-943.

White, M., F. Hoffman, W. Hargrove, and R. Nemani, 2005. A global framework for monitoring phenological responses to climate change. *Geophysical Research Letters,* **32**, Art. No. L04705 (Feb 04718).

Whittaker, R.H., 1975. *Communities and Ecosystems.* Macmillan, London : New York.

Wilcox, B.P., 2002. Shrub control and streamflow on rangelands: a process-based viewpoint. *Journal of Range Management*, **55**, 318-326.

Williams, D.G., and J.R. Ehleringer, 2000. Carbon isotope discrimination and water relations of oak hybrid populations in southwestern Utah. *Western North American Naturalist*, **60**,121-129.

Williams, J.W., and S.T. Jackson, 2007. Novel climates, no-analog communities, and ecological surprises. *Frontiers in Ecology and the Environment*, **5**, 475-482.

Williams, D.G., and Z. Baruch, 2000. African grass invasion in the Americas: ecosystem consequences and the role of ecophysiology. *Biological Invasions,* **2**, 123-140.

Wilmking, M., G.P. Juday, V.A. Barber, and H.S.J. Zald, 2004. Recent climate warming forces contrasting growth responses of white spruce at treeline in Alaska through temperature thresholds. *Global Change Biology,* **10**, 1724-1736.

Wisdom, M.J., M.M. Rowland, and L.H. Suring, (eds.), 2005. *Habitat threats in the sagebrush ecosytem: methods of regional assessment and applications in the Great Basin.* Allen Press/ Alliance Communicaton Group Publishing, Lawrence, KS 66044.

Wittig, V.E., C.J. Bernacchi, X.G. Zhu, C. Calfapietra, R. Ceulemans, P. Deangelis, B. Gielen, F. Miglietta, P.B. Morgan, and S.P. Long, 2005. Gross primary production is stimulated for three Populus species grown under free-air CO_2 enrichment from planting through canopy closure. *Global Change Biology,* **11**, 644-656.

Wondzell, S.M., G.L. Cunningham, and D. Bachelet, 1996. Relationships between landforms, geomorphic processes, and plant communities on a watershed in the northern Chihuahuan Desert. *Landscape Ecology,* **1,** 351-362.

Wood, Y.A., T. Meixner, P.J. Shouse, and E.B. Allen, 2006. Altered ecohydrologic response drives native shrub loss under conditions of elevated nitrogen deposition. *Journal of Environmental Quality*, **35**, 76-92.

Woodward, F.I., 1987. *Climate and Plant Distribution.* Cambridge University Press, Cambridge.

Wullschleger, S.D., P.J. Hanson, and D.E. Todd, 2001. Transpiration from a multi-species deciduous forest as estimated by xylem sap flow techniques. *Forest Ecology and Management,* **143**, 205-213.

Wurzler, S., T.G. Reisin, and Z. Levin, 2000. Modification of mineral dust particles by cloud processing and subsequent effects on drop size distributions. *Journal of Geophysical Research*, **105**, 4501-4512.

Wythers, K.R., P.B. Reich, M.G. Tjoelker, and P.B. Bolstad, 2005. Foliar respiration acclimation to temperature and temperature variable Q_{10} alter ecosystem carbon balance. *Global Change Biology*, **11**, 435-449.

Yao, J., D. Peters, K. Havstad, R. Gibbens, and J. Herrick, 2006. Multi-scale factors and long-term responses of Chihuahuan Desert grasses to drought. *Landscape Ecology,* **21**, 1217-1231.

Zender, C.S., and E.Y. Kwon, 2005. Regional contrasts in dust emission responses to climate. *Journal of Geophysical Research-Atmospheres,* **110**, D13201.

Ziska, L.H., 2003. Evaluation of the growth esponse of six invasive species to past, present and future atmospheric carbon dioxide. *Journal of Experimental Botany,* **54**, 395-404.

Zhu, Z.I., and D.L. Evans, 1994. United States forest types and predicted percent forest cover from AVHRR data. *Photogrammetric Engineering and Remote Sensing,* **60**, 525-531.

CHAPTER 4 REFERENCES

Alexander, R. B., and R.A. Smith, 2006: Trends in the nutrient enrichment of U.S. rivers during the late 20th century and their relation to changes in probable stream trophic conditions, *Limnology and Oceanography,* **51**, 639-654.

Andreadis, K.M., and D.P. Lettenmaier, 2006: Trends in 20th century drought over the continental United States, *Geophysical Research Letters,* **33**, doi:10.1029/2006GL025711.

Arnell, N., and C. Liu, 2001: Hydrology and water resources, pp. 191-233 in *Climate Change 2001: Impacts, Adaptation, and Vulnerability*, Cambridge University Press.

Arnell, N., 2002: *Hydrology and Global Environmental Change*, Pearson Education Ltd, Edinburgh, 346 p.

Barber, V.A., G.P. Juday, and B.P. Finney, 2000: Reduced growth of Alaskan white spruce in the twentieth century from temperature-induced drought stress, *Nature,* **405**, 668-673.

Bartholow, J.M., 2005: Recent water temperature trends in the Lower Klamath River, California, *Journal of Fisheries Management,* **25**, 152-162.

Bowling, L.C., P. Storck, and D.P. Lettenmaier, 2000: Hydrologic effects of logging in Western Washington, United States, *Water Resources Research*, **36**, 3223-3240.

Bowling, L.C. and D.P. Lettenmaier, 2001: The effect of forest roads and harvest on catchment hydrology in a mountainous maritime environment, *In*: *Land Use and Watersheds: Human Influence on Hydrology and Geomorphology in Urban and Forest Areas*, Water Science and Application, The American Geophysical Union, **2**, 145-164.

Brutsaert, W., and M.B. Parlange, 1998: Hydrologic cycle explains the evaporation paradox, *Nature,* **396**, 30.

Brutsaert, W., 2006: Indications of increasing land surface evaporation during the second half of the 20th century, *Geophysical Research Letters*, **33**, doi:10.1029/2006GL027532.

Burns, D.A., J. Klaus, and M.R. McHale, 2007: Recent climate trends and implications for water resources in the Catskill Mountain region, New York, USA, *Journal of Hydrology* **336**, 155-170.

Carroll, A.L., S.W. Taylor, J. Régnière, and L. Safranyik, 2003: Effects of climate change on range expansion by the mountain pine beetle in British Columbia, *In:* T.L. Shore, J.E. Brookes, and J.E. Stone, eds *Information Report BC-X-399, Mountain Pine Beetle Symposium: Challenges and Solutions* (pp. 223-232, Natural Resources Canada, Victoria, British Columbia.

Caspersen, J., S. Pacala, J. Jenkins, G. Hurtt, P. Moorcroft, and R. Birdsey, 2000: Contributions of land-use history to carbon accumulation in U.S. forests, *Science,* **290**, 1148-1151.

Cayan, D.R., S.A. Kammerdiener, M.D. Dettinger, J.M. Caprio, and D.L. Peterson, 2001: Changes in the onset of spring in the western United States, *Bulletin of the American Meteorological Society,* **82**, 299-415.

Chang, H.J. 2004: Water quality impacts of climate and land use changes in southeastern Pennsylvania, *Professional Geographer,* **56**, 240-257.

Christensen, N.S., and D.P. Lettenmaier, 2007: A multimodel ensemble approach to assessment of climate change impacts on the hydrology and water resources of the Colorado River basin, *Hydrology and Earth System Science,* **11**, 1417-1434.

Cohn, T.A., and H.F. Lins, 2005: Nature's style: Naturally trendy, *Geophysical Research Letters* **32**, doi:10.1029/2005GL024476.

Crozier, L., and R.W. Zabel, 2006. Climate impacts at multiple scales: evidence for differential population responses in juvenile Chinook salmon, *Journal of Animal Ecology*, **75**, 1100-1109.

Curriero, F.C., J.A. Patz, J.B. Rose, and S. Lele, 2001: The association between extreme precipitation and waterborne disease outbreaks in the United States, 1948-1994, *American Journal of Public Health,* **91**, 1194-1199.

Czikowsky, M.J., and D.R. Fitzjarrald, 2004: Evidence of seasonal changes in evapotranspiration in eastern U.S. hydrological records, *Journal of Hydrometeorology.,* **5**, 974-988.

Dai, A., K.E. Trenberth, and T. Qian, 2004. A global data set of Palmer Drought Severity Index for 1870-2002: Relationship with soil moisture and effects of surface warming, *Journal of Hydrometeorology,* **5**, 1117-1130.

Dressler, K.A., S.R. Fassnacht, and R.C. Bales, 2006: A Comparison of Snow Telemetry and Snow Course Measurements in the Colorado River Basin, *Journal of Hydrometeorology,* **7**, 705-712.

Easterling, D. R., and T. R. Karl, 2001: "Potential Consequences of Climate Variability and Change for the Midwestern United States, chapter 6 in National Assessment Team, U.S. Global Change Research Program, *Climate Change Impacts on the United States: The potential consequences of climate variability and change.*

Easterling, D.R., 2002: Recent changes in frost days and the frost-free season in the United States, *Bulletin of the American Meteorological Society*, **83**, doi: 10.1175/1520-0477.

Eaton, J.G., and R.M. Scheller, 1996: Effects of climate warming on fish thermal habitat in streams of the United States, *Limnology and Oceanography*, **41**, 1109-1115.

Elliott, J.A., I.D. Jones, and S.J. Thackeray, 2006: Testing the sensitivity of phytoplankton communities to changes in water temperature and nutrient load in a temperate lake, *Hydrobiolgica,* **559**, 401-411.

Feng, S. and Q. Hu, 2004: Changes in agro-meteorological indicators in the contiguous United States: 1951-2000. *Theoretical and App. Climatology,* **78**, 247-264.

Garbrecht, J., M. van Liew, and G.O. Brown, 2004: Trends in precipitation, streamflow, and evapotranspiration in the Great Plains of the United States, *Journal of Hydrologic Engineering,* **9**, 360-367.

GAO. 2004: The General Accounting Office (GAO) Report: *Watershed Management: Better Coordination of Data Collection Efforts Needed to Support Key Decisions*, http://www.gao.gov/new.items/d04382.pdf

Gleick, P.H., 1999: Introduction: Studies for the water sector of the National Assessment, *Journal of Water Resources Asso*ciation, **35**, 1297-1300.

Gleick, P.H., and D.B. Adams (ed.), 2000: Water: The potential consequences of climate variability and change for the water resources of the United States, U.S. Geological Survey, 151 p. (available from pistaff@pacinst.org).

Gleick, P.H., 1996: Basic water requirements for human activities: meeting basic needs, *Water International,* **21**, 83-92.

Golubev, V.S., J.H. Lawrimore, P.Y. Groisman, N.A. Speranskaya, S.A. Zhuravin, M.J. Menne, T.C. Peterson, and R.W. Malone, 2001: Evaporation changes over the contiguous United States and the former USSR: a reassessment, *Geophysical Research Letters*, **28**, 2665-2668.

Graf, W.L., 1999: Dam nation: A geographic census of American dams and their large-scale hydrologic impacts, *Water Resources Research*, **35**, 1305-1311.

Green, T.R., Taniguchi, M., Kooi, H. 2007: Potential impacts of climate change and human activity on subsurface water resources. *Vadose Zone Journal*, **6**, 531-532. doi: 10.2136/vzj2007.0098.

Groisman, P.Y., R.W. Knight, T.R. Karl, D.R. Easterling, B. Sun, and J.M. Lawrimore, 2004: Contemporary changes of the hydrological cycle over the contiguous United States: Trends derived from in-situ observations, *Journal of Hydrometeorology*, **5**, 64-85.

Groisman, P.Y., R.W. Knight, and T.R. Karl, 2001: Heavy precipitation and high streamflow in the contiguous United States: Trends in the twentieth century, *Bulletin of the American Meteorological. Society*, **82**, 219-246.

Groisman, P.Y., T.R. Karl, D.R. Easterling, R.W. Knight, P.F. Jamason, K. J. Hennessy, R. Suppiah, C. M. Page, J. Wibig, K. Fortuniak, V.N. Razuvaev, A. Douglas, E. Førland, and P.M. Zhai, 1999; Changes in the probability of heavy precipitation: Important indicators of climatic change. *Climatic Change*, **42**, 243-283.

Gurdak, J.J., R.T. Hanson, P.B. McMahon, B.W. Bruce, J.E. McCray, G.D. Thyne, and R.C. Reedy, 2007: Climate variability controls on unsaturated water and chemical movement, High Plains Aquifer, USA, *Vadose Zone Journal*, **6**, 533-547.

Hamlet A.F., and D.P. Lettenmaier, 2007: Effects of 20th Century Warming and Climate Variability on Flood Risk in the Western U.S., *Water Resources Research*, **43**, W06427, doi:10.1029/2006WR005099.

Hamlet A.F., Mote P.W, Clark M.P., Lettenmaier D.P., 2007: 20th Century Trends in Runoff, Evapotranspiration, and Soil Moisture in the Western U.S., *Journal of Climate*, **20** (8), 1468-1486.

Hamlet, A.F., P.W. Mote, M.P. Clark, and D.P. Lettenmaier, 2005. Effects of temperature and precipitation variability on snowpack trends in the western U.S., *Journal of Climate.*, **18**, 4545-4561.

Hanson, R,T., and M.D. Dettinger, 2005. Ground water/surface water responses to global climate simulations, Santa Clara-Calleguas basin, Ventura, California, *Water Resources Bulletin*, **41**, 517-536.

Hinzman, L.D., N.D. Bettez, W.R. bolton, F.S. Chapin, M.B. Dyurgerov, C.L. Fastie, B. Griffith, R.D. Hollister, A. Hope, H.P. Huntington, A.M. Jensen, G.J. Jia, T. Jorgenson, D.L. **Kane**, D.R. Klein, G. Kofinas, A.H. Lynch, A.H. Lloyd, A.D. Mcguire, F.E. Nelson, W.C. Oechel, T.E. Osterkamp, C.H. Racine, V.E. Romanovsky, R.S. Stone, D.A. Stow, M. Sturm, C.E. Tweedie, G.L. Vourlitis, M.D. Walker, D.A. Walker, P.J. Webber, J.M. Welker, K.S. Winker, and K. Yoshikawa, 2005. Evidence and implications of recent climate change in northern Alaska and other Arctic regions, *Climatic Change*, **72**, 251-298.

Hobbins, M.T., J.A. Ramirez, and T. Brown, 2004. Trends in pan evaporation and actual evapotranspiration across the conterminous U.S.: paradoxical or complementary? *Geophysical Research Letters*, **31**, doi: 10.1029/2004GL019846.

Hodgkins, G.A., and R.W. Dudley. 2006a: Changes in the timing of winter-spring streamflows in eastern North America, 1913-2002, *Geophysical Research Letters*, **33**, L06402, doi:10.1029/2005GL025593.

Hodgkins, G.A., and R.W. Dudley, 2006b. Changes in late-winter snowpack depth, water equivalent, and density in Maine, 1926-2004, *Hydrological Processes*, **20**, 741-751.

Hodgkins, G.A., R.W. Dudley, and T.G. Huntington, 2005. Changes in the number and timing of ice-affected flow days on New England rivers, 1930-2000, *Climatic Change*, **71**, 319-340.

Hodgkins, G.A., R.W. Dudley, and T.G. Huntington, 2003. Changes in the timing of high river flows in New England over the 20th century, *Journal of Hydrology*, **278**, 244-252.

Hodgkins, G.A., I.C. James, and T.G. Huntington, 2002. Historical changes in lake ice-out dates as indicators of climate change in New England, *International. Journal of Climatology*, **22**, 1819-1827.

Huntington, T. G., G.A. Hodgkins, B.D. Keim, and R.W. Dudley, 2004. Changes in the proportion of precipitation occurring as snow in New England (1949 to 2000), *Journal of Climate*, **17**, 2626-2636.

Hutson, S.S., N.L. Barber, J.F. Kenny, K.S. Linsey, D.S. Lumia, and M.A. Maupin, 2004. Estimated use of water in the United States in 2000, U.S. Geological Survey Circular 1268, 46 p. (available from www.usgs.gov).

Intergovernmental Panel on Climate Change (IPCC), 2000. *Special Report on Emission Scenarios*, Cambridge University Press, New York.

Jain, S., and V.P. Singh, 2003. *Water resource systems planning and management*, Elsevier, New York, 882 p.

Jha, M., Z. Pan, E.S. Takle, and R. Gu, 2004. Impacts of climate change on streamflow in the upper Mississippi River basin: A regional climate model perspective, *J Geophys. Res.*, **109**, doi:10.1029/2003JD003686.

Jolly, W.M., R. Nemani, and S.W. Running, 2005. A generalized, bioclimatic index to predict foliar phenology in response to climate, *Global Change Biology*, **11**, 619-632.

Jones, J. A., and G. E. Grant, 1996. Peak flow responses to clear-cutting and roads in small and large basins, western Cascades, Oregon, *Water Resources Research*, **32**, 959-974.

Joos, F., I.C. Prentice, and J.I. House, 2002. Growth enhancement due to global atmospheric change as predicted by terrestrial ecosystem models: consistent with U.S. forest inventory data, *Global Change Biology*, **8**, 299-303.

Keleher, C.J., and F.J. Rahel, 1996. Thermal limits to salmonid distributions in the Rocky Mountain region and potential habitat loss due to global warming: a Geographic Information System (GIS) Approach, *Trans Am. Fisheries Soc.*, **125**, 1-13.

Langbein, W.B. and Slack, J.R., 1982. Yearly variations in runoff and frequency of dry years for the conterminous United States, 1911-79: U.S. Geological Survey Open-File Report 82-751, 85 pp.

Lettenmaier, D.P., 2003. The role of climate in water resources planning and management, pp. 247-266 *In*: R. Lawford, D. Fort, H. Hartmann, and S. Eden, eds., *Water: Science, policy, and management*, Water Resources Monograph 16, American Geophysical Union.

Lettenmaier, D.P., E.F. Wood, and J.R. Wallis, 1994. Hydro-climatological trends in the continental U.S., 1948-88, *Journal of Climate*, **7**, 586-607.

Liepert, B.G., 2002. Observed reductions of surface solar radiation at sites in the United States and worldwide from 1961 to 1990, *Geophysical Research Letters,* **29**, 10.1029/2002GL014910.

Liang, X., D.P. Lettenmaier, E.F. Wood, and S.J. Burges, 1994: A Simple Hydrologically Based Model of Land and Energy Fluxes for General Circulation Models, *Journal of Geophysical Research*, **99**, 14,415-14,428.

Lins, H.F., 2007: Observed trends in hydrological cycle components, Section 197 in *Encyclopedia of Hydrological Sciences*, V. 5, part 17, p. 3035-3044., J. Wiley & Sons, London.

Lins, H.F., and J.R. Slack, 2005: Seasonal and regional characteristics of U.S. streamflow trends in the United States from 1940 to 1999, *Physical Geography,* **26**, 489-501.

Lins, H.F., and J.R. Slack, 1999: Streamflow trends in the United States, *Geophysical Research Letters,* **26**, 227-230.

Liu, A.J., S.T.Y. Tong, and J.A. Goodrich, 2000. Land use as a mitigation strategy for the water-quality impacts of global warming: a scenario analysis on two watersheds in the Ohio River Basin, *Environmental. Engineering and Policy,* **2**, 65-76.

Loaiciga, H.A., D.R. Maidment, and J.B. Valdes, 2000. Climate change impacts on a regional Karst aquifer, Texas, USA, *Journal of Hydrology* **227**, 173-194.

Logan, J.A., J. Regniere, and J.A. Powell: 2003. Assessing the impacts of global warming on forest pest dynamics, *Frontiers in Ecology and the Environment,* **1**, 130-137.

Lucht, W., I.C. Prentice, R.B. Myneni, S. Sitch, P. Friedlingstein, W. Cramer, P. Bousquet, W. Buermann, and B. Smith, 2002. Climate control of the high-latitude vegetation greening trend and Pinatubo effect, *Science,* **296**, 1687-1689.

Maass, A., M.M. Hufschmidt, R. Dorfman, H.A. Thomas, Jr., S.A. Marglin, and G.M. Fair, 1962. Design of water resources systems: New Techniques for Relating Economic Objectives, Engineering Analysis, and Governmental Planning. Harvard University Press, Cambridge, Mass.

Matheussen B., R.L. Kirschbaum, I.A. Goodman, G.M. O'Donnell, and D.P. Lettenmaier, 2000. Effects of land cover change on streamflow in the interior Columbia basin, *Hydrological Processes,* **14**, 867-885.

Mauget, S.A., 2003. Multidecadal regime shifts in U.S. streamflow, precipitation, and temperature at the end of the Twentieth Century, *Journal of Climate,* **16**, 3905-3916.

Mauget, S.A., 2004. Low frequency streamflow regimes of the central United States: 1939-1998, *Climatic Change,* **63**, 121-144.

Maurer, E.P., 2007. Uncertainty in hydrologic impacts of climate change in the Sierra Nevada, California under two emissions scenarios, *Climatic Change*, **82**, 309-325, doi: 10.1007/s10584-006-9180-9.

Maurer, E.P., A.W. Wood, J.C. Adam, D.P. Lettenmaier, and B. Nijssen, 2002. A long-term hydrologically-based data set of land surface fluxes and states for the conterminous United States, *Journal of Climate,* **15**, 3237-3251.

McCabe, G.J., and D.M. Wolock, 2002a. A step increase in streamflow in the conterminous United States, *Geophysical Research Letters,* **29**, doi:10.1029/2002GL015999.

McCabe, G.J., and D.M. Wolock, 2002b. Trends and temperature sensitivity of moisture conditions in the conterminous United States, *Climate Research,* **20**, 19-29.

McKenzie, D., A.E. Hessl, and D.L. Peterson, 2001. Recent growth of conifer species of western North American: Assessing spatial patterns of radial growth trends. *Canadian Journal Forest Research,* **31**, 526-538.

Milly, P.C.D., J. Betancourt, M. Falkenmark, R.M. Hirsch, Z. Kundzewicz, D.P. Lettenmaier, and R.J. Stouffer, 2008. Stationarity is Dead: Whither Water Management. *Science,* **319**, 573-574.

Milly, P.C.D., K.A. Dunne, and A.V. Vecchia, 2005. Global pattern of trends in streamflow and water availability in a changing climate, *Nature,* **438**, 347-350.

Milly, P.C.D., and K.A. Dunne, 2001. Trends in evaporation and surface cooling in the Mississippi River basin. *Geophysical Research Letters,* **28**, 1219-1222.

Moog, D.B., and P.J. Whiting, 2002. Climatic and agricultural contributions to changing loads in two watershed in Ohio, *Journal of. Environmental Quality,* **31**, 83-89.

Mote, P.W., A.F. Hamlet, M.P. Clark, and D.P. Lettenmaier, 2005. Declining mountain snowpack in western North America, *Bulletin of the American Meteorological Society.*, **86**, 39-49.

Mote, P.W., 2003. Trends in snow water equivalent in the Pacific Northwest and their climatic causes. *Geophysical Research Letters,* **30**, doi:10.1029/2003GL017258.

Murdoch, P.S., J.S. Baron, and T.L. Miller, 2000. Potential effects of climate change on surface-water quality in North America. *Journal of the American Water Resources Association,* **36**, 357-366.

Myneni, R.B., C.D. Keeling, C.J. Tucker, G. Asrar, and R.R. Nemani, 1997. Increased plant growth in the northern high latitudes from 1981-1991. *Nature,* **386**, 698-701, doi:10.1038/386698a0.

National Research Council, Global Water and Energy Experiment (GEWEX) panel, 1998. Global Water and Energy Experiment (GEWEX) Continental-Scale International Project: A review of progress and opportunities. National Academy Press, Washington D.C., 93 pp.

Nemani, R.R., C.D. Keeling, H. Hashimoto, W.M. Jolly, S.C. Piper, C.J. Tucker, R.B. Myneni, and S.W. Running, 2003. Climate-driven increases in global terrestrial net primary production from 1982 to 1999, *Science,* **300**, 1560-1563.

NRC (National Research Council of the National Academies), 2004; *Confronting the Nation's Water Problems: The role of research*: The National Academies Press, Washington, DC, http://books.nap.edu/books/0309092582/html/index.html

Oki, D.S., 2004. Trends in streamflow characteristics at long-term gaging stations, Hawaii. U.S. Geological Survey Scientific Investigations Report 2004-5080, 116 pp.

Pagano, T., and D. Garen, 2005. A recent increase in western U.S. streamflow variability and persistence. *Journal of Hydrometeorology,* **6**, 173-179.

Pagano, T., D. Garen, and S. Sorooshian, 2004. Evaluation of official western U.S. seasonal water supply outlooks, 1922-2002. *Journal of Hydrometeorology*, **5**, 896-909.

Petersen, J.H., and J.F. Kitchell, 2001. Climate regimes and water temperature changes in the Columbia River: bioenergetic implications for predators of juvenile salmon. *Canadian Journal of Fisheries and Aquatic Sciences,* **58**, 1831-1841.

Peterson, T.C., V.S. Golubev, and P.V. Groisman, 1995: Evaporation losing its strength, *Nature,* **377**, 687-688.

Poff, N.L., M. Brinson, and J.B. Day, 2002 Freshwater and coastal ecosystems and global climate change: A review of projected impacts for the United States. Pew Center on Global Climate Change, Arlington, VA. 44 pp. Available at www.pewclimate.org/global-warming-in-depth/all_reports/aquatic_ecosystems/index.cfm

Potter, K.W., 1991: Hydrologic impacts of changing land management practices in a moderate-sized agricultural catchment, *Water Resources Research*, **27**, 845-856.

Ramstack, J.M., S.C. Fritz, and D.R. Engstrom, 2004: Twentieth century water quality trends in Minnesota lakes compared with presettlement variability, *Canadian Journal of Fisheries and Aquatic Sciences.*, **61**, 561-576.

Roderick, M.L., and G.D. Farquhar, 2002: The cause of decreased pan evaporation over the past 50 years, *Science,* **298**, 1410-1411.

Rosenzweig, C., D.C. Major, K. Demong, C. Stanton, R. Horton, and M. Stults, 2007: Managing climate change risks in New York City's water system: Assessment and adaptation planning, *Mitigation and Adaptation Strategies for Global Change*, DOI 10.1007/s11027-006-9070-5.

Schaefer, G.L., M.H. Cosh, and T.L. Jackson, 2007: The USDA Natural Resources Conservation Service Soil Climate Analysis Network (SCAN), *Journal of Atmospheric and Oceanic Technology* **24**, 2073-2077.

Schoennagel, T., T.T. Veblen, and W.H. Romme, 2004: The interaction of fire, fuels, and climate across Rocky Mountain forests, *BioScience,* **54**, 661-676.

Scibek, J., and D.M. Allen, 2006: Comparing modeled responses to two high-permeability, unconfined aquifers to predicted climate change, *Global and Planetary Change,* **50**, 50-62.

Senhorst, H.A.J., and J.J.G. Zwolsman, 2005: Climate change and effects on water quality: a first impression, *Water Science Technology,* **51**, 53-59.

Schindler, D.W., S.E. Bayley, B.R. Parker, K.G. Beaty, D.R. Cruikshank, E.J. Fee, E.U. Schindler, and M.P. Stainton, 1996: The effects of climatic warming on the properties of boreal lakes and streams at the experimental lakes area, northwestern Ontario, *Limnology & Oceanography,* **41**, 1004-1017.

Schwartz, R.C., P.J. Deadman, D.J. Scott, and L.D. Mortsch, 2004: Modeling the impacts of water level changes on a Great Lakes community, *Journal of the American Water Resources Association*, **40**, 647-662.

Seager, R., M. Ting, I. Held, Y. Kushnir, J. Lu, G. Vecchi, H.-P. Huang, N. Harnik, A. Leetmaa, N.-C. Lau, C. Li, J. Velez, and N. Naik, 2007: Model projections of an imminent transition to a more arid climate in southwestern North America, *Science*, **316**, 1181-1184.

Shuttleworth, W.J., 1993: Evaporation, Chapter 4 in *Handbook of Hydrology*, D.R. Maidment, ed., McGraw Hill, New York.

Slack, J.R., A.M. Lumb, and J.M. Landwehr, 1993: Hydroclimatic data network (HCDN): A U.S. Geological Survey streamflow data set for the United States for the study of climate variation, 1874-1988. *Water Resources Investigations Reports*, **93**, 4076.

Stednick, J.D., 1996: The effects of timber harvest on annual water yield, *Journal of Hydrology,* **176**, 79-93.

Stefan, H.G., X. Fang, and J.G. Eaton, 2001: Simulated fish habitat changes in North American lakes in response to projected climate warming, *Transactions of the American Fisheries Society,* **130**, 459-477.

Stewart, I.T., D.R. Cayan, and M.D. Dettinger, 2005: Changes toward earlier streamflow timing across western North America, *Journal of Climate*, **18**, 1136-1155.

Sudler, C. E., 1927: Storage required for the regulation of streamflow, *Transactions of the American Society of Civil Engineers*, **91**, 622-660.

Szilagyi, J., G.G. Katul, and M.B. Parlange, 2001: Evapotranspiration intensifies over the conterminous United States, *Journal of Water Resources Planning and Management,* **127**, 354-362.

Takle, E.S., C. Anderson, M. Jha, and P.W. Gassman, 2006: Upper Mississippi River basin modeling system Part 4: Climate change impacts on flow and water quality, *In*: V.P. Singh and Y.J. Xu, eds.,*Coastal Hydrology and Processes* , 135-142, Water Resources Publications.

U.S. Geological Survey, 1998: A new evaluation of the USGS stream gauging network, Report to Congress, Nov. 30, 1998, 20 pp.

Vaccaro, J., 1992: Sensitivity of groundwater recharge estimates to climate variability and change, Columbia Plateau, Washington, *Journal of Geophysical Research,* **97**, 2821-2833.

Vogel, R.M., T. Yushiou, and J.F. Limbrunner, 1998: The regional persistence and variability of annual streamflow in the United States, *Water Resources Research*, **34**, 3445-3459.

Volney, W.J.A. and R.A. Flemming, 2000: Climate change impacts of boreal forest insects, *Agriculture, Ecosystems and Environment,* **82**, 283-294.

Walter, M.T., D.S. Wilks, J.Y. Parlange, and B.L. Schneider, 2004: Increasing evapotranspiration from the conterminous United States. *Journal of Hydrometeorology,* **5**, 405-408.

Westerling, A.L., A. Gershunov, T.J. Brown, D.R. Cayan, and M.D. Dettinger, 2003: Climate and wildfire in the western United States. *Bulletin of the American Meteorological Society,* **48**, doi:10.1175/BAMS-84-5-595.

Westerling, A.L., H.G. Hidalgo, D.R. Cayan, and T.W. Swetnam, 2006: Warming and earlier Spring increases western U.S. forest wildfire activity, *Science,* **313**, 940-943.

Williams, D.W. and A.M. Liebhold, 2002: Climate change and the outbreak ranges of two North American bark beetles, *Agricultural and Forest Entomology,* **4**, 87-99.

Woodhouse, C.A., and J.T. Overpeack, 1998: 2000 years of drought variability in the central United States, *Bulletin of the American Meteorological Society*, **79**, 2693-2714.

Wolfe, D.W., M.D. Schwartz, A.N. Lakso, Y. Otsuke, R.M. Pool, and N.J. Shaulis, 2004: Climate change and shifts in spring phenology of three horticultural woody perennials in northeastern USA, *International Journal of Biometeorology* **10**.1007/s00484-004-0248-9.

Zreda, M., and D. Desilets, 2005: Cosmic-ray neutron probe: Non-invasive measurement of soil water content. *American Geophysical Union*, Abstract U21B-0810.

CHAPTER 5 REFERENCES

Alexander, V., and H. J. Niebauer, 1981. Oceanography of the eastern Bering Sea ice-edge zone in spring. *Limnology and Oceanography*, **26**, 1111-1125.

Alongi, D.M., 2002. Present state and future of the world's mangrove forests. *Environmental Conservation,* **29,** 331-349.

Amstrup, S.C., B.G. Marcot, and D.C. Douglas, 2007. *Forecasting the range-wide status of polar bears at selected times in the 21st century.* Administrative Report, U.S. Department of the Interior. U.S. Geological Survey. 126pp.

Amstrup, S.C. and D.P. DeMaster, 1988. Polar bear – *Ursus maritimus*. Pages 39-56 *In:* J. W. Lentfer (ed.) *Selected Marine Mammals of Alaska: species accounts with research and management recommendations.* Marine Mammal Commission, Washington, D.C.

Amstrup, S.C., and C. Gardner, 1994. Polar bear maternity denning in the Beaufort Sea. *Journal of Wildlife Management,* **58**, 1-10.

Amstrup, S.C., G. Durner, I. Stirling, N.J. Lunn, and F. Messier, 2000. Movements and distribution of polar bears in the Beaufort Sea. *Canadian Journal of Zoology,* **78**, 948-966.

Atkinson, S.N. and M.A. Ramsay, 1995. The effects of prolonged fasting on the body composition and reproductive success of female polar bears. *Functional Ecology*, **9**, 559-567.

Baker, J.D., C.L. Littnan, and D.W. Johnston, 2006. Potential effects of sea level rise on the terrestrial habitats of endangered and endemic megafauna in the Northwestern Hawaiian Islands. *Endangered Species Research,* **4**, 1-10.

Bakun, A., 1990 Global climate change and intensification of global ocean upwelling. *Science*, **247**, 198-201.

Barnett, T.P., J.C. Adam, and D.P. Lettenmaier, 2005. Potential impacts of a warming climate on water availability in snow-dominated regions. *Nature*, **438**, 303-309.

Barry, J.P., C.H. Baxter, R.D. Sagarin, S.E. Gilman. 1995. Climate-related, long-term faunal changes in a California rocky intertidal community. Science **267**, 672-675.

Beaubien, E.G. and H.J. Freeland, 2000. Spring phenology trends in Alberta, Canada: Links to ocean temperature. *International Journal of Biometeorology,* **44**(2), 53-59.

Beaugrand, G., P.C. Reid, F. Ibanez, J.A. Lindley and M. Edwards, 2002. Reorganization of North Atlantic marine copepod biodiversity and climate. *Science*, **296**, 1692-1694.

Beaugrand, G. 2004. The North Sea regime shift: evidence, causes, mechanisms and consequences. *Progress in Oceanography,* **60**, 245-262.

Beebee, T.J.C., 2002. Amphibian Phenology and Climate Change. *Conservation Biology,* **16** (6), 1454-1454, doi:10.1046/j.1523-1739.2002.02102.x

Beebee, T.J.C., 1995. Amphibian Breeding and Climate. *Nature*, **374**, 219-220.

Beever, E.A., P.F. Brussard, and J. Berger, 2003. Patterns of apparent extirpation among isolated populations of pikas (Ochotona princeps) in the Great Basin. *Journal of Mammology,* **84**, 37-54.

Behrenfeld, M.J., R. O'Malley, D.Siegel, C. McClain, J. Sarmiento, G. Feldman, A. Milligan, P. Falkowski, R. Letelier, E. Boss, 2006. Climate-driven trends in contemporary ocean productivity. *Nature,* **444**, 752-755.

Belchansky, G. I., D. C. Douglas, I. N. Mordvintsev, and N. G. Platonov, 2004. Estimating the time of melt onset and freeze onset over Arctic sea-ice area using active and passive microwave data. *Remote Sensing of the Environment*, **92**(1), 21-39.

Bell, J.L., L.C. Sloan, and M.A. Snyder. 2004. Regional Changes in Extreme Climatic Events: A Future Climate Scenario. *Journal of Climate*, **17**(1), 81-87.

Beniston, M., and D.G. Fox, 1996. Impacts of climate change on mountain regions. Pages 191-213 *In:* R. T. Watson, M. C. Zinyowera, and R. H. Moss (eds.), *Climate change 1995 - Impacts, adaptations and mitigation of climate change.* Contribution of Working Group II to the Second Assessment Report of the IPCC. Cambridge University Press, New York, NY.

Bertness, M.D., G.H. Leonard, J.M. Levine, J.F. Bruno. 1999. Climate-driven interactions among rocky intertidal organisms caught between a rock and a hard place. *Oecologia* **120**, 446-450

Blaustein, A.R., Belden, L.K., Olson, D.H., Green, D.M., Root, T.L., and J.M. Kiesecker, 2001. Amphibian breeding and climate change. *Conservation Biology,* **15**(6), 1804-1809

Blaustein, A.R., T.L. Root, J.M. Kiesecker, L.K. Belden, D.H. Olson, and D.M. Green, 2002. Amphibian phenology and climate change. *Conservation Biology,* **16**(6), 1454-1455.

Blix, A.S. and J.W. Lentfer, 1979. Modes of thermal protection in polar bear cubs: at birth and on emergence from the den. *American Journal of Physiology*, **236**, 67-74.

Boisvenue, C. and S.W. Running, 2006. Impacts of climate change on natural forest productivity – evidence since the middle of the 20th century. *Global Change Biology,* **12**, 862-882.

Both, C. and M.E. Visser, 2005. The effect of climate change on the correlation between avian life-history traits. *Global Change Biology,* **11**(10), 1606-1613.

Both, C., 2006. Climate change and adaptation of annual cycles of migratory birds. *Journal of Ornithology,* **147**(5, Suppl. 1), 68-68.

Brandt, M. In Press. Coral disease and bleaching relationships in South Florida. *Further citation to come.*

Breshears, D.D., N.S. Cobb, P.M. Rich, K.P. Price, C.D. Allen, R.G. Balice, W.H. Romme, J.H. Kastens, M.L. Floyd, J. Belnap, J.J. Anderson, O.B. Myers, and C.W. Meyer, 2005. Regional vegetation die-off in response to global-change-type drought. *Proceedings of the National Academy of Sciences,* **102**, 15144-15148

Brooks M.L., 2003. Effects of increased soil nitrogen on the dominance of alien annual plants in the Mojave Desert. *Journal of Applied Ecology*, **40**, 344-353.

Bryant, D., L. Burke, J. McManus, M. Spalding. 1998. Reefs at risk: a map-based indicator of threats to the world's coral reefs. World Resources Institute. Washington, DC Brown, B.E., 1997. Coral bleaching: causes and consequences. *Coral Reefs,* **16** (Supplement 1), S129-138.

Buckley, J. R., T. Gammelsrød, A. Johannessen, 0. M. Johannessen, and L. P. Røed. 1979. Upwelling: Oceanic structure at the edge of the Arctic ice pack in winter. *Science,* **203**, 165-167.

Burkett, Virginia R., D.A. Wilcox, R. Stottlemeyer, W. Barrow, D. Fagre, J. Baron, J. Price, J.L. Nielsen, C.D. Allen, D.L. Peterson, G. Ruggerone, T. Doyle. 2005. Nonlinear dynamics in ecosystem response to climatic change: case studies and policy implications. *Ecological Complexity,* **2**, 357-394.

Butler, C., 2003. The disproportionate effect of climate change on the arrival dates of short-distance migrant birds. *Ibis,* **145**, 484-495.

Caldeira, K. and M.E. Wickett, 2003. Anthropogenic carbon and ocean pH. *Nature*, **425**, 365.

Calvert, W. and I. Stirling, 1990. Interactions between polar bears and overwintering walruses in the central Canadian high arctic. *International Conference on Bear Research and Management*, **8**, 351-356.

Carlton, J.T., 2000. Global Change and Biological Invasions in the Oceans. *In:* Mooney, H.A. and R.J. Hobbs (eds.). *Invasive Species in a Changing World.* Island Press. pp 31-54

Carr, M.E, M.A. Friedrichs, M. Schmeltz, M.N. Aita, D. Antoine, K.R. Arrigo, I. Asanuma, O. Aumont, R. Barber, M. Behrenfeld, R. Bidigare, E.T. Buitenhuis, J. Campbell, A. Ciotti, H. Dierssen, M. Dowell, J. Dunne, W. Esaias, B. Gentili, W. Gregg, S. Groom, N. Hoepner, J. Ishizaka, T. Kameda, C. Le Quere, S. Lohrenz, J. Marra, F. Melin, K. Moore, A. Morel, T.E. Reddy, J. Ryan, M. Scardi, T. Smyth, K. Turpie , G. Tilstone, K. Waters, Y. Yamanaka, 2006. A comparison of global estimates of marine primary production from ocean color. *Deep Sea Research*, **53**, 741-770

Carey, C., W.R. Heyer, J. Wilkinson, R.A. Alford, J.W. Arntzen, T. Halliday, L. Hungerford, K.R. Lips, E.M. Middleton, S.A. Orchard, A.S. Rand. 2001. Amphibian declines and environmental change: use of remote sensing data to identify environmental correlates. *Conservation Biology*, **15**, 903-913.

Caspersen, J.P., S.W. Pacala, J. Jenkins, G.C. Hurtt, P.R. Moorcroft and R.A. Birdsey, 2000. Contributions of land-use history to carbon accumulation in U.S. forests. *Science* **290**, 1148-1151.

Cayan, D.R., S. Kammerdiener, M.D. Dettinger, J.M. Caprio, and D.H. Peterson, 2001. Changes in the onset of spring in the western United States. *Bulletin of the American Meteorological Society,* **82**, 399-415.

Cesar, H. 2000. Coral Reefs: Their Functions, Threats and Economic Value, in H. Cesar (ed.) Collected Essays on the Economics of Coral Reefs, CORDIO, Kalmar University, Kalmar, Sweden.

Cesar, H., L. Burke and L. Pet-Soede, 2003. *The Economics of Worldwide Coral Reef Degradation.* Cesar Environmental Economic Consulting, Arnhem, The Netherlands.

Chavez, F., L. Ryan, S.E. Lluch-Cota and M. Ñiguen, 2003. From anchovies to sardines and back: Multidecadal change in the Pacific Ocean. *Science*, **229**(5604), 217-221.

Clarkson, P.L. and D. Irish, 1991. Den collapse kills female polar bears and two newborn cubs. *Arctic*, **44**, 83-84.

Coley, P.D., and T.M. Aide, 1991. Comparison of plant defenses in temperate and tropical broad-leaved forests. Pages 25-49 *In:* P.W. Price, T.M. Lewinsohn, G.W. Fernandes and W.W. Benson (eds.). *Plant-Animal Interactions: Evolutionary Ecology in Tropical and Temperate Regions.* John Wiley & Sons, Inc., New York.

Coley, P.D., and J.A. Barone. 1996. Herbivory and plant defenses in tropical forests. *Annual Review of Ecology and Systematics,* **27**, 305-335.

Comiso, J.C., C. L. Parkinson, R. Gersten, and L. Stock, 2008. Accelerated decline in the Arctic sea ice cover. *Geophysical Research Letters*, **35**, L01703

Corn, P.S. 2003. Amphibian breeding and climate change: Importance of snow in the mountains. *Conservation Biology*, **17**, 622-625.

Cotton, P., 2003. Avian migration phenology and global climate change, *Proceedings of the National Academy of Sciences*, **100**, 12219-12222.

Cronin, M.A., S.C. Amstrup, G.W. Garner, and E.R. Vyse, 1991. Interspecific and intraspecific mitochondrial DNA variation in North American bears (Ursus). *Canadian Journal of Zoology*, **69**, 2985-2992.

Crozier, L., 2004. Warmer winters drive butterfly range expansion by increasing survivorship. *Ecology,* **85**, 231-241.

Crozier, L., 2003. Winter warming facilitates range expansion: cold tolerance of the butterfly *Atalopedes campestris*. *Oecologia,* **135**, 648-656.

D'Antonio C.M. and L.A. Meyerson, 2002. Exotic plant species as problems and solutions in ecological restoration: a synthesis. *Restoration Ecology*, **10**, 703-13.

Daszak, P., A.A. Cunningham, A.D. Hyatt, 2000. Emerging Infectious Diseases of Wildlife: Threats to Biodiversity and Human Health. *Science,* **287**, 443- 448.

D'Elia, C.F., R.W. Buddemeier and S.V. Smith, 1991. Workshop on coral bleaching. Coral Reef Ecosystem and Global Change: Report of Proceedings. College Park, University of Maryland, Maryland Sea Grant UM-SG-TS-91-03.

Deméré, T.A., A. Berta, and P.J. Adam, 2003. Pinnipedimorph evolutionary biogeography. *Bulletin of the American Museum of Natural History,* **279**, 32-76.

Derocher, A.E., D. Andriashek, and I. Stirling, 1993. Terrestrial foraging by polar bears during the icefree period in western Hudson Bay. *Arctic*, **4,** 251-254.

Derocher, A.E., R.A. Nelson, I. Stirling, M.A. Ramsay, 1990. Effects of fasting and feeding on serum urea creatinine levels in polar bears. *Marine Mammal Science*, **6**, 196-203.

Derocher, A.E., Ø. Wiig, and G. Bangjord, 2000. Predation of Svalbard reindeer by polar bears. *Polar Biology*, **23**, 675-678.

Derocher, A.E., N.J. Lunn and I. Stirling. 2004. Polar bears in a warming climate. *Integrative and Comparative Biology*, **44**, 163-176.

Diaz, H.F., J.K. Eischeid, C. Duncan, and R.S. Bradley, 2003. Variability of freezing levels, melting season indicators, and snow cover for selected high-elevation and continental regions in the last 50 years. *Climatic Change*, **59**, 33-52.

Dippner, J.W., G. Ottersen. 2001. Cod and climate variability in the Barents Sea. *Climate Research*, **17**, 73-82.

Dirnbock, T., S. Dullinger, G. Grabherr. 2003. A regional impact assessment of climate and land-use change on alpine vegetation. *Journal of Biogeography*, **30**, 401-417.

Donner, S.D., W.J. Skirving, C.M. Little, M. Oppenheimer and O. Hoegh-Guldberg, 2005. Global assessment of coral bleaching and required rates of adaptation under climate change. *Global Change Biology*, **11**, 1-15.

Drinkwater, K.F., 2005. The response of Atlantic cod (*Gadus morhua*) to future climate change. *ICES Journal of Marine Science* **62**,1327-1337.

Dukes, J.S., 2000: Will the rising atmospheric CO_2 concentration affect biological invaders? *Invasive Species in a Changing World*, H. Mooney and R. Hobbs, eds. Island Press, Washington, D.C., 95-113.

Dukes, J.S. and H.A. Mooney, 1999. Does global change increase the success of biological invaders? *Trends in Ecology and Evolution*, **14**(4), 135-139.

Dunn, P.O. and D. Winkler, 1999. Climate change has affected the breeding date of tree swallows throughout North America. *Proceedings of the Royal Society of London Bulletin*, **266**, 2487-2490.

Durner, G.M. and S.C. Amstrup, 1995. Movements of a polar bear from northern Alaska to northern Greenland. *Arctic*, **48**, 338-341.

Durner, G.M., S.C. Amstrup, and A.S. Fischbach, 2003. Habitat characteristics of polar bear terrestrial maternal den sites in northern Alaska. *Arctic*, **56**, 55-62.

Eakin et al. In Press.Caribbean Corals in Hot Water: Record-Setting Thermal Stress and Coral Bleaching in 2005. *Citation details to come.*

Ehrlich, P.R., D.D. Murphy, M.C. Singer, C.B. Sherwood, RR. White and I.L. Brown, 1980. Extinction, reduction, stability and increase: The responses of checkerspot butterfly populations to the California drought. *Oecologia*, **46**, 101-105.

Elsner, J.B., 2006. Evidence in support of the climate change – Atlantic hurricane hypothesis. *Geophysical Research Letters*, **33**(16), L16705.

Emanuel, K., 2005. Increasing destructiveness of tropical cyclones over the past 30 years. *Nature*, **436**, 686-688.

Fay, F.H., 1982. Ecology and biology of the Pacific walrus, *Odobenus rosmarus divergens* Illiger. *North American Fauna Series*. U.S. Fish and Wildlife Service. Washington, D.C. 275 pp.

Federal Register, 2006. Rules and Regulations. Endangered and Threatened Species: Final Listing Determination for Elkhorn Coral and Staghorn Coral, 71 Fed. Reg. 26852.

Ferguson, S.H., M.K. Taylor, E.W. Born, A. Rosing-Asvid, and F. Messier, 1999. Determinants of home range size for polar bears (*Ursus maritimus*). *Ecology Letters*, **2**, 311-318.

Ferguson, S.H., I. Stirling and P. McLoughlin, 2005. Climate change and ringed seal (*Phoca hispida*) recruitment in western Hudson Bay. *Marine Mammal Science*, **21**, 121-135.

Field, J.C., Boesch, D.F. Scavia, D. Buddemeier, R. Burkett, V.R. Cayan, D. Fogerty, M. Harwell, M. Howarth, R. Mason, C. Pietrafesa, L.J. Reed, D. Royaer, T. Sallenger, A. Spranger, M. and J.G. Titus, 2001. Potential consequences of climate variability and change on coastal and marine resources. *In*: *Climate Change Impacts in the United States: Potential Consequences of Climate Change and Variability and Change.* Foundation Document. U.S. Global Change Research Program: Cambridge, UK, Cambridge University Press.

Field, M.E., J.V. Gardner and D.B. Prior, 1999: Geometry and significance of stacked gullies on the northern California slope. *Marine Geology*, **154**, 271-286.

Fields, P.A., J.B. Graham, R.H. Rosenblatt, G.N. Somero. 1993. Effects of expected global climate change on marine faunas. TREE 8: 361-367.

Fischer, AG., 1960. Latitudinal variation in organic diversity. *Evolution*, **14**, 64-81.

Fitt, W.K. and M.E. Warner, 1995. Bleaching patterns of four species of Caribbean reef corals. *Biological Bulletin*, **189**, 298-307.

Forister, M.L., and A.M. Shapiro, 2003. Climatic trends and advancing spring flight of butterflies in lowland California. *Global Change Biology*, **9**, 1130-1135.

Franco, A.M.A., J.K. Hill, C. Kitschke, Y.C. Collingham, D.B. Roy, R. Fox, B. Huntley, and C.D. Thomas, 2006. Impacts of climate warming and habitat loss on extinctions at species' lowlatitude range boundaries. *Global Change Biology*, **12**, 1545-1553.

Furnell, D.J., and D. Oolooyuk 1980. Polar bear predation on ringed seals in ice-free water. *Canadian Field-Naturalist*, **94**, 88-89.

Garner, G.W., S.C. Amstrup, I. Stirling, and S.E. Belikov, 1994. Habitat considerations for polar bears in the North Pacific Rim. *Transactions of the North American Wildlife and Natural Resources Conference*, **29**, 111-120.

Gibbs, J.P., and A.R. Breisch, 2001. Climate warming and calling phenology of frogs near Ithaca, New York, 1900-1999. *Conservation Biology*, **15**, 1175-1178.

Glynn, P.W., 1984. Widespread coral mortality and the 1982-83 El Niño warming event. *Environmental Conservation*, **11**, 133-146.

Glynn, P.W., 1993. Coral reef bleaching: ecological perspectives. *Coral Reefs*, **12**, 1-17.

Gobbi, M., D. Fontaneto, and F. De Bernardi, 2006. Influence of climate changes on animal communities in space and time: the case of spider assemblages along an alpine glacier foreland. *Global Change Biology*, **12**, 1985-1992.

Goreau, T.J. and R.M. Hayes, 1994. Coral bleaching and ocean "Hot spots." *Ambio*, **23**, 176-180

Government Accountability Office, 2007. Climate Change: Agencies should develop guidance for addressing the effects on Federal land and water resources. Government Accountability Office 07-863. Washington, DC.

Grabherr, G., M. Gottfried, and H. Pauli. 1994. Climate effects on mountain plants. *Nature*, **369**, 448.

Grebmeier, J.M., J.E. Overland, S.E. Moore, E.V. Farley, E.C. Carmack, L.W. Cooper, K.E. Frey, J.H. Helle, F.A. McLaughlin, and S.L. McNutt, 2006. A major ecosystem shift in the Northern Bering Sea. *Science*, **311**, 1461-1464.

Groffman, P.M., J.S. Baron, T. Blett, A.J. Gold, I. Goodman, L.H. Gunderson, B.M. Levinson, M.A. Palmer, H.W. Paerl, G.D. Peterson, N.L. Poff, D.W. Rejeski, J.F. Reynolds, M.G. Turner, K.C. Weathers, J. Wiens, 2006. Ecological thresholds: the key to successful environmental management or an important concept with no practical application? *Ecosystems*, **9**, 1-13.

Guralnick, R., 2007. Differential effects of past climate warming on mountain and flatland species distributions; a multispecies North American mammal assessment. *Global Ecol. Biogeogr.* **16**, 14-23.

Hansen, J., 2006. Expert report submitted to the United States District Court, District of Vermont in regard to Case No. 2:05-CV-302 and 2:05-CV-304, Green Mountain Chrysler-Plymouth-Dodge-Jeep et al. v. Thomas W. Torti, Secretary of Vermont Agency of Natural Resources, et al.

Hansen, J., L. Nazarenko, R. Ruedy, M. Sato, J. Willis, A. Del Genio, D. Koch, A. Lacis, K. Lo, S. Menon, T. Novakov, J. Perlwitz, G. Russell, G. A. Schmidt, and N. Tausnev. 2005. Earth's energy imbalance: confirmation and implications. Science 308:1431-1435.

Harrington, C.R. 1968. Denning habits of the polar bear (*Ursus maritimus* Phipps). *Canadian Wildlife Service Report Series*, Number **5**. Ottawa.

Harvell, C.D., C.E. Mitchell, J.R. Ward, S. Altizer, A.P. Dobson, R.S. Ostfeld, Hay, M.E., and W. Fenical, 1988. Marine plant-herbivore interactions: the ecology of chemical defense. *Annual Review of Ecology and Systematics*, **19**, 111-145.

Hayhoe, K., C. Wake, T.G. Huntington, L, Luo, M.D. Schwartz, J. Sheffield, E.F. Wood, B. Anderson, J. Bradbury, T.T. DeGaetano, and D. Wolfe, 2006: Past and future changes in climate and hydrological indicators in the U.S. Northeast. *Climate Dynamics*, **10**, doi:1007/s00382-006-0187-8.

Hayhoe, K. D. Cayan, C.B. Field, P.C. Frumhoff, E.P. Maurer, N.L. Miller, S.C. Moser, S.H. Schneider, K.N. Cahill, E.E. Cleland, L. Dale, R. Drapek, R.M. Hanemann, L.S. Kalkstein, J. Lenihan, C.K. Lunch, R.P. Neilson, S.C. Sheridan, and J.H. Verville, 2004: Emissions pathways, climate change, and impacts on California. *Proceedings of the National Academy of Sciences*, 101(34): 12422-12427.

Hays, G.C., A.J. Richardson and C. Robinson, 2005. Climate change and marine plankton. *Trends in Ecology and Evolution*, **20**, 337-344.

Helmuth, B., J.G. Kingsolver, E. Carrington, 2005. Biophysics, physiological ecology, and climate change: Does mechanism matter? *Annual Review of Physiology*, **67**, 177-201.

Helmuth, Brian, N. Mieszkowska, P. Moore, S.J. Hawkins. 2006. Living on the edge of two changing worlds: forecasting the responses of rocky intertidal ecosystems to climate change. *Annual Review of Ecology and Systematics*, **37**, 373-404.

Hicke, J. A., and D.B. Lobell, 2004. Spatiotemporal patterns of cropland area and net primary production in the central United States estimated from USDA agricultural information. *Geophysical Research Letters*, **31**, L20502, doi:10.1029/2004GL020927.

Hierro J.L., D. Villarreal, O. Eren, *et al.*, 2006. Disturbance facilitates invasion: the effects are stronger abroad than at home. *American Naturalist*, **168**, 144-56.

Hill, J.K., C.D. Thomas, R. Fox, M.G. Telfer, S.G. Willis, J. Asher, and B. Huntley, 2002. Responses of butterflies to twentieth century climate warming: Implications for future ranges. *Proceedings of the Royal Society Biological Sciences Series B*, **269**(1505): 2163-2171.

Hoegh-Guldberg, O., 1999. Climate change, coral bleaching and the future of the world's coral reefs. *Marine Freshwater Research*, **50**, 839-866.

Hoegh-Guldberg, O., R. Berkelmans, and J. Oliver, 1997: Coral bleaching: implications for the Great Barrier Reef Marine Park. *Proceedings of the Cooperative Research Centre Conference in Research and Reef Management*. CRC for the sustainable use of the Great Barrier Reef, 21-43.

Hoegh-Guldberg, O.; P.J. Mumby, A.J. Hooten, R.S. Stenek, P. Greenfield, E. Gomez, C.D. Harvell, P.F. Sale, A.J. Edwards, K. Caldeira, N. Knowlton, C.M. Eakin, R. Iglesias-Prieto, N. Muthiga, R.H. Bradbury, A. Dubi, and M.E. Hatziolos. 2007. Coral reefs under rapid climate change and acidification. *Science*, **318**, 1737-1742.

Holland, M.M., C.M. Bitz, and B. Tremblay, 2006: Future abrupt reductions in the summer Arctic sea ice. *Geophyical. Research Letters*, **33**, L23503, doi:10.1029/2006GL028024.

Hooff, R.C. and W.T. Peterson, 2006. Copepod biodiversity as an indicator of changes in ocean and climate conditions of the northern California Current. *Limnological and Oceangraphy*, **51**, 2607-2620.

Hoyos, C.D., P.A. Agudelo, P.J. Webster, and J.A. Curry, 2006. Deconvolution of the factors contributing to the increase in global hurricane intensity. *Science*, **312**, 94-97.

Hsieh, C., S.M. Glaser, A.J. Lucas, G. Sugihara. 2005. Distinguishing random environmental fluctuations from ecological catastrophes in the north Pacific Ocean. *Nature*, **435**, 336-340.

Huenneke L.F., S.P.Hamburg, R. Koide, H.A.Mooney, P.M.Vitousek, 1990. Effects of soil resources on plant invasion and community structure in California serpentine grassland. *Ecology*, **71**, 478-491.

Hughes, C.L., J.K. Hill, and C. Dytham, 2003. Evolutionary trade-offs between reproduction and dispersal in populations at expanding range boundaries. *Proceedings of the Royal Society Biological Sciences Series B,* **270**(Supplement 2), S147-S150.

Hughes, T.P.; A.H. Baird; D.R. Bellwood; M. Card; S.R. Connolly; C. Folke; R. Grosberg; O. Hoegh-Guldberg; J.B.C. Jackson; J. Kleypas; J.M. Lough; P. Marshall; M. Nystrom; S.R. Palumbi; J.M. Pandolfi; B. Rosen; J. Roughgarden, 2003. Climate change, human impacts and the resilience of coral reefs. *Science* **301**, 929-933.

Humphries, M.M., J. Umbanhowar, K.S. McCann, 2004. Bioenergetic prediction of climate change impacts on northern mammals. *Integrative and Comparative Biology,* **44**, 152-162

Hunt Jr., G. A., P. Stabeno, G. Walters, E. Sinclair, R. D. Brodeur, J. M. Napp, N. A. Bond, 2002. Climate change and control of the southeastern Bering Sea pelagic ecosystem. *Deep Sea Research II*, **49,** 5821-5853.

Inouye, D.W., 2007. Consequences of climate change for phenology, frost damage, and floral abundance of sub-alpine wildflowers. *Ecology, In press.*

Inouye, D.W., 2007. Impacts of global warming on pollinators. *Wings.* **30**(2), 24-27.

Inouye, D.W., B. Barr, K.B. Armitage, and B.D. Inouye, 2000. Climate change is affecting altitudinal migrants and hibernating species. *Proceedings of the National Academy of Sciences*, **97**, 1630-1633.

Inouye, D.W., W.A. Calder, and N.M. Waser, 1991. The effect of floral abundance on feeder censuses of hummingbird abundance. *Condor*, **93**, 279-285.

Inouye, D.W., M. Morales, and G. Dodge, 2002. Variation in timing and abundance of flowering by *Delphinium barbeyi* Huth (Ranunculaceae): the roles of snowpack, frost, and La Niña, in the context of climate change. *Oecologia,* **139**, 543-550.

Inouye, D.W., F. Saavedra, and W. Lee, 2003. Environmental influences on the phenology and abundance of flowering by *Androsace septentrionalis* L. (Primulaceae). *American Journal of Botany*, **90**, 905-910.

Inouye, D.W., and F.E. Wielgolaski, 2003. High altitude climates. Pages 195-214 *In:* M. D. Schwartz (ed.). *Phenology: an Integrative Environmental Science*. Kluwer Academic Publ, PO Box 17/3300 AA Dordrecht/Netherlands.

IPCC, 2007: *Climate Change 2007: The Physical Science Basis, Contribution from Working Group 1 to the Fourth Assessment Report, Policy Maker Summary*. Intergovernmental Panel on Climate Change. Cambridge University Press, Cambridge, UK.

IPCC, 2001: *Climate Change 2001: The Scientific Basis, Contribution from Working Group I to the Third Assessment Report.* Intergovernmental Panel for Climate Change. Cambridge University Press, Cambridge, UK.

IPCC, 1990: *Climate Change 1990: The Scientific Basis, Contribution from Working Group I to the First Assessment Report.* Intergovernmental Panel for Climate Change. Cambridge University Press, Cambridge, UK.

Iverson, L.R. and A.M. Prasad, 2001. Potential changes in tree species richness and forest community types following climate change. *Ecosystems*, **4**, 186-199.

Jablonski, D., 1993. The tropics as a source of evolutionary novelty through geological time. *Nature,* **364**, 142-144.

Janetos, A.C., R. Kasperson, T. Agardy, J. Alder, N. Ashe, R. Defries, and G. Nelson. 2005. *Chapter 28: Synthesis: Conditions and trends in systems and services, tradeoffs for human well-being, and implications for the future*. Conditions and Trends Volume. Millennium Ecosystem Assessment. R.J. Scholes, R. Hassan and N. Ashe (Eds). Island Press. Washington, DC. 823-834.

Jenni, L. and M. Kéry, 2003. Timing of autumn bird migration under climate change: advances in long-distance migrants, delays in short-distance migrants. Proceedings of the Royal Society of Biological Science, **270**(1523), 1467-1471, doi: 0.1098/rspb.2003.2394.

Johannessen, O. M., E. V. Shalina, and M. W. Miles, 1999. Satellite evidence for an Arctic sea ice cover in transformation. *Science,* **286**, 1937-1939.

Johnson, T.R., 1998. Climate change and Sierra Nevada snowpack. M.S. Thesis. University of California, Santa Barbara, Santa Barbara.

Jokiel, P.L. and S.L. Coles, 1990. Response of Hawaiian and other Indo-Pacific reef corals to elevated temperature. *Coral Reefs*, **8**, 155-162.

Joos, F., I. C. Prentice, J. I. House, 2002. Growth enhancement due to global atmospheric change as predicted by terrestrial ecosystem models: consistent with US forest inventory data. *Global Change Biology, 8/4*, 299-303.

Jonzen, N., A. Lindén, T. Ergon, E. Knudsen, J.O. Vik, D. Rubolini, D. Piacentini, C. Brinch, F. Spina, L. Karlsson, M. Stervander, A. Andersson, J. Waldenström, A. Lehikoinen, E. Edvardsen, R. Solvang, andN.C. Stenseth, 2006. Rapid advance of spring arrival dates in long-distance migratory birds. *Science,* **312**, 1959-1961.

Karamouz, M., and B. Zahraie, 2004. Seasonal streamflow forecasting using snow budget and El Niño-Southern Oscillation climate signals: Application to the salt river basin in Arizona. *Journal of Hydrologic Engineering*, **9**, 523-533.

Keeling, C.D., J.F.S. Chin, and T.P. Whorf, 1996. Increased activity of northern vegetation inferred from atmospheric CO_2 measurements. *Nature,* **382**(6587), 146-149.

Kelly, B.P., 2001. Climate change and ice breeding pinnipeds. Pages 43-55 *In:* G.R. Walther, C.A. Burga and P.J. Edwards (eds). *"Fingerprints" of climate change: adapted behaviour and shifting species' ranges.* Kluwer Academic/Plenum Publishers, New York and London.

Kelly, B. P., O. H. Badajos, M. Kunnasranta, and J. R. Moran, 2006. Timing and re-interpretation of ringed seal surveys. *Final Report OCS Study MMS 2006-013*. Coastal Marine Institute, University of Alaska Fairbanks.

Kendall, M.A., M.T. Burrows, A.J. Southward, S.J. Hawkins, 2004. Predicting the effects of marine climate change on the invertebrate prey of birds of rocky shores. *Ibis*, **146**, 40-47.

Kennedy V.A., R.R. Twilley, J.A. Kleypas, J.H. Cowan, Jr. and S.R. Hare, 2002. Coastal and Marine Ecosystems and Global Climate Change: Potential Effects on U.S. Resources. Pew Center for Global Climate Change, Arlington, VA. 52pp.

Kenny, A. and C. Mollmann, 2006. Towards intergrated ecosystem assessments for the North and Baltic Seas: synthesizing GLOBEC research. *GLOBEC International Newsletter*, **12**(2), 64-65.

Kiesecker, Joseph M., A.R. Blaustein, and L.K. Belden, 2001. Complex causes of amphibian population declines. *Nature*, **410**, 681-684.

King, J.E., 1983. *Seals of the world, 2nd Edition.* Comstock Publishing Associates, Ithaca, NY.

Kleypas, J.A., R.W. Buddemeier, D. Archer, J.P. Gattuso, C. Langdon, and B.N. Opdyke, 1999. Geochemical Consequences of Increased Atmospheric Carbon Dioxide on Coral Reefs. *Science*, **284**, 5411, 118.

Kowalska, Z., 1965. Cross-breeding between a female European brown bear (*Ursus arctos*) and a male polar bear (*U. maritimus*) in the Logzkin Zoo. *Przegi Zool,*. **9**, 313-319.

Kühl. M., R. N. Glud, J. Borum, R. Roberts, and S. Rysgaard, 2001. Photosynthetic performance of surface-associated algae below sea ice as measured with a pulse-amplitude-modulated (PAM) fluorometer and O_2 microsensors. Marine Ecology Progress Series 223:1-14.

Legendre, L., S.F. Ackley, G.S. Dieckmann, B. Gulliksen, R. Horner, T. Hoshiai, I.A. Melnikov, W.S. Reeburgh, M. Spindler, and C.W. Sullivan, 1992. Ecology of sea ice biota. *Polar Biology*, **12**, 429-444.

Lehikoinen, E., T.H. Sparks, M. Zalakevicius, 2004. Arrival and departure dates. *Advanced Ecological Research*, **35**, 1-31.

Lesica, P., and B. McCune, 2004. Decline of arctic-alpine plants at the southern margin of their range following a decade of climatic warming. *Journal of Vegetation Science*, **15**, 679-690.

Lesser, M.P., W.R. Stochaj, D.W. Tapley and J.M. Shick, 1990. Bleaching in coral reef anthozoans: effects of irradiance, ultraviolet radiation and temperature on the activities of protective enzymes against active oxygen. *Coral Reefs*, **8**, 225-232.

Li, W. and A.T. Smith, 2005. Dramatic decline of the threatened Ili pika, *Ochotona iliensis* (Lagomorpha: Ochotonida) in Xinjiang, China. *Oryx*, **39**, 30-34.

Lister, A.M., 2004. The impact of Quaternary ice ages on mammalian evolution. *Philosophical Transactions of the Royal Society of London*, **359**, 221-241.

Lobell, D.B., Ortiz-Monasterio, J. Ivan, Addams, C. Lee, and G.P. Asner, 2002. Soil, climate and management impacts on regional wheat productivity in Mexico from remote sensing. *Agricultural and Forest Meteorology*, **114**, 31-43.

Logan, J.A., J. Régnière, J.A. Powell, 2003. Assessing the impacts of global warming on forest pest dynamics. *Frontiers in Ecology and Environment*, **1**(3): 130-137.

Lovejoy, T.E. and L.J. Hannah, Eds., 2005. Climate change and biodiversity. Yale University Press, New Haven.

Lucht, W., I.C. Prentice, R.B. Myneni, S. Sitch, P. Friedlingstein, W. Cramer, P. Bousquet, W. Buermann, and B. Smith, 2002. Climatic control of the high-latitude vegetation greening trend and Pinatubo effect. *Science*, **296**(5573), 1687-1689.

Lumsden, S.E., T.F. Hourigan, A.W. Bruckner, G. Dorr (eds.), 2007: *The State of Deep Coral Ecosystems of the United States.* NOAA Technical Memorandum CRCP-3. Silver Spring MD. 365 pp.

Lunn, N. J. and I. Stirling, 1985. The significance of supplemental food to polar bears during the ice-free period of Hudson Bay. *Canadian Journal of Zoology*, **63**, 2291-2297.

Lunn, N. J. and I. Stirling, 2001. Climate change and polar bears: long-term ecological trends observed in Wapusk National Park. *Research Links*, **9**, 5-6.

Lydersen, C., and T.G. Smith, 1989. Avian predation on ringed seal *Phoca hispida* pups. *Polar Biology*, **9**, 489-490.

MacArthur, R.H., 1972. Geographical ecology: patterns in the distribution of species. Harper and Row, New York, New York, USA.

MacCracken, M., E. Barron, D. Easterling, B. Fetzer, and T. Karl, 2001. Scenarios for climate variability and change. Pages 13-71 *In:* N.A.S. Team, (ed.). *Climate change impacts on the United States: the potential consequences of climate variability and change.* Cambridge University Press, Cambridge.

Mackas, D.L., W.T. Peterson, M.D. Ohman, and B.E. Lavaniegos, 2006. Zooplankton anomalies in the California Current system beforeandduringthewarmoceanconditionsof2005. *Geophysical Research Letters*, **33**, L22S07, doi:10.1029/2006GL027930.

MacMynowski, D. and T. Root, 2007. Climate and the Complexity of Migratory Phenology: Sexes, Migratory Distance, and Arrival Distributions. *Biometeorology*, **51**, 361-373

Mann, M.E., and K.A. Emanuel, 2006. Atlantic hurricane trends linked to climate change. *Eos: Transactions of the American Geophysical Union*, **87**, 233-244.

Mantua, N.J, R.H. Hare, Yuan Zhang, J.M. Wallace, and R.C. Francis, 1997. A Pacific interdecadal climate oscillation with impacts on salmon production. *Bulletin of the American Meteorological Society*,**78**, 1069-1079)

Martinez-Meyer, E., Townsend, P.A., Hargrove, W.W. 2004. Ecological niches as stable distributional constraints on mammal species, with implications for Pleistocene extinctions and climate change projections for biodiversity. *Global Ecology and Biogeography*, **13**, 305-314.

McGowan, J.E., D.R.Cayan and L.M.Dorman. 1998. Climate-ocean variability and ecosystem response in the northeast Pacific. *Science*, **281**, 210-217.

McKee, K.L., I.A. Mendelssohn, and M.D. Materne, 2004. Acute salt marsh dieback in the Mississippi River deltaic plain: a drought-induced phenomenon? *Global Ecology and Biogeography*, **13**, 65-73.

McKenzie C., S. Schiff , R. Aravena, C. Kelly, V.S. Louis VS, 1998. Effect of temperature on production of CH_4 and CO_2 from peat in a natural and flooded boreal forest wetland. *Climate Change*, **40**, 247-66.

McLaughlin, J.F., J.J. Hellmann, C.L. Boggs, and P.R. Ehrlich, 2002. Climate change hastens population extinction. *Proceedings of the National Academy of Sciences,* **99**, 6070-6074.

Meehl, G.A., T.F. Stocker, W.D. Collins, P. Friedlingstein, A.T. Gaye, J.M. Gregory, A. Kitoh, R. Knutti, J.M. Murphy, A. Noda, S.C.B. Raper, I.G. Watterson, A.J. Weaver, and Z.C. Zhao, 2007. 2007: Global Climate Projections. *In:* S. Solomon, D. Qin, M. Manning, Z. Chen, M. Marquis, K.B. Averyt, M. Tignor, and G.H. Miller, (eds.). *Climate Change 2007: The Physical Science Basis.* Contribution of Working Group I to the Fourth Assessment Report of the Intergovernmental Panel on Climate Change. Cambridge University Press, Cambridge University Press, Cambridge, UK, and New York, NY, USA.

Menzel, A., G. Jakobi, R. Ahas, H. Scheifinger, and N. Estrella, 2003. Variations of the climatological growing season (1951-2000) in Germany compared with other countries. *International Journal of Climatology*, **23**(7), 793-812.

Messier, F., M.K. Taylor, and M.A. Ramsay, 1994. Denning ecology of polar bears in the Canadian Arctic archipelago. *Journal of Mammalogy,* **75**, 420-430.

Mieszkowska, N., S.J. Hawkins, M.T. Burrows, M.A. Kendall, 2007. Long-term changes in the geographic distribution and population structures of Osilinus lineatus (Gastropoda: Trochidae) in Britain and Ireland. *Journal of the Marine Biological Assocation of the United Kingdom,* **87**, 537-545.

Millennium Ecosystem Assessment, 2005. *Ecosystems and Human Well-being: Biodiversity Synthesis.* World Resources Institute, Washington, DC.

Mieszkowska, N. M.A. Kendall, S.J. Hawkins, R. Leaper, P. Williamson, N.J. Hardman-Mountford, A.J. Southward, 2006. Changes in the range of some common rocky shore species in Britain – a response to climate change? *Hydrobiologia* **555**, 241-251.

Mieszkowska, N., R. Leaper, P. Moore, M.A. Kendall, M.T. Burrows, D. Lear, E. Poloczanska, K. Hiscock, P.S. Moschella, R.C. Thompson, R.J. Herbert, D. Laffoley, J. Baxter, A.J. Southward, S.J. Hawkins. 2005. Marine biodiversity and climate change: Assessing and predicting the influence of climate change using intertidal rocky shore biota: Final report for United Kingdom funders. Marine Biological Association Occasional Publ. No. 20.

Moore, P., 2004. Favoured aliens for the future. *Nature*, 427:594.

Monson R.K., J.P. Sparks, T.N. Rosenstiel, L.E. Scott-Denton, T.E. Huxman, P.C. Harley, A.A. Turnipseed, S.P. Burns, B. Backlund, J. Hu, 2005. Climatic influences on net ecosystem CO_2 exchange during the transition from wintertime carbon source to springtime carbon sink in a high-elevation, subalpine forest. *Oecologia,* **146**, 130-147.

Morris, J.T., P.V. Sundareshwar, C.T. Nietsch, B. Kjerfve, D.R. Cahoon, 2002. Responses of coastal wetlands to rising sea-levels. *Ecology,* **83**, 2869-2877

Muller, E.M., C.S. Rogers, A.S. Spitzack and R. van Woesik, 2007. Bleaching increases likelihood of disease on Acropora palmate (Lamarck) in Hawksnest Bay, St John, U.S. Virgin Islands. *Coral Reefs.* **27**(1):191-195.

National Research Council, 2007. *Colorado River Basin Water Management: Evaluating and Adjusting to Hydroclimatic Variability*. The National Academies Press, Washington, D.C.

National Research Council, 2006. *Status of Pollinators in North America.* National Academies Press, Washington, D.C.

Nelson, G.C., E. Bennett, A. Asefaw Berhe, K. Cassman, R. DeFries, T. Dietz, A. Dobermann, A. Dobson, A. Janetos, M. Levy, D. Marco, N. Nakicenovic, B. O'Neill, R. Norgaard, G. Petschel-Held, D. Ojima, P. Pingali, R. Watson and M. Zurek, 2006: Anthropogenic drivers of ecosystem change: an overview. *Ecology and Society* 11 (2): 29.

Nemani, R.R., M.A. White, P.E. Thornton, K. Nishida, S. Reddy, J. Jenkins, and S. Running, 2002. Recent trends in hydrologic balance have enhanced the terrestrial carbon sink in the United States. *Geophysical Research Letters*, **29**, 1468, doi:10.1029/2002GL014867.

Orr, James C., V.J. Fabry, O. Aumont, L. Bopp, S.C. Doney, R.A. Feeley, A. Gnanadesikan, N. Gruber, A. Ishida, F. Joos, R.M. Key, K. Lindsay, E. Meier-Reimer, R. Matear, P. Monfray, A. Mouchet, R.G. Najjar, G.K. Plattner, K.B. Rodgers, C.L. Sabine, J.L. Sarmiento, R. Schlitzer, R.D. Slater, I.J. Totterdell, M-F. Weirig, Y. Yamanaka, A. Yool, 2005. Anthropogenic ocean acidification over the twenty-first century and its impact on calcifying organisms. *Nature,* **437**, 681-686.

Overland, J.E., and M. Wang, 2007. Future regional sea ice declines. *Geophysical Research Letters,* **34**, L17705.

Overpeck, J.T., B.L. Otto-Bliesner, G.H. Miller, D.R. Muhs, R.B. Alley, and J.T. Kiehl, 2006. Paleoclimatic evidence for future ice-sheet instability and rapid sea-level rise. *Science,* **311**, 1747-1750.

Overpeck, J., K. Hughen, D. Hardy, R. Bradley, R. Case, M. Douglas, B. Finney, K. Gajewsky, G. Jacoby, A. Jennings, S. Lamoureux, A. Lasca, G. MacDonald, J. Moore, M. Retelle, S. Smith, A. Wolfe, and G. Zielinski, 1997. Arctic environmental change of the last four centuries. *Science,* **278** (5341), 1251-1256.

Overpeck, J.T., M. Sturm, J.A. Francis, D.K. Perovich, M.C. Serreze, R. Benner, E.C. Carmack, S. Chapin III, S.C. Gerlach, L.C. Hamilton, L.D. Hinzman, M. Holland, H.P. Huntington, J.R. Key, A.H. Lloyd, G.M. MacDonald, J.McFadden, D. Noone, T.D. Prowse, P. Schlosser, C. Vörösmarty, 2005. Arctic System on Trajectory to New, Seasonally Ice-Free State. *Eos: Transactions of the American Geophysical Union,* **86**(34), 309, 312-313.

Pandolfi, J.M., R.H. Bradbury, E. Sala, T.P. Hughes, K.A. Bjorndal, R.G. Cooke, D. McArdle, L. McClenachan, M.J.H. Newman, G. Paredes, R.R. Warner, J.B.C. Jackson, 2003. Global trajectories of the long-term decline of coral reef ecosystems. *Science,* **301**, 955-958.

Park, R.A., M.S. Trehan, P.W. Mausel and R.C. Howe, 1989. *The Effects of Sea Level Rise on U.S. Coastal Wetlands.* U.S. EPA, Offices of Policy, Planning and Evaluations.

Parmesan, C., 2006. Ecological and Evolutionary Responses to Recent Climate Change. *Annual Review of Ecology, Evolution and Systematics,* **37**, 637-669.

Parmesan, C., 1996: Climate and species' range. *Nature* **382**, 765-766.

Parmesan, C. and G. Yohe, 2003. A Globally Coherent Fingerprint of Climate Change Impacts across Natural Systems. *Nature*, **421**, 37.

Parmesan, C., N. Ryrholm, C. Stefanescu, J.K. Hill, C.D. Thomas, H. Descimon, B. Huntley, L. Kaila, J. Kullberg, T. Tammaru, W.J. Tennent, J.A. Thomas, and M. Warren. 1999. Poleward shifts in geographical ranges of butterfly species associated with regional warming. *Nature,* **399**(6736), 579-583.

Pearcy, W.G., 1991. *Ocean ecology of north Pacific salmonids.* Washington State Sea Grant Program, The University of Washington Press, Seattle. 179 pp.

Pelejero, C., E. Calvo, M.T. McCulloch, J.F. Marshall, M.K. Gagan, J.M. Lough, B.N. Opdyke, 2005. *Science,* **309**, 2204-2207.

Peters, R.L. and T.E. Lovejoy. 1992. *Global Warming and Biological Diversity.* Yale University Press, New Haven, CT.

Peterson, W.T. and F.B. Schwing, 2003. A new climate regime in northeast Pacific ecosystems. *Geophysical Research Letters*, **38** (17), 1896, doi 10.1029/2003GL017528.

Petes, L.E., B.A. Menge, G.D. Murphy, 2007. Environmental stress decreases survival, growth, and reproduction in New Zealand mussels. *Journal of Experimental Marine Biology and Ecologyl,*: **351**, 83-91.

Poff, N.L., M.M. Brinson, J.W. Day, Jr. 2002. Aquatic ecosystems and global climate change: Potential impacts on inland freshwater and coastal wetland ecosystems in the United States. Pew Center on Global Climate Change. Washington, DC.

Polar Bear Specialist Group. 2006. Status of the polar bear. N. J. Lunn, and A. E. Derocher (eds.). Polar Bears: proceedings of the 14th working meeting of the IUCN/SSC Polar Bear Specialist Group. IUCN, Gland, Switzerland and Cambridge, UK. 35-55.

Pounds, J.A., 2001. Climate and amphibian declines. *Nature,* **410,** 639-640

Pounds, J. Alan and M. Crump, 1994. Amphibian declines and climate disturbances: the case of the Golden Toad and the Harlequin Frog. *Conservation Biology*, **8**, 72-85.

Pounds, J.A., M.P.L. Fogden, and J.H. Campbell, 1999. Biological response to climate change on a tropical mountain. *Nature,* **398**, 611-615.

Pounds, J.A., M.R. Bustamante, L.A. Coloma, J.A. Consuegra, M.P.L. Fogden, P.N. Foster, E. La Marca, K.L. Masters, A. Merino-Viteri, R. Puschendorf, S.R. Ron, G.A. Sanchez-Azofeifa, C.J. Still, and B.E. Young, 2006. Widespread amphibian extinctions from epidemic disease driven by global warming. *Nature,* **439**(7073), 161-167.

Powell, J. A., and J.A. Logan, 2005. Insect seasonality: circle map analysis of temperature-driven life cycles. *Theoretical Population Biology*, 67(3), 161-179.

Powell, J.A. and J.A. Logan, 2001. Ghost Forests, Global Warming, and the Mountain Pine Beetle (Coleoptera: Scolytidae). *American Entomologist*, **3**, 160-172.

Powell, J.J., J. Jenkins, J. Logan, and B. Bentz, 2000. Seasonal temperature alone can synchronize life cycles. *Bulletin Mathematical Biology*, **62**, 977-998.

Rahmstorf, S, 2007. A Semi-Empirical Approach to Projecting Future Sea-Level Rise. *Science*, **315** (5810), 368.

Ramsay, M. A. and K. A. Hobson. 1991. Polar bears make little use of terrestrial food webs: evidence from stable isotope analysis. *Oecologia*, **86**, 598-600.

Ramsay, M.A. and I. Stirling, 1988. Reproductive biology and ecology of female polar bears (*Ursus maritimus*). *Journal of Zoology* (London), **214**, 601-634.

Ramsay, M.A. and I. Stirling, 1990. Fidelity of female polar bears to winter-den sites. *Journal of Mammalogy*, **71**, 233-236.

Ravens, J., K. Caldeira, H. Elderfield, O. Hoegh-Guldberg, P. Liss, U. Riebessell, J. Shepard, C. Turley and A. Watson, 2005. *Ocean Acidification due to Increasing Carbon Dioxide.* The Royal Society, London, England.

Reading, C.J., 1998. The effect of winter temperatures on the timing of breeding activity in the common toad Bufo bufo. *Oecologia,* **117**, 469-475.

Rhymer, J.M. and D. Simberloff, 1996. Extinction by hybridization and introgression. *Annual Review of Ecology and Systemantics*, **27**, 83-109.

Richardson, A.J. and D.S. Shoeman, 2004. Climate impact on plankton ecosystems in the Northeast Atlantic. *Science*, **305**, 1609-1612.

Rignot, E., and P. Kangaratnam, 2006. Changes in the velocity structure of the Greenland Ice Sheet. *Science,* **311**, 986-990.

Roessig, J.M., C.M. Woodley, J.J. Cech and L.J. Hansen, 2004. Effects of global climate change on marine and estuarine fishes. *Reviews in Fish Biology and Fisheries*, **14**, 215-275.

Roman, J., 2006. Diluting the founder effect: cryptic invasions expand a marine invader's range *Proceedings of the Royal Society*: **273**, 2453-2459

Romme, W.H., J. Clement, J. Hicke, D. Kulakowski, L.H. MacDonald, T.L. Schoennagel, and T.T. Veblen, 2006. *Recent Forest Insect Outbreaks and Fire Risk in Colorado Forests: A Brief Synthesis of Relevant Research.*

Root, T.L. and S.H. Schneider, 2006. Conservation and climate change: The challenges ahead. *Conservation Biology,* **20**(3), 706-708.

Root, T.L., J.T. Price, K.R. Hall, S.H. Schneider, C. Rosenzweig and J.A. Pounds, 2003. Fingerprints of global warming on wild animals and plants. *Nature*, **421**, 57-60.

Root, T.L., D.P. MacMynowski, M. Mastrandrea, and S.H. Schneider. 2005. Human-modified temperatures induce species' changes: joint attribution. *Proceedings of the National Academy of Sciences,* **21**, 7465-7469.

Roots, E.F., 1989. Climate change: high latitude regions. *Climate Change*, **15**, 223-253.

Rothrock, D.A., J. Zhang and Y. Yu, 2003. The arctic ice thickness anomaly of the 1990s: A consistent view from observations and models. *Journal of Geophysical Research*, **108**(C3), 3083, doi:10.1029/2001JC001208.

Roy, D.B. and T.H. Sparks, 2000. Phenology of British butterflies and climate change. *Global Change Biology*, **6**, 407-416.

Saavedra, F., D.W. Inouye, M.V. Price, and J. Harte, 2003. Changes in flowering and abundance of *Delphinium nuttallianum* (Ranunculaceae) in response to a subalpine climate warming experiment. *Global Change Biology*, **9**, 885-894.

Sacks, W., D.Schimel, and R.Monson, 2007. Coupling between carbon cycling and climate ina high elevation, subalpine forest: a model-data fusion analysis. *Oecologia,* **151**(1), 54-68, doi:10.1007/s00442-006-0565-2.

Sæther, B.E., S. Engen, A.P. Møller, H. Weimerskirch, M.E. Visser, W. Fiedler, E. Matthysen, M.M. Lambrechts, R. Freckleton, A. Badyaev, P.H. Becker, J.E. Brommer, D. Bukacinski, M. Bukacinska, H. Christensen, J. Dickinson, C. du Fau, F. R. Gelbach, D. Heg, H. Hötker, J. Merilä, J.T. Nielsen, W. Rendell, D.L. Thomson, J. Török, and P. Van Hecke, 2005. Life history variation predicts the effect of demographic stochasticity on avian population dynamics. *American Naturalist,* **164**, 793-802.

Sagarin, R., J.P. Barry, S.E. Gilman and C.H. Baxter, 1999. Climate-related change in an intertidal community over short and long time scales. *Ecological Monographs*, **69**, 465-490.

Salathé , E. 2005. Downscaling simulations of future global climate with application to hydrologic modeling. *International Journal of Climate*, **25**, 419-436.

Samue, M. D., 2002: Climate Warming and Disease Risks for Terrestrial and Marine Biota. *Science*, **296**, 2158-2162.

Scavia, D., J.C. Field, D.F. Boesch, R.W. Buddemeier, V. Burkett, D.R. Cayan, M. Fogarty, M.A. Harwell, R.W. Howarth, C. Mason, D.J. Reed, T.C. Royer, A.H. Sallenger, and J.G. Titus, 2002. Climate change impacts on US coastal and marine ecosystems. *Estuaries,* **25**, 149-164.

Scheffer, V.B., 1958. *Seals, sea lions and walruses: A review of the pinnipedia.* Stanford University Press, Stanford, CA.

Schneider, S.H. and T.L. Root, 2002. Introduction: the Rationale for the National Wildlife Federation Cohort of Young Scientists Studying Wildlife Responses to Climate Change. *In*: *Wildlife Responses to Climate Change: North American Case Studies*, Island Press Washington, DC. p. xi-xv.

Schneider, S.H. and T.L. Root, 1996. Ecological implications of climate change will include surprises. *Biodiversity and Conservation,* **5**(9), 1109-1119.

Scholze, M., W. Knorr, N.W. Arnell, I.C. Prentice, 2006. A climate-change risk analysis for world ecosystems. *Proceedings of the National Academy Of Sciences*, **103**, 13116-13120.

Schwing, F.B., N.A. Bond, S.J. Bograd, T. Mitchell, M.A. Alexander and N. Mantua, 2006. Delayed coastal upwelling along the U.S. west coast in 2005: a historical perspective. *Geophysical Reseach Letters*, **33**, L22S01, doi:10.1029/2006GL026911.

Schwartz M.D. and B.E. Reiter, 2000. Changes in North American spring. *International Journal of Climatology,* **20**, 929-932.

Serreze, M. C., M. M. Holland, and J. Stroeve. 2007. Perspectives on the Arctic's shrinking sea-ice cover. *Science,* **315**,1533-1536.

Serreze, M.C., J.E. Walsh, F.S. Chapin III, T. Osterkamp, M. Dyergerov, V. Romanovsky, W.C. Oechel, J. Morison, T. Zhang, and R.G. Barry, 2000. Observational evidence of recent change in the northern high latitude environment. *Climate Change*, **46**, 159-207.

Sher, A.A. and L.A. Hyatt, 1999. The disturbed resource-flux invasion matrix: a new framework for patterns of plant invasion. *Biological Invasions*, **1**, 107-14.

Short, F.T. and H. Neckles, 1999. The effects of global climate change on seagrasses. *Aquatic Botany,* **63**, 169-196.

Singer, F.J. and K. Harter, 1996. Comparative effects of elk herbivory and 1988 fires on northern Yellowstone National Park grasslands. *Ecological Applications,* **6**(1), 185-200.

Singer, M.C. and P.R. Ehrlich, 1979. Population dynamics of the checkerspot butterfly *Euphydryas editha. Fortschritte der Zoologie,* **25**, 53-60.

Smith, T.G., 1985. Polar bears, *Ursus maritimus*, as predators of belugas, *Delphinapterus* leucas. *Canadian Field-Naturalist*, **99**, 71-75.

Smith, T.G., 1980. Polar bear predation of ringed and bearded seals in the land-fast sea ice habitat. *Canadian Journal of Zoology*, **58**, 2201-2209.

Snyder, M.A., L.C. Sloan, N.S. Diffenbaugh, and J.L. Bell, 2003. Future Climate Change and Upwelling in the California Current, *Geophysical Research Letters*, **30**(15), 1823-1827, 10.1029/2003GL017647.

Southward, A.J., S.J. Hawkins and M.T. Burrows, 1995. Seventy years' observations of changes in distribution and abundance of zooplankton and intertidal organisms in the western English Channel in relation to rising sea temperature. *Journal of Thermal Biology*, **20**, 127-155.

Sparks, T.H., D.B. Roy, and R.L.H. Dennis, 2005. The influence of temperature on migration of Lepidoptera into Britain. *Global Change Biology*, **11**(3), 507-514. doi:10.1111/j.1365-2486.2005.00910.x.

Soto, C.G., 2002. The potential impacts of global climate change on marine protected areas. *Reviews in Fish Biology and Fisheries*, **11**, 181-195.

Sriver, R. and M. Huber, 2006. Low frequency variability in globally integrated tropical cyclone power dissipation. *Geophysical Research Letters*, **33**, L11705, doi:10.1029/2006GL026167.

Stanley, S.M., 1979. *Macroevolution, pattern and process.* W.H. Freeman, San Francisco.

Stefanescu C., J. Peñuelas, and I. Filella, 2003. Effects of climatic change on the phenology of butterflies in the northwest Mediterranean Basin. *Global Change Biology*, **9** (10), 1494-1506.

Stenseth, N.C. and A. Mysterud, 2002. Climate, changing phenology, and other life history traits: nonlinearity and match-mismatch to the environment. *Proceedings of the National Academy of Sciences,* **99**, 13379-13381.

Stenseth, N.C., A. Mysterud, G. Ottersen, J.W. Hurrell, K.S. Chan, and M. Lima, 2002. Ecological effects of climate fluctuations. *Science*, **297**, 1292-1296.

Stirling, I., 1974. Midsummer observations on the behavior of wild polar bears (*Ursus maritimus*). *Canadian Journal of Zoology*, **52**, 1191-1198.

Stirling, I. and E.H. McEwan, 1975. The caloric value of whole ringed seals (*Phoca hispida*) in relation to polar bear (*Ursus martimus*) ecology and hunting behavior. *Canadian Journal of Zoology,* **53**, 102-127.

Stirling, I., and T.G. Smith, 2004. Implications of warm temperatures and an unusual rain event for the survival of ringed seals on the coast of Southeastern Baffin Island. *Arctic*, **57**, 59-67.

Stirling, I. and T.G. Smith, 1975. Interrelationships of Arctic Ocean mammals in the sea ice habitat. *Circumpolar Conference on Northern Ecology*, **2**, 129-136.

Stirling, I. and D. Andriashek, 1992. Terrestrial maternity denning of polar bears in the eastern Beaufort Sea area. *Arctic*, **45**, 363-366.

Stirling, I. and W.R. Archibald, 1977. Aspects of predation of seals by polar bears. *Journal of Fisheries Research Board of Canada*, **34**, 1126-1129.

Stirling, I., and A.E. Derocher, 1993. Possible impacts of climate warming on polar bears. *Arctic*, **46**, 240-245.

Stirling, I., N.J. Lunn and J. Iacozza, 1999. Long-term trends in the population ecology of polar bears in western Hudson Bay in relation to climate change. *Arctic*, **52**, 294-306.

Stirling, I. and N.A. Øristland, 1995. Relationships between estimates of ringed seal and polar bear populations in the Canadian Arctic. *Canadian Journal of Fisheries and Aquatic Sciences*, **52**, 2595-2612.

Stone, R. S., E. G. Dutton, J. M. Harris and D. Longenecker, 2002. Earlier spring snowmelt in northern Alaska as an indicator of climate change. *Journal of Geophysical Research – Atmoshpere*, **107**(D9&10/ACL10):1-15. doi: 10.1029/2000JD000286.

Stroeve, J.C., M.C. Serreze. F. Fetterer. T. Aretter, W. Meier, J. Maslanik and K. Knowles, 2005. Tracking the Arcitc's shrinking ice cover: Another extreme September minimum in 2004. *Geophysical Research Letters*, **32**, L04501, doi:10.1029/2400GL021810.

Stroeve, J., M.M. Holland, W. Meier, T. Scambos, and M. Serreze, 2007. Arctic sea ice decline: Faster than forecast, *Geophyical. Research Letters*, **34**, L09501, doi:10.1029/2007GL029703.

Stuart, S.N., J.S. Chanson, N.A. Cox, B.E. Young, A.S.L. Rodrigues, D.L. Fischman, R.W. Waller, 2004. Status and trends of amphibian declines and extinctions worldwide. *Science*, **306**, 1783-1786.

Swetnam, T.W., and J.L. Betancourt, 1998. Mesoscale disturbance and ecological response to decadal climatic variability in the American Southwest. *Journal of Climate*, **11**, 3128-3147.

Sydeman, W.J., R.W. Bradley, P. Warzybok, C.L. Abraham, J. Jahncke, J.D. Hyrenbach, V.Kousky, J.M. Hipfner and M.D. Ohman, 2006. Planktivorous auklet Ptychoramphus aleuticus responses to ocean climate, 2005: unusual atmospheric blocking? Geophysical Research Letters, **33**, L22S09, doi:10.1029/2006GL026736.

Talbot, S.L. and G.F. Shields, 1996. Phylogeography of brown bears (*Ursus arctos*) of Alaska and paraphyly within the Ursidae. *Molecular Phylogenetics and Evolution*, **5**, 477-494.

Taugbøl, G. 1984. Ringed seal thermoregulation, energy balance and development in early life, a study of *Pusa hispida* in Kongsfj., Svalbard. Zoology Thesis. University of Oslo, Norway.

Tebaldi, C., K. Hayhoe, J.M. Arblaster, and G. A. Meehl, 2006: Going to the extremes; An intercomparison of model-simulated historical and futre changes in extreme events, *Climatic Change*, **79**, 185-211.

Thomas, C.D., M.C. Singer and D. Boughton, 1996. Catastrophic extinction of population sources in a butterfly metapopulation. *American Naturalist*, **148**, 957-975.

Thomas, D.N., and G. S. Dieckmann, 2002. Antarctic Sea Ice – a Habitat for Extremophiles. *Science*, **295**, 641-644.

Thompson, R.C., T.P. Crowe and S.J. Hawkins, 2002. Rocky intertidal communities: past environmental changes, present status and predictions for the next 25 years. *Environmental Conservation*, **29**, 168-191.

Tynan, C.T. and D.P. DeMaster, 1997. Observations and predictions of arctic climatic change: potential effects on marine mammals. *Arctic*, **50**, 308-322.

Vermeij, G.J., 1978. *Biogeography and adaptation.* Harvard University Press, Cambridge, Massachusetts, USA.

Vila, M., J.D. Corbin, J.S. Dukes, J. Pino, and S.D. Smith, S.D. In press. Linking plant invasions to global environmental change. *In*: J. Canadell, D. Pataki, L. Pitelka, (eds.). *Terrestrial* Ecosystems in a Changing World, Springer, New York.

Visser, M.E. and C. Both, 2005. Shifts in phenology due to global climate change: the need for a yardstick. *Proceedings of the Royal Society*, **272**, 2561-2569.

Visser M.E., L.J.M. Holleman and P. Gienapp, 2006. Shifts in caterpillar biomass phenology due to climate change and its impact on the breeding biology of an insectivorous bird. *Oecologia*, **147**, 164-172.

Visser M.E., C. Both and M.M. Lambrechts, 2004. Global climate change leads to mistimed avian reproduction. *Advances in Ecological Research*, **35**, 89-110.

Von Holle, B., and G. Motzkin, 2007. Historical land use and environmental determinants of nonnative plant distribution in coastal southern New England. *Biological Conservation*, **136**, 33-43.

Waits, L.P., S.L. Talbot, R.H. Ward, and G.F. Shields, 1998. Mitochondrial DNA phylogeography of the North American brown bear and implications for conservation. *Conservation Biology*, **12**, 408-417.

Wake, D.B., 2007. Climate change implicated in amphibian and lizard declines. *Proceedings of the National Academy Of Science*, **104**, 8201-8202.

Walther, G.R., E. Post, P. Convey, A. Menzel, C. Parmesan, T.J.C. Beebee, J.M. Fromentin, O. Hoegh-Guldberg, and F. Bairlein, 2002. Ecological responses to recent climate change. *Nature*, **416**, 389-395.

Ward, J. and K. Lafferty. Biology The Elusive Baseline of Marine Disease: Are Diseases in Ocean Ecosystems Increasing? *Plos Biology*, **2**, 0542-0547.

Wielgolaski, F.E., and D.W. Inouye, 2003. High latitude climates. Pages 175-194 *In:* M. D. Schwartz (ed.). *Phenology: an Integrative Environmental Science*. Kluwer Academic Publ, PO Box 17/3300 AA Dordrecht/Netherlands.

Wilson, R.J., D. Gutiérrez, J. Gutiérrez, and V.J. Monserrat, 2007. An elevational shift in butterfly species richness and composition accompanying recent climate change. *Global Change Biology* **13**: 1873-1887.

Winkler, D.W., P.O. Dunn, and C.E. McCulloch. Predicting the effects of climate change on avian life-history traits. *Proceedings, National Academy of Sciences,* **99**, 13595-13599.

Wolfe, D.W., M.D., Schwartz, A.N. Lakso, Y. Otsuki, R.M. Pool, N. Shaulis, 2005. Climate change and shifts in spring phenology of three horticultural woody perennials in northeastern USA. *International Journal of Biometeorology*, **49**(5), 303-309.

Yates, K.K. and R.B. Halley, 2006. CO_3^{2-} concentration and pCO_2 thresholds for calcification and dissolution on the Molokai reef flat, Hawaii. *Biogeosciences,* **3**, 1-13.

Zhou, L. M., C.J. Tucker, R.K. Kaufmann, D. Slayback, N.V. Shabanov, and R.B. Myneni. 2001. Variations in northern vegetation activity inferred from satellite data of vegetation index during 1981 to 1999. *Journal of Geophysical Research – Atmospheres*, **106**(D17), 20069-20083.

Ziska, LH and K. George, 2004: Rising carbon dioxide and invasive, noxious plants: potential threats and consequences. *Water Resources Review* **16**: 427-46.

www.ingramcontent.com/pod-product-compliance
Lightning Source LLC
Chambersburg PA
CBHW080636180526
45168CB00008B/3188

* 9 7 8 1 5 0 7 8 7 3 9 4 6 *